Beam Optics of Quadrupole Probe-Forming Systems

Beam Optics of Quadrupole Probe-Forming Systems

G W Grime
F Watt
Department of Nuclear Physics, University of Oxford

Adam Hilger Ltd, Bristol

British Library Cataloguing in Publication Data

Grime, G. W.
Beam optics of quadrupole probe–forming systems
1. Beam optics
I. Title II. Watt, F.
530.1′41 QC665.B/

ISBN 0-85274-546-X

Consultant Editor: **Professor G Dearnaley**

Published by Adam Hilger Ltd, Techno House, Redcliffe Way, Bristol BS1 6NX.
The Adam Hilger book-publishing imprint is owned by The Institute of Physics.

Printed in Great Britain by J W Arrowsmith Limited, Bristol

For Amanda and Bradley

For Simon, Helen and John

Contents

Preface

During the last decade, the use of high-energy ion beams for the purpose of elemental analysis has increased enormously, for by using the three techniques of particle-induced x-ray emission (PIXE), Rutherford backscattering (RBS) and nuclear reaction analysis (PRA), the elemental composition of a sample may be determined with sensitivities in the parts per million region.

The use of focused high-energy ion beams or microprobes to provide information on the distribution of elements in a sample is an obvious development of the technique, and since the first microprobe was constructed at Harwell in the early 1970s there has been a slow but steady increase in the number of these instruments throughout the world. There is no doubt that the ion microprobe will prove to be of great importance in a broad range of analytical applications.

That the field of ion microprobes has not expanded faster is primarily due to two factors. The first is that the production of ions with energies suitable for PIXE, RBS and PRA requires at least a million-volt particle accelerator, and thus the majority of microprobes tend to be found only in the nuclear structure or physics laboratories which have such accelerators available. The second factor concerns the relative complexity of focusing such high-energy ions to form a probe. The high rigidity of ions with energies of several MeV means that the successful technology developed for the electron microscope cannot be used and the focusing can only be achieved using stronger focusing elements such as quadrupole lenses.

Although quadrupole lenses have been used for many years as beam transport devices, their application to probe formation is largely unknown. The available literature on quadrupole lenses tends to be theoretical in nature, and although a few configurations with simplifying symmetry have been investigated, the results are not directly applicable to the design of practical ion microprobe systems.

This book is an attempt to present the theory of quadrupole focusing with the practical development of microprobe systems in mind. The introductory chapter outlines the analytical processes mentioned above and provides a background to the problems of designing quadrupole probe-forming systems. Following a description of the quadrupole lens and the mathematical models used to describe the fields within both magnetic and electrostatic quadrupole lenses (chapter 2), we describe two techniques for calculating the path of a charged particle through a beam-optical system; matrix methods (chapter 3) and numerical raytracing (chapter 4). Both of these techniques are derived from the equation of motion, but rather than attempting to extract general results directly from the theory, the emphasis in these chapters is on demonstrating how the theory may be used to develop computer programs for the design of microprobe systems.

As a guide to the design of the quadrupole microprobe systems, the intrinsic beam-optical parameters of a large number of quadrupole configurations have been calculated and are presented graphically and in tabular form in Appendix 1. Chapter 5 describes how the appendix may be used to estimate the resulting spot size for any of the wide range of doublet, triplet and quadruplet systems investigated.

The minimum spot size achieved by a quadrupole system is influenced strongly by parasitic aberrations due to imperfections in the construction and alignment of the instrument. Due to the large number of parameters involved, it is not possible to investigate parasitic aberrations systematically as was done for the intrinsic properties and so in chapter 6 we have attempted to assess the importance of the various effects by reference to their influence in three existing microprobe systems.

Finally, in chapter 7, we describe briefly some novel and interesting alternatives to the quadrupole lens as the focusing element and also discuss ways in which the focused ion beam may be further reduced in size.

Unfortunately, due to limitations of space we have not been able to consider the hardware associated with high-energy ion microprobes or any of the many applications which have been reported. It is hoped, however, that sufficient references to the literature have been included to allow the reader to pursue these aspects further.

It is of course impossible and futile to attempt to acknowledge the many individuals who have helped either directly or indirectly in the production of this book. The authors however wish especially to acknowledge members of the computer groups of the particle physics and nuclear structure divisions of the Nuclear Physics Laboratory at the University of Oxford for their invaluable help and cooperation. The authors are also grateful to the two other members of the Oxford microprobe group, G D Blower and J Takacs for their encouragements and comments on the text. We also wish to thank Professor K W Allen for providing the environment in which the work carried out for this book was stimulated.

G W Grime
F Watt
Oxford 1983

1

High-energy ion microprobes

1.1 Introduction

Several powerful analytical techniques are available when energetic ions impinge upon the surface of a specimen. If the energy of the incident ions is sufficiently high, then both atomic and nuclear processes can take place, which between them can provide simultaneous information on the elemental content of the sample over the whole periodic table, down to the parts per million level of sensitivity. Of the many reactions initiated by an incident ion beam of a few MeV in energy (figure 1.1), there are four mechanisms which are of primary interest:

(1) Proton- (or particle-) induced x-ray emission (PIXE)
(2) Elastic (or Rutherford) backscattering (RBS)
(3) Nuclear reactions
(4) Ion microscopy.

All four techniques will be discussed in sections 1.2 to 1.5, with the main emphasis on the processes

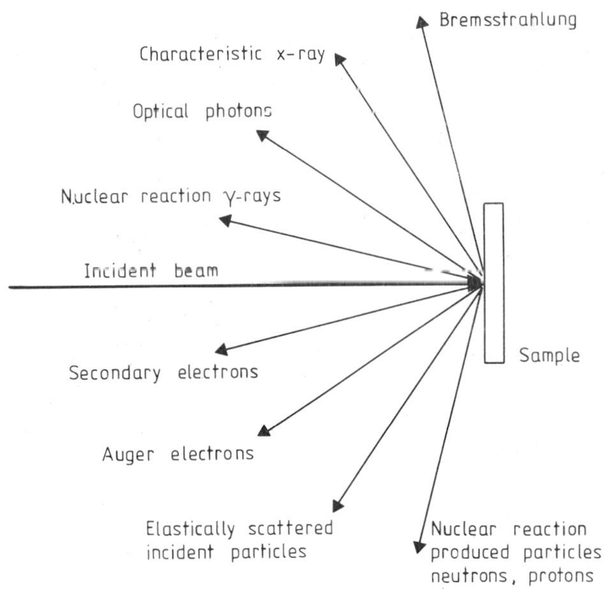

Figure 1.1 Schematic diagram of possible reactions initiated by an ion beam of MeV energy.

which have been shown to have major advantages over existing techniques, i.e. PIXE, RBS and nuclear reaction analysis techniques. Section 1.6 summarises the beam energy requirements for the analytical techniques considered. The advantages of focused ion microprobes should by then be apparent and subsequent sections will deal with the focusing of high-energy ion beams.

Section 1.7 deals with the concepts of brightness and emittance in beam optics and discusses the suitability of current ion sources for the production of ions in a form suitable for probes.

Section 1.8 considers the problem of ion-probe formation at the energies required by the techniques of PIXE, PBS and nuclear reaction analysis.

Section 1.9 describes very briefly and in simple terms, the action of the strong focusing quadrupole lens used in the probe formation.

Section 1.10 is a brief review of the beam optics terminology used extensively throughout the book.

Section 1.11 is a review of some microprobe systems in use at the time of writing. Since the subject matter of the book is concerned with beam optics, examples of the usage of microprobes are not considered. For information on specific applications and hardware, the reader should consult the respective references given in the section.

Lastly, section 1.12 is a brief review of the theoretical work carried out into quadrupole probe-forming systems. Again, the references listed in that section should be consulted for further information.

1.2 Proton-induced x-ray emission (PIXE)

X-ray emission used for analytical purposes has been known for some time, mainly in the form of x-ray fluorescence analysis, i.e. the induction of elemental characteristic x-rays by incident x-rays of higher energy. It was not until 1970 that Johansson *et al* [JO70a] showed that characteristic x-ray excitation by energetic protons, combined with detection of the x-rays by a solid-state detector cooled to liquid nitrogen temperatures constituted a powerful, multi-elemental analytical method of high sensitivity. For the reader who wishes a comprehensive review on PIXE there are several excellent review articles [JO76, FO78, FO75, ZI76, FO76a, FO76b] and published conference proceedings available [IO79, PI80, PI76].

Proton-induced x-ray emission is an atomic process (shown schematically in figure 1.2) in which there is a high probability of an inner-shell electron being ejected from an atom by an incoming proton.

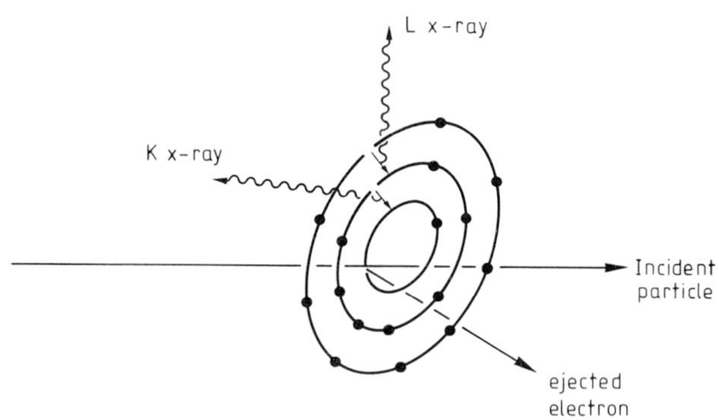

Figure 1.2 Schematic diagram showing the production of characteristic K and L x-rays by MeV ion collision.

The filling of the vacancy thus created by an electron from a higher shell causes an x-ray quantum to be emitted, the energy of which is characteristic of the atom under bombardment. The ability of a solid-state detector to resolve the energies of adjacent characteristic x-rays with high efficiency and to detect simultaneously a wide range of x-rays gives us information on which elements are present in the analysed sample and in what quantities.

As a brief example to illustrate the power and versatility of this technique, we have chosen the analysis of a sample of environmental dust collected in the City of Oxford over a period of three days. Figure 1.3 shows an x-ray energy spectrum of the dust, induced by the bombardment of 4MeV protons from a nuclear accelerator. As a guide to the speed and sensitivity of PIXE for such environmental samples, elemental lead levels of about 0.1 μg were obtained from this spectrum, which took only a few minutes to accumulate.

In laboratories where pollution studies are carried out routinely and require a fast sample

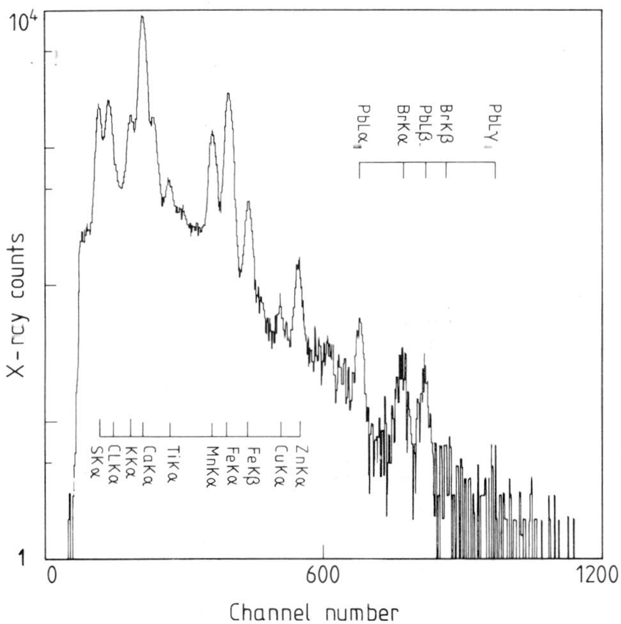

Figure 1.3 Proton induced x-ray energy spectrum from environmental dust. Proton energy = 4 MeV. Accumulated charge = 0.3 μC. Run times = 5 min.

turn-round time, the use of computer software [KA77, KA81, CL81] to extract absolute elemental quantities enables the composition of a sample to be obtained minutes after the end of a run.

The elemental constituents of a sample are identified through their characteristic x-ray energies and some of the prominent x-ray emission energies [JO70b] that are encountered in x-ray spectroscopy are listed in Appendix 2. From this table it is seen that the characteristic K_α x-ray energies corresponding to elements of low atomic number are small, and in practice the induced characteristic x-rays from the elements hydrogen to about sodium, (K_α x-ray energy = 1041 eV) are absorbed in the window of the detector assembly and any surface dead layer of the solid-state detector. Detection of low A elements is possible using a crystal spectrometer which has excellent energy resolution but which unfortunately has poor efficiency compared with that of solid-state detectors. Because it is a wavelength dispersive device, the analysis is not multi-elemental and a broad range of x-ray energies cannot be recorded simultaneously. From Appendix 2 it is also apparent that for elements above about Ga in the periodic table, elemental analysis can be achieved by observing the characteristic L x-rays (Ga L_α, x-ray energy = 1098 keV). This is useful (as will be seen below) in order to maintain high elemental x-ray yields throughout the higher-mass region of the periodic table.

The dependence of inner-shell ionisation cross sections on incident particle energy has been extensively studied both theoretically and experimentally, with comparisons between the available theories and experiments in the energy and Z regions of interest good to within 10–30%. Johansson *et al* [JO76] have plotted data from K and L x-ray ionisation measurements on thin targets and fitted to the results a fifth degree polynomial of the form

$$\ln(\sigma u_i^2) = \sum_{n=0}^{5} b_n x^n \quad (i = \text{K or L})$$

where

$$x = \ln(E/\lambda u_i) .$$

u_i is the ionisation energy in eV

E is the proton energy in eV

σ is the ionisation cross section in units of 10^{-14} cm^2

λ is the ratio of the proton mass to the electron mass

u_L was calculated from

$$u_L = 1/4 \ (u_{L_1} + u_{L_2} + 2\mu L_3)$$

and the b_n coefficients are given by

Shell	b_0	b_1	b_2
K	2.0471	−0.00659	−0.47448
L	3.6082	0.37123	−0.36971
Shell	b_3	b_4	b_5
K	0.09919	0.04606	0.00608
L	−0.00008	0.00251	0.00126

The relevant parameter for elemental analysis is the x-ray production cross section σ_p, related to σ by $\sigma_p = \sigma\omega k$ where ω is the fluorescent yield and k the relative intensity of possible transitions to fill an inner shell vacancy.

Figures 1.4 and 1.5 show plots of K_α and L x-ray production against proton energy for some common elements taken from values calculated by Johansson and Johansson [JO76] using the expressions above. Comparison of figures 1.4 and 1.5 show that due to the larger L x-ray cross sections for elements in the higher mass regions in the periodic table, observation of the characteristic L x-rays rather than the K x-rays for high-mass elements will result in more rapid analysis times.

It is seen from these graphs that, in general, the higher the proton energy the greater the x-ray production cross sections. The analytical sensitivity, however, is not only determined by the cross section but also by the x-ray background radiation, and the dominant part of this background radiation at low energies is the radiation caused by the deceleration of the secondary electrons ejected by incident particle impact (bremsstrahlung).

If E_e^m is the maximum energy transferred from an ion of mass M and energy E_p to a free electron of mass m then the bremsstrahlung background radiation decreases rapidly above the classical limit for free electron scattering given by

$$E_e^m = (4m/A_p M)\, E_p .$$

For example, E_e^m for a 2 MeV proton is 4.36 keV, corresponding approximately to the Sc K_α x-ray energy. By increasing the proton energy, therefore, although the x-ray production cross sections increase, so does the energy at which the bremsstrahlung background becomes significant.

The background caused by direct bremsstrahlung from the projectile is apparent only at higher radiation energies, being masked by the secondary electron bremsstrahlung up to the E_e^m cut-off energy. The cross section for this process is given by [JO76]:

$$\frac{d\sigma}{dE_x} = C\,\frac{AZ^2 Z_1^2}{EE_x}\left(\frac{Z}{A} - \frac{Z_1}{A_1}\right)$$

where Z, A, and E are the charge, mass and energy respectively of the incident particle, E_x is the x-ray energy and Z_1 and A_1 the charge and mass of the target atoms. Note that the yield, unlike that of the secondary electron bremsstrahlung, increases with decreasing particle energy.

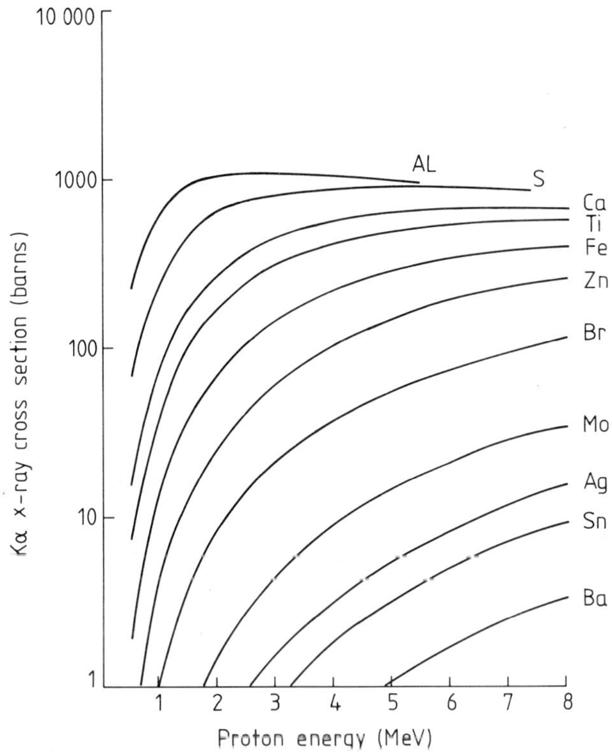

Figure 1.4 Plot of K_α x-ray production cross sections against incident proton energy for several common elements.

A third important source of background, which again is only noticeable at higher x-ray energies, is when the incident particles have a sufficiently high energy to produce nuclear reaction γ-radiation in the target. This nuclear background produces a high-energy tail in the spectrum due to Compton scattering in the detector.

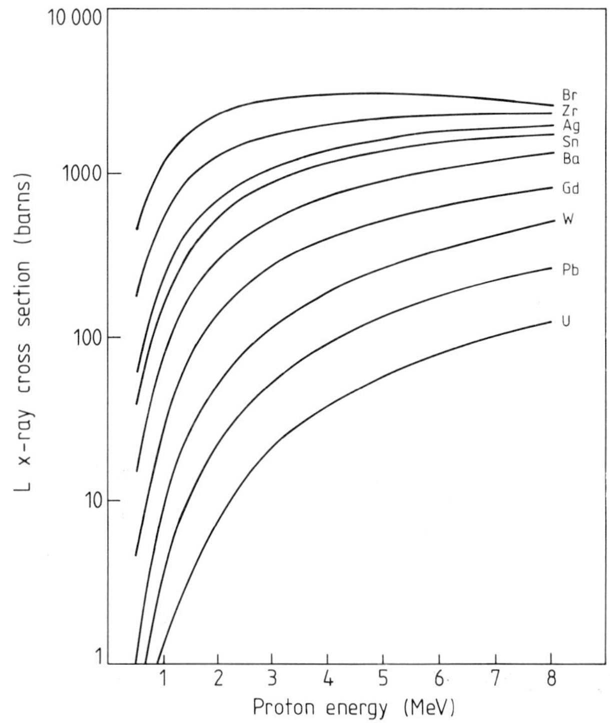

Figure 1.5 Plot of L x-ray production cross-sections against incident proton energy for several common elements.

The optimum incident particle energy for maximum sensitivity in PIXE is therefore balanced between higher incident energies for increased x-ray production cross sections and reduced incident energies to minimise background. In general, most PIXE work is carried out with proton energies between 2 and 6 MeV. The figures for optimum concentration sensitivities and absolute detection limits of an element in a sample are dependent on several factors. Sensitivity, and therefore absolute detection limits are degraded by:

(1) The target thickness: in general, the thicker the target the greater the secondary electron bremsstrahlung radiation.

(2) Low x-ray production cross sections: although for elements in the high Z range, the more prolific L x-rays can be utilised.
(3) Target constituents: multi-elemental targets can be difficult to analyse because of overlapping K_α, K_β and L x-ray lines.
(4) Electronic pile-up: spurious counts due to overlapping or coincident pulses at high count rates. This is a common occurrence for trace element analysis in medium to high Z matrices where the prolific x-rays from the matrix material cannot be filtered out. This problem, however, is not so great for elemental analysis of biological material, where the energies of the x-rays from the main bulk of the sample are below the threshold of detection.
(5) Detector resolution: currently available resolutions of 140 eV at 5.9 keV x-ray energy are adequate for separating K x-ray peaks down to the detection threshold. Poorer resolution due to electronic noise, large area silicon crystals, inferior pulse processing etc may result in overlapping peaks.
(6) Sample charging: internal or surface charging of a sample caused by beam bombardment may cause localised intense electric fields which can increase energy of any secondary electrons above the E_e^m cut-off thereby extending the range of the secondary electron bremsstrahlung background.
(7) Primary particle bremsstrahlung.
(8) Nuclear reaction Compton scattering.
(9) Associated with any intense x-ray peak from a major elemental constituent is a low-energy tail caused by incomplete charge collection within the detector and Compton scattering of the primary x-ray.

Folkmann [FO78] has made an attempt to calculate concentration sensitivities and absolute detection limits using, as a model, trace elements embedded in several different elemental matrices of thickness 1 mg cm^{-2}. Using x-ray production cross section data, together with calculated secondary electron and incident particle bremsstrahlung backgrounds, he concluded that the optimum concentration sensitivity is in the region 10^{-6}–10^{-7} within a Z range that can be moved with beam energy. The absolute detection limit is easily deduced by multiplying the concentration sensitivity by the matrix thickness (mass/area) and the beam area. For a typical concentration sensitivity of 10^{-6}, target thickness of 1–0.1 mg cm^{-2} and incident beam areas of 1 cm^2–1 mm^2, then trace elements in the range 10^{-9}–10^{-12} g are detectable. It is perhaps interesting to note at this stage that for a beam focused down to 1 μm while maintaining the 10^{-6} sensitivity and assuming no target damage, the detection limit is 10^{-18} g!

In summarising the analytical power of proton-induced x-ray emission we have:

(1) Fast.
(2) High concentration sensitivity for elements above $Z = 11$ (10^{-6}–10^{-7}).
(3) Low minimum detectable concentrations (routinely 10^{-9}–10^{-12} g).
(4) Multi-elemental: information on elemental composition of the sample is obtained simultaneously.
(5) Elements can be identified easily by their characteristic x-ray energies.
(6) Microbeams can be used. This not only decreases the minimum detectable concentration but as we shall see later, enables two-dimensional elemental mapping to be carried out.
(7) Essentially non-destructive (apart from possible damage in some applications, e.g. intense microprobe beam-mapping of biological material).

The comparison between proton-induced x-ray emission and other analytical techniques has been carried out recently by several groups. Deconninck *et al* [DE81] have investigated the trace element content of albumin using PIXE and compared the results with those obtained by atomic absorption spectroscopy (AAS), a standard medical laboratory technique, for which commercial instruments are available. The conclusions drawn from the comparisons were the same as we have experienced in Oxford in monitoring rubidium levels in blood cells and plasma using PIXE and AAS: AAS is subject to practical difficulties in which inaccuracies occur due to varying flame intensity, background instabilities etc. Nevertheless, under the control of an experienced operator, the technique is as good as, if not superior to, PIXE as an analytical technique in the detection of a single element only.

Several groups [BU81, CO73, GO73, GI73] have measured trace element concentrations in samples using both PIXE and x-ray fluorescence (XRF). The techniques are basically similar in that PIXE uses energetic ions to induce the characteristic elemental x-rays whereas XRF uses x-rays either from an x-ray tube or a radioactive source. A comparison between the sensitivities of these two multi-elemental techniques across the periodic table is difficult because of the differences in the x-ray background of the two processes. As discussed above, in PIXE the major source of background is of low-energy character, corresponding to the bremsstrahlung generated by secondary electrons. In XRF the problem is with the Compton scattering of the primary x-ray radiation which is essentially high energy in character [WO73]. In general, the concentration sensitivities in both processes seem to be similar, providing the energy of the excitation source is optimised. When considering minimum detection limits on an absolute scale, however, the superiority of PIXE is pronounced. When using tube-produced x-rays, and x-rays from a radioactive source, the incident photon flux is much smaller than that of accelerator-produced protons over a given area. XRF then requires thicker samples, which also introduce matrix effects (absorption of the emitted x-rays in the bulk of the material) into the technique, thereby complicating the analysis. In cases where the sample mass is small (e.g. pollution targets, biological tissue, semiconductor material etc) the lower detectable limits of PIXE are necessary, although there are many materials analysis problems for which the relatively less complex XRF is more than adequate.

Neutron activation analysis (NAA) is a powerful technique with sensitivities superior to PIXE for certain elements. The sample is bombarded with neutrons and the trace element identification is carried out by observing characteristic γ-decay of short-lived product radionuclides. Maenhaut *et al* [MA81a] have investigated the trace element content of serum, serum albumin and living tissues with both PIXE and NAA. They conclude that the two techniques are complementary, in that more trace elements could be determined using both PIXE and NAA than by using one technique alone.

The three techniques described above, AAS, XRF and NAA, have been compared to PIXE because they achieve more or less equivalent concentration sensitivities although they are inferior in the minimum detectable concentration limits. There are obviously many more analytical techniques available, most of which involve low-energy electrons, ions or photons as the incident or analysed radiation and are hence surface techniques, and it is outside the scope of this book to consider them all.

As we have seen, the ability to focus protons down to small spot sizes improves the minimum detectable concentration drastically and it is because of this that PIXE offers more potential for small sample analysis than AAS, XRF and NAA. This, however, is not the only advantage. The ability to focus a proton beam to a small spot and scan a sample while observing characteristic x-rays enables two-dimensional elemental mapping to be carried out.

The technique of electron-induced characteristic x-ray emission has this capability and for many years electron microscopists have utilised the detection of x-rays in conjunction with secondary electrons to enable mapping of elemental concentrations in samples. Unfortunately, the large unavoidable x-ray background caused by the bremsstrahlung of the primary electrons degrades the concentration sensitivity to only one part in 10^3–10^4, thereby allowing only the major elemental constituents to be detected. A comparison of x-ray spectra of a pollen tube taken using both 2.5 MeV protons and 10 keV electrons is given in work carried out by the Heidelberg group [BO80a] showing the differences in sensitivity of the two techniques to trace element analysis.

The proton probe therefore seems to offer a unique service, that of two-dimensional trace element mapping for elements with $Z>11$ at concentration levels of parts per million.

1.3 Particle-induced nuclear reactions

The use of nuclear reactions for elemental analysis is an expanding field and covers a broad range of reaction mechanisms. A nuclear reaction is usually written in the form A(B,C)D where $A+B\rightarrow C+D + Q$, with A the target nucleus, B the incident particle, C the reaction product and D the residual nucleus. Q is the energy released during the reaction as particle kinetic energy or γ-radiation, or absorbed from the incident particle energies. The elements under investigation are identifiable through the energy of the reaction products in the same way that an excited atom is identified through its characteristic x-ray radiation. Because the Q-values of nuclear reactions vary so dramatically even for elements adjacent in the periodic table [MA66] and the lowest excited states of the light nuclei are in general widely spaced and well defined [LE76a], the reaction products have distinctive energies and are therefore easily recognisable. An added advantage is that nuclear characteristics differ markedly between isotopes of the same element and so the reaction products are specific to a particular isotope of that element. Figure 1.6 shows a schematic diagram of the possible nuclear reactions that can take place and for an excellent review of the subject the reader is referred to an article by Wolicki [WO75].

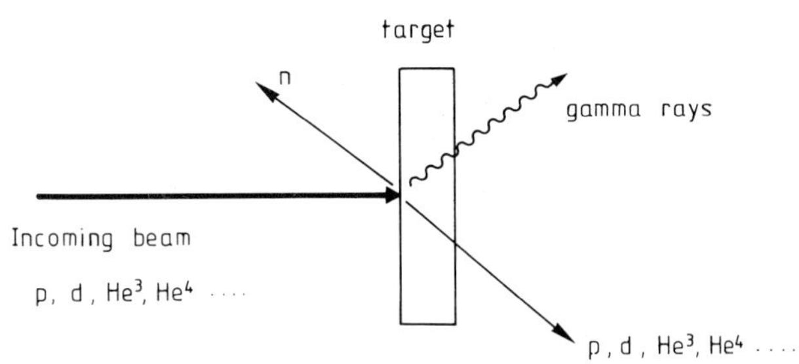

Figure 1.6 Schematic diagram of the production of nuclear reaction particles induced by an ion beam of MeV energy.

For particle-induced nuclear reaction cross sections to be high enough for useful elemental analysis, the incident particle must have sufficient energy to overcome the electrostatic potential energy which builds up as it approaches a target nucleus. The height of this Coulomb barrier for a particle of charge Z_1e and radius R_1 impinging on a target nucleus of charge Z_2e and radius R_2 is given by

$$E_C = \frac{Z_1 Z_2\, e^2}{(R_1 + R_2)4\pi\varepsilon_0}$$

where

$R_1 + R_2 = R$, the interaction radius
$= R_0\,(A_1^{1/3} + A_2^{1/3})$

where A_1, A_2 are the mass numbers of the nucleus and the incident particle. In general, R_0 lies between 1.0 and 1.4 fm. (An approximate but useful estimate of the Coulomb barrier height per incident nuclear charge is $E_C = ZA^{-1/3}$MeV.) For a given incident beam particle, the barrier height is greater as the atomic number of the target nucleus increases. In general, if the most readily available lightest particles (p,d, ^{3}He and ^{4}He) are used, then, at incident particle energies of about 4MeV, the Coulomb barrier will only be overcome for target atoms of atomic number up to about 20, and therefore at these comparatively economic energies it is only for these light elements that the nuclear reaction cross sections can be high.

It is interesting to point out at this stage the remarkable complementary relationship between analysis using nuclear reactions and PIXE. As we have already seen, in PIXE, elements below $Z = 11$ are in general not detected because their characteristic x-ray energies are so low that they are absorbed before detection. The PIXE technique, therefore, is most efficient in detecting heavy trace elements of $Z > 11$ embedded in a light matrix of $Z < 11$. In nuclear reaction analysis, however, the cross sections are small for elements of $Z > 20$, due to the effects of the nuclear Coulomb potential, and so this analytical technique is most sensitive for light elements embedded in a heavy matrix.

Analysis using nuclear reactions can be divided into two distinct types; charged particle activation analysis (CPAA) and prompt radiation analysis (PRA).

In charged particle activation analysis, the residual nucleus is left in a state of radioactivity and the decay products of the excited nucleus are measured some time after the reaction. The identifying characteristics for the radioactive nucleus can be the half-life for decay, the types of radiation emitted and the characteristic γ-rays emitted from the daughter nucleus. The concentration sensitivities for this technique can be better than one part per million for many light elements [TI66, MA62, DE71, AA72]. For obvious reasons, it is not a technique where elemental mapping can be carried out, and therefore the interest in using the method in conjunction with a focused beam is limited. The reader is therefore referred to a comprehensive bibliography [LU71] on activation analysis if he wishes to pursue the subject further.

Prompt radiation analysis (PRA) is a technique in which an element in a sample is detected through the nuclear radiations emitted instantaneously from nuclear reactions induced by the incident particle beam. In many respects, the experiments resemble those of early nuclear structure research although the objectives are clearly different.

Nuclear reactions can be broadly categorised into three basic mechanisms. The first of these is termed a capture reaction in which the incident particle is captured by the target nucleus to form a compound nucleus which subsequently decays by the emission of one or more γ-rays, e.g. ^{20}Ne $(\alpha, \gamma)^{24}$Mg. The second is the inelastic scattering of the incoming particle in which some of the kinetic energy of the particle is absorbed by the target nucleus, which subsequently decays by γ-emission, e.g. ^{12}C(p,pγ)^{12}C. The third category is when the particles which separate after the reaction are not the same as the incident particle and target nucleus, e.g. ^{19}F(p,α)^{16}O.

The great variety of ways in which information about a target sample can be obtained through nuclear reactions is indicated by the vast quantity of research carried out into nuclear properties and it would be difficult in any discussion to list all the available reaction mechanisms useful to the experimenter, especially since the full range of useful reactions has yet to be defined and the compilation of relevant cross section data is incomplete. Much of the data required is available from the nuclear physics literature, especially from nuclear structure and reaction compilations of light nuclei [AJ79, AJ80, AJ76, AJ71, AJ78, EN78]. An excellent guide to the most widely used reactions in PRA together with relevant cross section data has been attempted [BI74, BI78, BI80], although the authors state that the compilations do not constitute an exhaustive survey of all the work in such a complex and varied field and therefore must only be taken as illustrative of current trends. The reader is also directed towards a comprehensive three-volume compilation by R A Jarjis and J M Calvert [JA79b].

Considered below is an example which illustrates the diversity and sensitivity of the PRA technique. At Oxford in 1976 an experiment was undertaken to investigate the level structure of ^{16}O through the α-capture reaction ^{12}C(α,γ) ^{16}O for alpha bombarding energies between 5.22 MeV and 6.9 MeV [HU80]. Unfortunately, however, when naturally occurring carbon is used (containing 1.1% abundance of ^{13}C isotope), the prolific contaminant reaction ^{13}C(α,nγ) ^{16}O causes a high γ-ray background up to $E\gamma = 6.13$ MeV. In order to eliminate all possible sources of the ^{13}C contaminant reaction, the Oxford cryogenically pumped gas target was used [WA78] with isotopically enriched carbon-12 methane as the target gas. Prior to the nuclear structure experiment, the ^{13}C content of the enriched carbon-12 methane was measured using the same reaction as that responsible for the contaminant γ background, i.e. ^{13}C(α,nγ) ^{16}O. The cross section for this reaction has been measured previously [SP63] and is shown schematically in figure 1.7 for α-particle energies between 5 and 7 MeV. With an incident α-beam energy of 6.107 MeV, corresponding to the energy at which the reaction cross section (and therefore analytical sensitivity) is at a maximum, the 6.13 MeV γ-rays characteristic of this contaminant reaction were searched for using an intrinsic germanium detector. After a total of 0.186 Coulombs of doubly charged helium had passed through the gas, no 6.13 MeV

Table 1.1 Microbeam analysis using nuclear reactions.

Element	Reaction	Beam energy (MeV)	Sensitivity (ppm)	Positional resolution (μm)
^{1}H	(^{11}B,α)2α	2.0	1000	90 × 30
^{2}H	(^{3}He, p)	0.75	40	5 × 30
^{7}Li	(p,α)α	2.8	100	10 × 30
^{9}Be	(α,n$_0$)	2.5	200	13
	(d,p$_0$)	13, 0.85	200	13
^{10}B	(p,α_0)	2.67	<2000	16
^{11}B	(p,α_0)	13	<1000	~10
^{10}B	(d,p$_0$)		<1000	
^{12}C	(d,p$_0$)	1.1–1.3	10	<5
	(d,p$_0$)	1.4	100	5 × 30
^{14}N	(d,p$_0$)+(d,α_0)	1.9	50	4–20
	(d,p$_0$)	2.0	1000	150
	(d,α_0)	1.4	100	5 × 30
^{16}O	(d,pγ)	2.5	1000	5.5 × 15
	(d,pγ)	1.1		5–20
	(d,pγ)	1.1		
	(d,p$_1$)	1.1	100	
	(d,p$_0$)	1.6	200	>12 × 200
^{18}O	(p,α_0)	0.87	<60	<250
	(p,α_0)	0.87		~60
	(p,α_0)	0.84	100	5 × 30
^{19}F	(p,α_0)	1.5	10	10
^{24}Mg	(d,p)	1.8	4000	20
^{27}Al	(p,p$_1\gamma$)	1.8		>660
^{28}Si	(d,p$_0$)	1.9	300	~5

A subscript $_{0,1}$ etc indicates the state of the final nucleus: ground, first excited, etc.

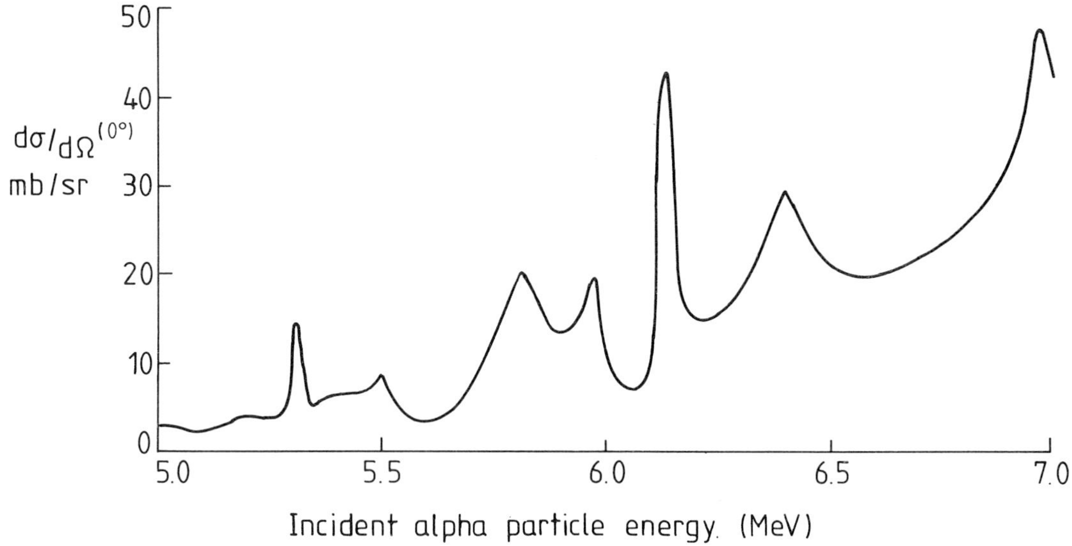

Figure 1.7 Schematic representation of the excitation function for the production of 6.13 MeV γ-rays from the reaction $^{13}C(\alpha,n\gamma)^{16}O$.

γ-rays from the gas had been observed, imposing a limit of less than 6 parts per million on the ^{13}C abundance in the isotopically enriched methane.

In prompt radiation analysis (PRA), by scanning a well-focused nuclear particle beam over a solid target and measuring instantaneously the energy of the nuclear reaction products, an elemental map of the sample can be obtained. It is a technique complementary to scanning microprobe PIXE in that the light element profiles are easiest to obtain. Table 1.1 shows a summary of the nuclear reactions that have been used in microprobe PRA taken from a review article by Cookson [CO79]. The reactions mainly involve charged particle detection methods, since the high efficiency obtainable with silicon detectors usually permits a more rapid analysis with microbeams than would γ or neutron detectors.

In general, materials analysis has been carried out at low beam energies, mainly due to the availability of low-energy accelerators. At these low incident beam energies, nuclear reaction cross sections may exhibit resonance behaviour and change rapidly with beam energy (for example see figure 1.7). The incident particle beam loses energy rapidly as it traverses a thick target, and therefore the analytical sensitivity may change drastically with target depth. In principle, for nuclear reactions which exhibit sharp well-defined resonance structures in cross section, a depth distribution for a thick target could be obtained by deconvolution techniques, although for particle reactions the interpretation of the results could be complex, due to the energy straggling of both the incident beam and the outgoing reaction products. In capture reactions, e.g. (p, γ), (^{4}He, γ), the situation is somewhat easier, since the resultant reaction γ-ray will, in general, have little self-absorption in the target. Successively deeper layers of the sample, therefore, may be probed by increasing the energy of the beam, thereby increasing the depth at which the incident particle has attained the optimum energy to induce the characteristic nuclear reaction products.

A focused nuclear particle beam has, therefore, the capability of three-dimensional single-element mapping, although the technique is complicated by the fact that any beam energy variation requires readjustment of the probe-forming elements to maintain spatial resolution. In general, the cross section for nuclear reactions is much lower (<100 mb sr^{-1}) than that of PIXE (see figures 1.4, 1.5), thereby limiting the analytical sensitivity in many cases.

1.4 Elastic or Rutherford backscattering (RBS)

For a nuclear particle beam incident on a sample, the cross section for a nuclear reaction to occur is large only when the beam energy is sufficient to overcome the electrostatic interaction between the incident particle and the target nucleus. When the incident particle energy is much smaller than that required to overcome the Coulomb barrier, then nuclear reactions become improbable and the primary mechanism is that of elastic scattering based on the conservation of momentum between the incoming particle and the target nucleus. Materials analysis based on backscattering spectrometry has been known to be a powerful technique for several years and several excellent review articles on the subject have been published [CH77, MA77, ME76, ZI75, MA75a, WE75].

The physics of the technique is relatively straightforward. If a light incident particle collides with a heavy target nucleus, then the target nucleus absorbs only a small amount of energy, consistent with the conservation of momentum, and the backscattered particle will retain most of its initial energy. If the incident particle backscatters from a very light target nucleus then the target nucleus will absorb a good proportion of the collision energy and the projectile will be left with only a fraction of its initial energy. Since the energy of the scattered-particle can be measured very accurately using surface barrier particle detectors, the mass of the target atom can be deduced from the energy of the backscattered projectile.

The incident ion beam energy is in general chosen to be high enough to allow maximum penetration of the beam, but with an energy sufficiently low so that the scattering is not complicated by nuclear resonance effects. The incident ion beam energy is typically 200–400 keV for $^1H^+$ or 1–3 MeV for $^4He^+$.

For elastic scattering, the ratio of the energy of the scattered particle E_1 to the incident energy E_0 is given in the non-relativistic limit by

$$\frac{E_1}{E_0} = \left(\frac{M_1 \cos\theta + (M_2^2 - M_1^2 \sin^2\theta)^{1/2}}{M_1 + M_2}\right)^2 \qquad (1.1)$$

where θ is the scattering angle in the laboratory frame of reference and M_1, M_2 are the masses of the incident particle and the target nucleus respectively. The ratio $K = E_1/E_0$ is commonly termed the kinematic factor and is the ratio of the particle energies before and after collisions. Target mass identification is then carried out by knowing E_0, m and θ and measuring E_1.

The elastic scattering process is a classical problem and was explained as early as 1911 by the Rutherford atomic model [RU11]. The laboratory frame differential cross section for the Rutherford scattering of particles of mass M_1, charge Z_1e and incident energy E(MeV) by a target consisting of particles of mass M_2 and charge Z_2e is given as

$$\left(\frac{d\sigma}{d\Omega}\right)_{\text{lab}} = 1.296\left(\frac{Z_1Z_2}{E}\right)^2\left[\left(\operatorname{cosec}^4(\theta/2) - 2\left(\frac{M_1}{M_2}\right)^2 + \ldots\right)\right] \text{mb sr}^{-1}.$$

Figure 1.8 shows the differential cross section for 2 MeV α-particles scattered at various angles and

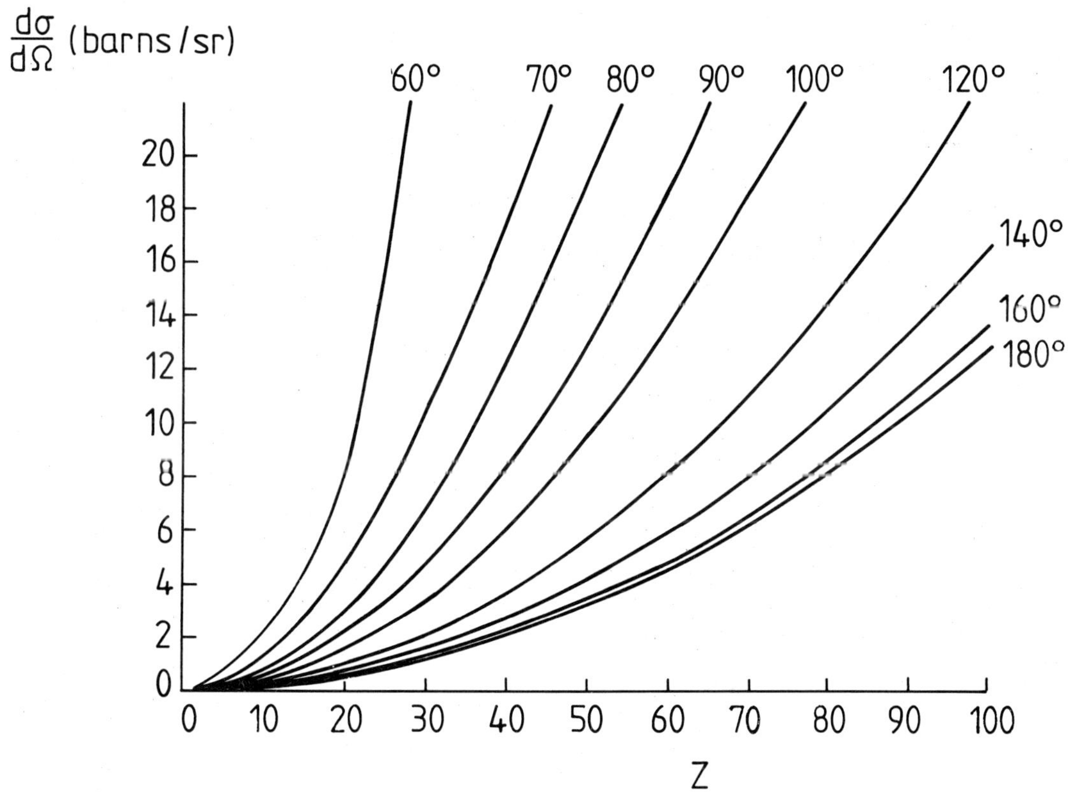

Figure 1.8 A plot of the differential cross section for elastic scattering at various angles of 2 MeV α-particles.

indicates the efficiency of the technique for high Z targets due to the Z^2 dependence of the cross section.

Since the Coulomb scattering cross section can be calculated exactly, a quantitative analysis is possible without a comparative standard. There can be deviations, however, from the classical Rutherford scattering formula in cases where the incident particle has a sufficiently high energy and impinges on a light target nucleus, thereby resulting in nuclear resonance interference, or when the incident beam is of low energy, thereby introducing complications caused by particle–electron interactions.

From the above equations some practical considerations arise which are worth commenting upon. Figure 1.9 shows the energy loss of incident ions scattered from a range of target atoms at various angles. The diagram shows that in order to separate elements of similar mass, the backscattered particles should be detected at an angle as near to 180° as possible, which unfortunately, from figure 1.8, is the angle for which the scattering cross section has the lowest value. Furthermore, figure 1.9 shows that the mass resolution is best for low-Z materials whereas unfortunately, because of the Z^2 dependence of the cross section, the analytical sensitivity at low Z is relatively poor.

Heavy-ion probes offer great potential since the mass separation between elements goes approximately as $2M_1/(M_1 + M_2)^2$ and the scattering cross section is proportional to Z_1^2. Unfortunately, poor detector resolution for heavy-ion probes (such as ^{12}C) nullifies the effect of the recoil energy separation and although magnetic analysis of the backscattered particles can be used, the complexity of the experimental technique increases and the collection efficiency is degraded.

The ideal probe seems to be, therefore, a ^4He beam at energies between 1–3 MeV, especially when one considers that the ^4He nucleus has nuclear properties analogous to a closed electron shell and is therefore relatively inert to nuclear resonance interference. From figure 1.8, differential cross sections for elastic scattering of α-particles at backward angles can be in the 1–10 b sr^{-1} range for middle-Z elements and therefore smaller than the alternative PIXE cross sections. For the light elements ($Z < 12$) the differential cross sections at backward angles are in the 1–100 mb sr^{-1} range and are in general smaller than the equivalent specific nuclear reaction cross sections.

The major advantage of material analysis using Rutherford backscattering over PRA and PIXE is that depth profiling can be carried out in a fast, simple and direct manner. When a projectile penetrates a target it loses energy due to collisions with atomic electrons and target nuclei. When the projectile encounters a hard collision (nuclear backscattering), it can change its trajectory into an outward direction and during its path back out of the target it will again lose energy to the target atoms. By measuring how much energy loss the particle has suffered, one can then estimate the depth to which the incoming projectile has penetrated. The energy loss of the projectile per unit path length depends on the energy of the particle, the nature of the projectile and the constituents of the target. Several excellent recent review articles [ZI80, SI79] and energy loss compilations [ZI79, AN77, ZI77, NO70] are available for information.

A very clear and precise account of the theory behind the extraction of depth profiles from backscattered particle energy spectra has been published [WE75] and so only a brief physical description is given here. Consider a schematic, idealised, backscattered energy spectrum (figure 1.10) taken from a mixture of element A and B laid thinly down on a substrate element C with the condition that $M_A > M_B > M_C$. It is clear from figure 1.10 that elements A and B lie at the surface of the sample since by using the kinematic equation (1.1) the leading edges

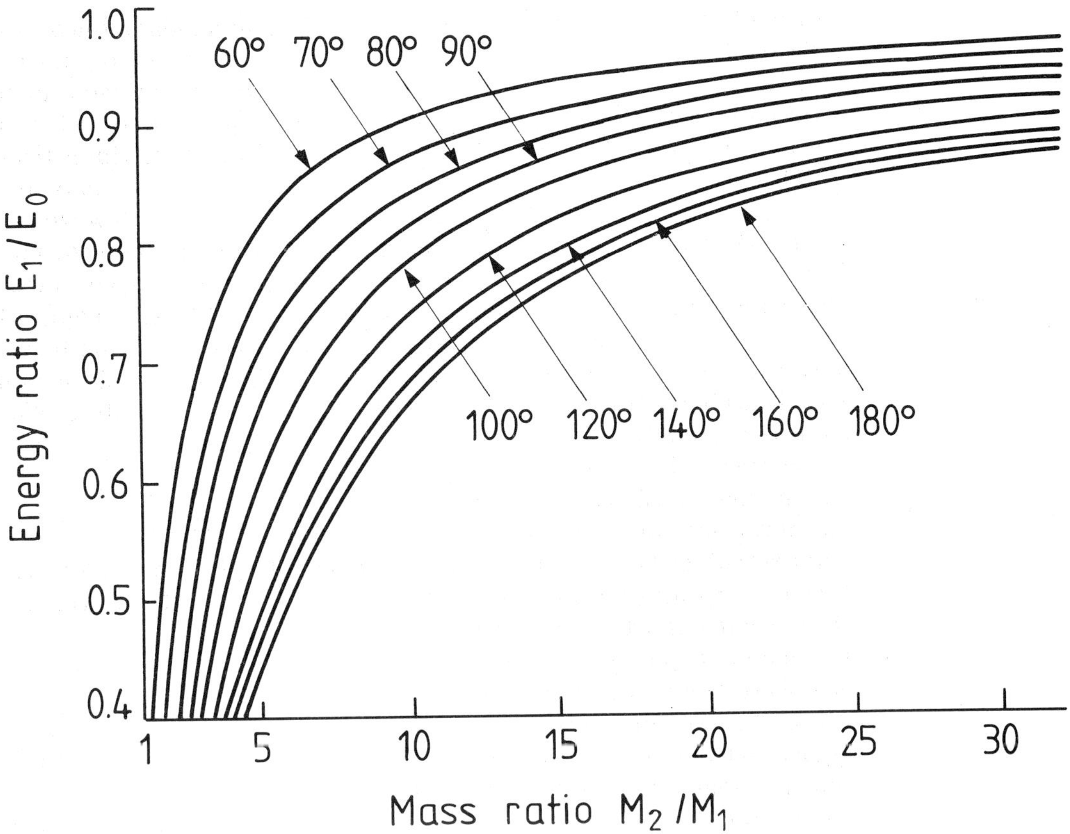

Figure 1.9 The energy loss of incident ions elastically scattered from a range of target atoms at various angles. E_1 and E_0 are the initial and final energies of the incident ion. M_1 and M_2 are the masses of the incident particle and target atom.

of the A and B peaks correspond to the energy calculated for the backscattered particles. However, because the energy of the backscattered particles from C lies lower than the energy expected from kinematic considerations alone, the element C clearly lies at a depth which can be calculated from the energy loss ΔE_C. Also, because the peaks in the energy spectrum due to backscattering from A and B are of similar width and are flat-topped, it can be deduced that elements A and B are uniformly distributed in the same surface layer. Following the notation in [WE75], the ratio of the elements A and B can be calculated from the equation

$$\frac{m}{n} = \frac{N_A}{N_B} = \frac{H_A}{H_B}\,\frac{\sigma_B}{\sigma_A}\,\frac{\varepsilon_A^{A_m B_n}}{\varepsilon_B^{A_m B_n}}$$

where H_A/H_B is the ratio of the heights of the backscattered particles from elements A and B, σ_B/σ_A is the ratio of the scattering cross sections for B and A and

$$\varepsilon_A^{A_m B_n} = K_A\,\varepsilon^{A_m B_n}(E_0) + \frac{1}{|\cos\theta|}\,\varepsilon^{A_m B_n}(K_A E_0)$$

$$\varepsilon_B^{A_m B_n} = K_B\,\varepsilon^{A_m B_n}(E_0) + \frac{1}{|\cos\theta|}\,\varepsilon^{A_m B_n}(K_B E_0)$$

where $\varepsilon^{A_m B_n}$ is the stopping cross section of the mixture A and B. From Bragg's rule, this can be written

$$\varepsilon^{A_m B_n} = m\varepsilon^A + n\varepsilon^B.$$

ε^A and ε^B are the stopping cross sections of A and B respectively, the values of which can be extracted from the compilations listed above.

The example considered above obviously represents a very favourable situation. In more unfavourable circumstances the peak heights of the backscattered particles will not be flat-topped and may overlap appreciably, thereby making the extraction of relevant parameters from the energy spectrum very complex.

There are obviously very many applications of backscattering spectrometry, especially in the determination of composition variation or impurity distribution with depth. Applications include (1) thin film studies, (2) ion implantation profiling, (3) surface contamination, (4) surface diffusion studies, etc. The technique is ideally suited for the detection of medium to heavy elements in a light matrix where depth resolutions of 20 nm in a surface layer of 1 μm thickness is possible without recourse to sample erosion or layer removal.

Very little work appears to have been carried out in backscattering spectrometry utilising focused microbeams, which is surprising considering the potential of the technique. A focused scanning microbeam can enable three-dimensional, non-destructive elemental mapping of the surface layer of a sample to be carried out.

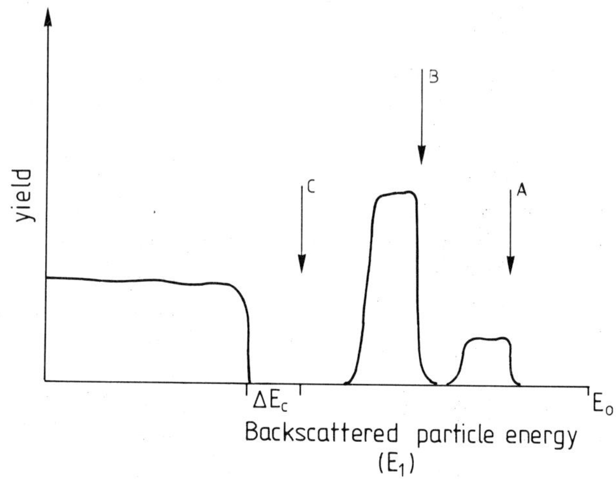

Figure 1.10 Schematic energy spectrum of a particle beam backscattered from a mixture of elements A and B laid down on a substrate C. E_0 is the energy of the incident ion energy. ΔE_c represents the energy loss suffered by the incident particles in traversing the A+B mixture.

1.5 High resolution scanning ion microscopy

For several decades, electron probes have been used almost universally for microscopy at the sub-micron level and such is the effort that has been put into electron microscopy over this period that resolutions of atomic proportions have been reached. However, the useful resolution limit for a probe may not necessarily correspond to the technical instrumental resolution, since the increase in specimen damage becomes a problem for small probe diameters. Consideration, therefore, of the dose minimisation, especially in the case of fragile organic material, becomes important when assessing the merits of different types of high resolution probes, and at present it seems that the scanning transmission electron microscope (STEM) is the instrument which seems to offer the minimum damage at the highest resolution [LA76].

The possibility and advantages of using ions rather than electrons in transmission microscopy has been investigated by several groups [GR78a, CH54] and most recently by the work of Levi-Setti [LE74] at the University of Chicago. The Chicago group have developed a first-generation scanning transmission ion microscope (STIM) based on a target-imaging low-energy spectrometer to separate the neutral from the charged component of the transmitted beam. With the development of high-performance field-ion sources, a 55 keV STIM with spatial resolutions at the 1 nm level with sufficient transmitted H^+ ion current for adequate imaging is envisaged [LE80a]. Comparisons between transmission microscopy using electrons and protons [LE80a] have shown that the STIM compares very favourably with the STEM, especially when considering information retrieval, radiation damage and interaction localisation.

At present, the Chicago STIM operates at the low energy of about 50 keV and focusing is achieved using relatively weak cylindrical lenses, as in conventional electron microscopy. Only if higher energy proton beams are used in transmission microscopy will stronger focusing lenses (quadrupoles) be required to focus the beam down to the required spatial dimensions. For further information, the reader is directed towards a recent article by R Levi-Setti [LE80c].

A second type of ion microscopy concerns the 'destructive' methods of microscopy or microanalysis and is based on the detection of secondary ions sputtered by the focused ion probe from the surface of the sample [CA62, LI67]. The development of the electrohydrodynamic (EHD) liquid metal ion source with parameters relevant to this type of microscopy (i.e. brightness, source size, sputtering characteristics and energy spread), outclassing any other ion source [SW79], has provided incentive for further development of a second generation high-resolution scanning ion-probe microanalyser (SIPM), [SE79] and [LE80c]. In this case, due to the high rigidity of the heavy-ion beam even at intermediate energies, it may well be that strong focusing could play an important part in the development of such a system.

Similarly, the advent of the heavy metal ion source has sparked off interest in micro ion implantation for use in the semiconductor industry. Since the required ion energy is in the range 50 keV–5 MeV to achieve the desired depths for the wide range of ion masses in all target materials, quadrupole lenses could well provide the strong focusing action required for such microbeam production. For a recent relevant discussion on ion implantation technology, the reader is asked to consult an excellent article by R G Wilson [WI80].

1.6 Beam energy requirements for PIXE, PRA and RBS

In the previous sections the three major techniques utilising high-energy ion beams, proton-induced x-ray emission (PIXE), Rutherford backscattering (RBS) and prompt analysis using nuclear reactions (PRA) were introduced as being important analytical techniques, especially when ion probes are used to enable two- or three-dimensional elemental mapping to be carried out.

In an ion probe, the type of focusing employed in the probe formation depends on the energy of the beam, and although the optimal beam energy requirements for PIXE, PRA and RBS have been discussed earlier, it is nevertheless worth summarising the conclusions.

In proton-induced x-ray emission, the optimum energy for maximum sensitivity occurs when the incident particle velocity is sufficiently well matched to the velocity of the inner core electron of the atom under investigation. Figures 1.4 and 1.5 indicate that the x-ray production cross section drops off dramatically for most of the more common elements for beam energies below 2 MeV. Depending on the element under investigation, for beam energies above 2 MeV the sensitivity for analysis is reduced, due to the increased background caused by secondary electron bremsstrahlung and scattered nuclear reaction γ-radiation. In general, the optimum energy for PIXE is taken to be about 2 MeV, varying up to about 6 MeV for the detection of middle-Z elements.

For analysis using nuclear reactions, the optimum energy for maximum sensitivity is dependent on the particular nuclear reaction used in the analysis. In the detection of light elements using nuclear reactions, a necessary criterion is that the incident particle has sufficient energy to overcome the nuclear Coulomb repulsion between the projectile and the target nucleus. In general, the nuclear reaction cross sections are adequate for analytical purposes only for incident beam energies of approximately 1 MeV and above.

When the incident particle has an energy lower than that required to cross the Coulomb barrier, then elastic collisions take place between the projectile and the target nucleus (RBS). Depth profiling can be achieved by comparing the energy of the backscattered particles to that of the incident beam. The ion-beam energy used is usually high enough to achieve good penetration of the sample, but with an energy sufficiently low so that the scattering is not complicated by nuclear resonance interferences. In general, RBS is carried out with a $^4He^+$ beam at energies between 1 and 4 MeV.

The beam requirements, therefore, for the three techniques are remarkably similar, and require a small accelerator such as a Van de Graaff electrostatic generator capable of delivering a beam of light ions of energy between 2 and 6 MeV. The rapidly growing interest in PIXE, PRA and RBS is due partly to the availability of such small accelerators; most of the accelerators used in early nuclear structure research have outlived their usefulness at low energies and have been replaced by higher energy machines. Even so, the interest in such applied techniques has been such that at least one company is manufacturing small accelerators for this very purpose†. For a detailed description of low-energy accelerators, the reader is referred to the standard work of Livingston and Blewett [LI62].

The nature of the incident beam depends upon which analytical technique is being used, although it is possible that the three techniques could be used simultaneously. It is well established experimentally that in particle-induced x-ray emission the use of heavy ions tends to limit the analytical

† General Ionex Corporation, Newburyport, Massachusetts, USA.

sensitivity due to an increase in background and hence most laboratories specialising in PIXE use protons as the incident beam. In analysis by nuclear reactions, it is the light ions $^1H^+$, $^2D^+$, $^3T^+$, $^3He^+$. $^4He^+$ . . . which are used most frequently with the choice of projectile depending on the light element or isotope under investigation. The relatively inert ^4He nucleus seems to be the ideal projectile in RBS, since the incident energy, and therefore the penetration depth, can be increased significantly without gross nuclear resonance interference.

The success of the techniques depends, therefore, on the production of hydrogen, helium and their associated isotopic ions, their subsequent acceleration to energies of approximately 2 MeV, and their transportation to the experimental area.

1.7 Ion source and related beam-optical parameters

The efficient injection of a beam of particles from an ion source into an accelerator and transportation of the accelerated beam to the experimental region requires the knowledge of several beam-optical parameters.

A particle is completely specified if its three position coordinates x, y, z and its three momentum coordinates p_x, p_y, p_z are known. This information can be represented by the position of a point in a six-dimensional space, called the phase space, with coordinates x, y, z, p_x, p_y, p_z. A beam of particles can therefore be represented by a group of points in a six-dimensional phase space, the density of which according to Liouville's theorem (see for example [LE80b]) is invariant as a function of time. If the three components of the motion are mutually independent in real space, which is the case in most beam transport devices, then Liouville's theorem simplifies to the statement that the beam of particles can be represented by a group of points in three planes (x, p_x), (y, p_y), and (z, p_z), the respective areas of which remain invariant.

In general, for transportation of high-energy beams in the absence of acceleration and beam degraders, the axial momentum p_z can be assumed to be constant and the transverse momenta p_x and p_y can be replaced by the angular divergence parameters

$$x' = p_x/p_z \quad \text{and} \quad y' = p_y/p_z.$$

The transverse motion of the beam can therefore be represented by the respective areas in the phase plane diagrams (x, x') and (y, y') and although the shapes may change in time, the areas remain constant.

As a simple example, consider a beam of particles represented by an ellipse [BA66] in the phase plane (x, x') (figure 1.11 (a)). If there are no external forces on the beam, i.e. the beam is in a drift space region, then after a time t the beam will be represented by the limiting ellipse as shown in figure 1.11 (b). Every point of the ellipse has been displaced parallel to the x-axis by an amount proportional to the initial divergence of the beam such that the ellipse has been sheared but its area conserved.

The beam can be characterised in one plane by the area A it occupies in phase space or by the emittance E defined by

$$E = A/\pi.$$

The inclusion of the factor π in the definition of emittance dates back from early work on narrow beams for which elliptical phase space contours are generally to be expected.

Analogously, the acceptance of a device can be characterised in one plane by the phase space area determined by its limiting geometry and is defined in a similar way.

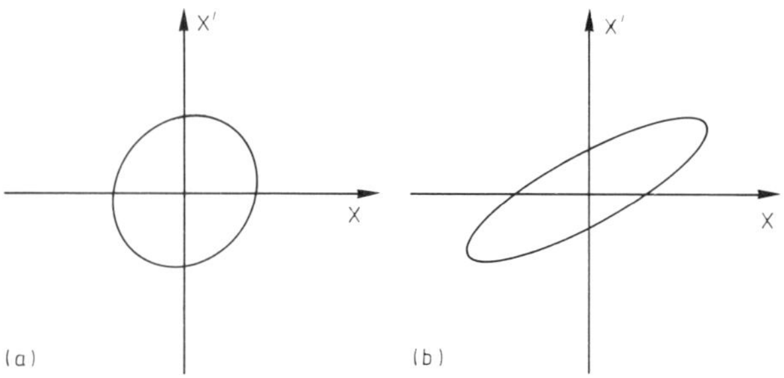

Figure 1.11 (a) A beam of particles represented by an ellipse in the phase plane (x,x'). (b) The phase plane representation of the particles after drifting through a field-free drift space after time t.

The transformation of the phase space ellipse in going through beam transport devices (e.g. lenses) indicates whether or not suitable conditions can be achieved for beam matching, i.e. the efficient transmission of the beam from the ion source to the experimental region. If, therefore, the acceptance of any device is greater or equal to the emittance of the beam presented to it, then the beam will achieve maximum transmission provided the emittance ellipse is adjusted to fit within the acceptance area. For matching an ion source to the accelerator and transport system, however, a parameter termed the normalised emittance needs to be introduced. This is necessary because the emittance is invariant only when the axial momentum p_z is constant. In an accelerating field, however, this is not the case and it becomes necessary to multiply the emittance by a factor proportional to the axial momentum to maintain invariance. There are two ways in common usage of expressing this new parameter, the normalised emittance:

$$E_n = E\, p/m_0c = E\beta\gamma \text{ (mm mrad)}$$

and

$$E_n = E(\text{KE})^{1/2} \qquad \text{(mm mrad (MeV}^{1/2}\text{)}$$

where $\beta = v/c$ and $\gamma = (1 - \beta^2)^{-1/2} = M/M_0$ and KE is the kinetic energy of the beam in units of MeV. E_n, the normalised emittance, is then a parameter useful in determining the type of source and transport mechanism for efficient matching of the beam to the experimental apparatus.

A parameter which is often used to specify the optical qualities of an ion source is the brightness, defined as $B = I/\Omega S$ where I is the total beam current, Ω the solid angle containing the beam particles emerging from a point on the beam axis and S the area of this source. The brightness is the current divided by the volume of the four-dimensional transverse phase space region occupied by the beam and is related to the emittance E by the relationship

$$B = 2I/\pi^2E^2.$$

For the comparison of ion sources, which may produce beams of different energies, the normalised brightness is more useful and is related to the normalised emittance by the relationship

$$B_n = 2I/\pi^2E_n^2.$$

For a more thorough account of the above discussions, the reader is referred to the excellent book by Banford [BA66] and articles by Septier [SE67] and Regenstreif [RE67]. For a more comprehensive and recent discussion of phase space, brightness and emittance the reader should also consult [LE80b].

There are many types of ion sources available for the production of positive and negative light ions suitable for work in PIXE, PRA and RBS. A recent review by Sidenius [SI78] briefly describes existing ion sources with their respective principles of operation and presents a useful survey of previous ion-source review articles. The most frequently used ion sources in the production of light ions are the radio frequency discharge and duoplasmatron sources.

The RF source is a very reliable and simply constructed instrument used for the production of ions from gases [BL61]. In this device a radio frequency field excites a secondary electron avalanche in a low-pressure gas contained in a glass or quartz container, and the positive ions produced are extracted by biasing the outlet relative to the plasma.

The duoplasmatron source is a very efficient, high-brightness source and is used for the production of hydrogen and helium ions [BR65]. A plasma is created by a discharge of primary electrons and is compressed by using a strong magnetic field to increase the ion density. Because the plasma density is high, normal electrostatic extraction is inefficient and so the ions are allowed to drift through a small outlet channel in the anode before being extracted from the plasma. Early work at the Los Alamos Scientific Laboratory [LA65] showed that surrounding the plasma core was a diffuse sheath containing negative ions, and by offsetting the anode channel, large currents (~100 μA) of negative hydrogen ions could be delivered without appreciable amounts of accompanying electrons. The off-axis direct-extraction duoplasmatron therefore allows tandem Van de Graaff accelerators (which require the preliminary injection of negative ions) to be used in the production of beams suitable for PIXE, RBS and PRA.

The question of whether these existing sources are able to provide a suitably intense beam for ion microprobe spectrometry is a very relevant issue and is worth further discussion. We shall take as an example the production of proton beams for the Oxford microprobe [WA82] using an off-axis direct-extraction duoplasmatron in conjunction with the EN tandem Van de Graaff accelerator of the Nuclear Physics Laboratory, University of Oxford.

Measurements of the emittance of this source [GR81a] are shown pictorially in a phase plane plot (figure 1.12). The use of current contours illustrates that most of the current is concentrated in the central portion of the beam and has relatively high brightness (or low emittance). The brightness of the source, estimated by measuring the beam current of divergence 1.4 mrad passing through a square collimator with dimensions 0.5 mm × 0.5 mm, yielded a result of $10^9\,\mathrm{A m^{-2} rad^{-2}}$ for the normalised brightness B_n. This figure represents the normalised brightness of the source measured within the 10nA contour (figure 1.12) and is consistent with quoted figures for brightness measurements in RF sources and duoplasmatrons of between 10^9 and $5 \times 10^9\,\mathrm{Am^{-2}\,rad^{-2}}$ [SE67].

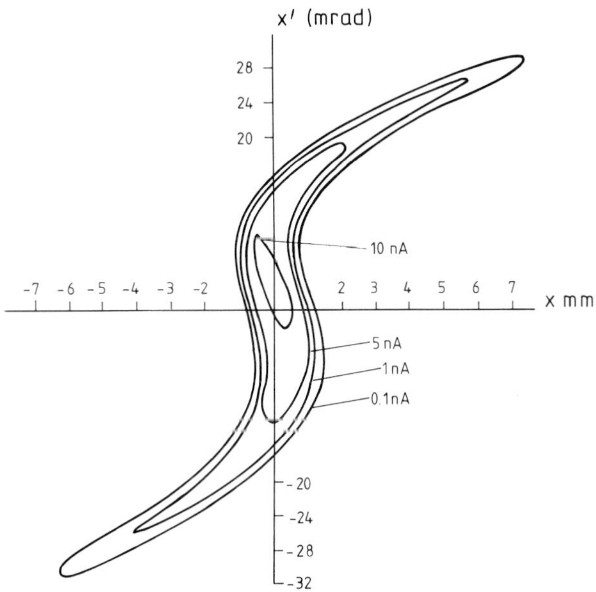

Figure 1.12 Phase plot of an off-axis direct-extraction duoplasmatron.

To calculate the amount of beam available for delivery to the Oxford microprobe, knowledge of the acceptance in phase space of the microbeam apparatus is required. If we assume for the moment that the probe entrance aperture is 25 μm × 25 μm, and the divergence of the beam in the x and y plane allowed through the aperture is limited to 0.3 mrad, to minimise aberration effects, then the four-dimensional normalised acceptance volume of the microbeam apparatus V_A can be calculated to be approximately 5×10^{-17} m^2 rad^2. The useful current available from the source is then $B_n \times V_A = 50$ nA.

An important point which has been ignored in the above discussion concerns the breakdown of Liouville's theorem for beams accelerated using a tandem Van de Graaff accelerator. As the beam of negative ions is stripped of electrons in the terminal of the accelerator, the phase space area occupied by the beam is increased due to multiple scattering. Measurements of the beam intensity from the EN tandem accelerator at Oxford have shown that with gas stripping, approximately 1 nA of beam with divergence 0.3 mrad is transmitted through a 25 μm × 25 μm aperture, showing that this effect may be significant. It is worth noting, however, that this figure falls to 0.1 nA when carbon stripper foils are used.

Whether beam currents of the order of 1 nA are of sufficient intensity for use in microprobe spectrometry can be estimated from the respective reaction cross sections for PIXE, PRA and RBS. For the simple case where the target is sufficiently thin so that the energy of the incident particle does not change appreciably with depth and therefore the cross section σ(E) can be assumed constant as the particle traverses the target, then $Y = QN_t\sigma\Omega\eta$, where Y is the number of counts detected, Q is the total charge incident on the sample, N_t is the number of target atoms per unit area, σ is the cross section, Ω the solid angle subtended by the detector and η the detection efficiency.

Consider the case of the detection by PIXE of 1 ng cm^{-2} of lead distributed in a 1 mg cm^{-2} organic matrix using a 4 MeV proton beam. The characteristic x-rays are detected by a 30 mm^2 Si(Li) detector 28 mm from the sample and the efficiency for detection is assumed to be one, i.e. there is no x-ray attenuation from the sample, detector windows etc. For a total charge of 1μC, corresponding to 1 nA for 1000 s, and using the value $\sigma_{PbL} = 93 \times 10^{-28}$ m^2 taken from figure 1.5, the number of lead L x-ray counts detected is calculated to be 60.

In the case of Rutherford backscattering and nuclear reaction spectrometry, where the reaction cross sections are in general lower than for PIXE, the detection limits are reduced accordingly because of the lower count rate.

1.8 Probe formation

There are two ways of producing a small diameter beam, either by simple collimation or by using a lens system which demagnifies a beam passing through an object aperture.

With collimation, the angular divergence of the beam passing through the collimating aperture must be sufficiently small so as to minimise the spreading of the beam as it hits the target. A beam of particles with angular divergence 1 mrad increases in size by 1 μm for each millimetre of travel and therefore it becomes necessary to position the target close to the collimation aperture. This presents problems, due to increased background radiation from the aperture material. Also, because of practical problems associated with the construction of microcollimators made from material thick enough to stop the incident beam, nuclear probes formed by collimation with diameters of less than 10 μm are difficult to construct. Several groups, however, have used collimated probes in microprobe analysis, [SC78, SH78, GE76, GR76, BA78, SC73, AR71, PR69, PI68], some of which have proved especially useful in experiments carried out externally to the vacuum envelope.

The production of nuclear probes of less than 10 μm in diameter requires a focusing system to demagnify the beam emerging from an object aperture. The focusing systems currently used in the production of electron or low-energy ion probes are based on a series of elements of cylindrical symmetry. The focusing action of these elements, i.e. the force acting such as to return the charged particles to the beam axis, is generated only by the fringing field, since the field within the element has only an axial component. To produce a focused ion beam of energies suitable for analytical work (several MeV) introduces problems when conventional cylindrical lenses are used because of the physical difficulty in bending such an energetic beam of massive particles. In general, for magnetic focusing, the ratio between the field strengths for singly charged ions and electrons is given by

$$B_I/B_e = (M_I\, E_I/M_e\, E_e)^{1/2}$$

where M_I, M_e are the masses and E_I, E_e are the energies of the ions and electrons respectively. Comparing the fields required for a 2 MeV proton and a 20 keV electron, it is seen that the strength required for focusing the ion beam is about 430 times that for an electron beam.

In general, the problem of focusing MeV ion beams can be tackled in two ways, either by using a superconducting cylindrical lens which is technologically difficult to construct but has the advantage that the beam optics of such a lens are relatively easy and well understood, or by using the strong focusing quadrupole lens which is easier to construct but has the disadvantage that the beam optics theory is difficult and sparsely documented in the literature. Several other new and novel techniques, which are being investigated at the time of writing, are discussed further in chapter 7.

1.9 Brief description of the quadrupole lens

Since the main part of this book is concerned with understanding and documenting ways of forming an energetic ion probe using the quadrupole lens as the building block, it is worth at this stage describing the action of a quadrupole lens on a charged particle using a simple physical description.

Magnetic quadrupole lenses, used since 1952 in ion-beam transport systems, are strong focusing lenses and differ from cylindrical lenses in that the field lines are at right angles to the direction of the ion motion and hence the action of the magnetic field force $\boldsymbol{F} = \boldsymbol{v} \times \boldsymbol{B}$ on the particle trajectory is necessarily greater. Figure 1.13 shows the cross section of a typical quadrupole magnet with its associated field lines. The field lines are rectangular hyperbolae, asymptotic to axes inclined 45° to the x and y axes marked in the diagram.

In the frame of reference shown, the field components are such that the x-component of the magnetic field is proportional to the distance from the central axis in the y-direction and the y-component of the magnetic field is proportional to the distance from the axis in the x-direction:

$$B_x = G_m y$$
$$B_y = G_m x$$

where G_m is the magnetic field gradient on the axis.

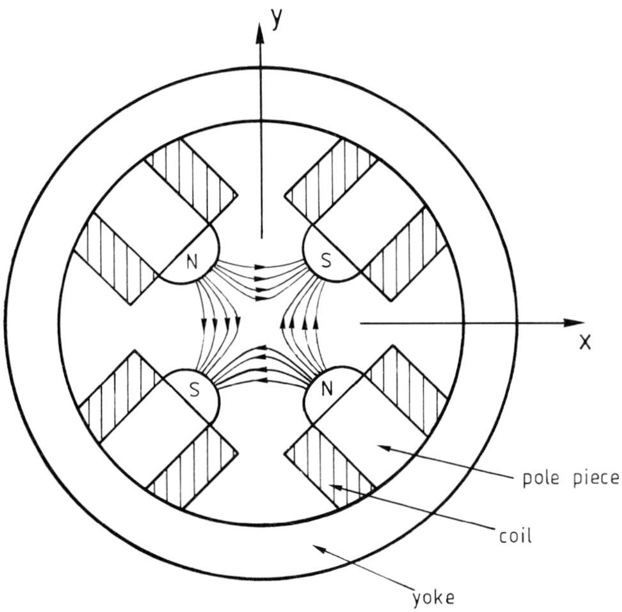

Figure 1.13 Cross section of a typical quadrupole lens showing the associated field lines.

The action of the quadrupole field on a charged particle passing through the lens is shown schematically in figure 1.14. The forces on a positively charged particle passing down into the paper at positions 1 and 2 are in a direction towards the central axis and so in the y-plane there is converging or focusing action. The forces on the charged particle at positions 3 and 4 are such that the particles are moved away from the central axis and hence in the x-plane there exists a diverging or defocusing action. For particles entering the quadrupole lens not in the x- or y-planes (position 5) then the acting force has both a defocusing x-component and a converging y-component. Moreover, the further the particle is from the central axis, the greater the magnitude of the respective converging or diverging forces. The quadrupole field is discussed in more detail in chapter 2.

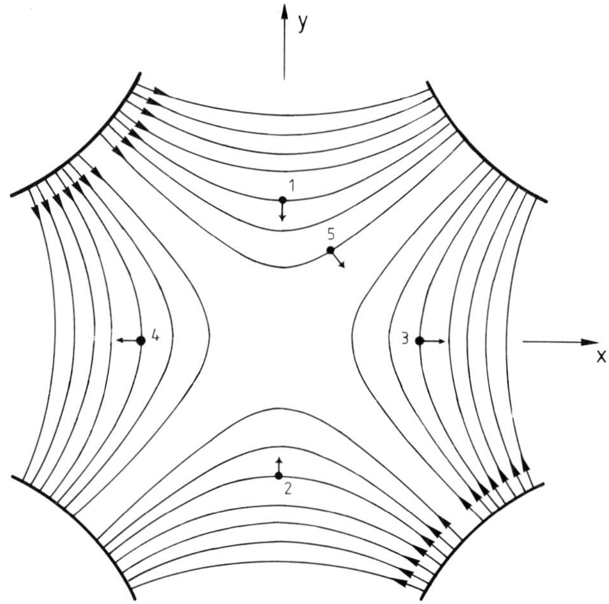

Figure 1.14 The action of the quadrupole field on a charged particle. The arrows represent the direction of the magnetic field force on a charged particle passing down into the lens for various positions in the cross section plane.

At this stage, it is clear that one quadrupole lens is insufficient to focus a charged particle beam in both planes simultaneously and it therefore requires a minimum of at least two such lenses of opposite polarity in order to focus an ion beam down to small spot sizes.

It is also possible to construct quadrupole lenses

using electrostatic fields. In this case the quadrupole field is created between four hyperbolic electrodes arranged symmetrically about the axis of the element. These electrodes are excited alternately $+V_p$, $-V_p$, $+V_p$ and $-V_p$, where V_p is the pole voltage (volts) and the created field profile is identical with that shown in figure 1.14 for the magnetic case. However, the force on a charged particle in an electrostatic field is parallel to the field lines. In order to obtain the same focusing action with respect to the coordinate system shown in figure 1.14 for magnetic quadrupoles, the electrostatic lens must be rotated through 45° so that the poles lie on the coordinate axes. In this case the electrostatic fields are given by

$$E_x = -G_e x$$
$$E_y = G_e y$$

where G_e is the electrostatic field gradient on the axis.

Electrostatic quadrupoles are not widely used because the pole voltages required are proportional to the particle energy and as a result, breakdown problems occur for high-energy ions. They may have advantages in low-energy machines, however, since the electrostatic focusing properties are independent of the mass of the particle.

1.10 Introduction to beam optics terminology

A typical quadrupole probe-forming system consists of a combination of two or more lenses which forms a demagnified image of a small object aperture located several metres from the lenses. Many factors influence the minimum spot size attainable and the aim of calculations is to determine the coordinates of a particle in the image plane (x_i,y_i) as a function of the system parameters and entrance coordinates. We take a set of cartesian coordinates with the origin at the object and the positive z-axis coinciding with the optical axis of the system: the x-axis is arbitrarily taken to be horizontal. It is assumed that when perfectly aligned, the axes of the lenses coincide with the z-axis and their planes of antisymmetry coincide with the xz- and yz-planes. A quadrupole is described as converging (C) if a particle moving in the xz-plane is deflected toward the axis and diverging (D) if the particle is deflected away from the axis.

The position and path of a particle in the system may be specified by the coordinates (x,y,z,θ,ϕ), where θ and ϕ are instantaneous slopes of the path of the particle projected onto the xz- and yz-planes respectively. The coordinates of the particle in the image plane may be written

$$x_i = F_x(x_o,y_o,\theta_o,\phi_o,\delta,P)$$
$$y_i = F_y(x_o,y_o,\theta_o,\phi_o,\delta,P)$$

where the suffix o denotes object coordinates, δ is the fractional momentum shift of the particle from the mean and P represents parameters such as quadrupole misalignment etc. We assume that the functions F are polynomial, so that

$$x_i = A_o + A_1 x_o + A_2 y_o + A_3 \theta_o + \ldots + A_n v_1^j v_2^k v_3^l + \ldots$$

where v_n represents a system variable.

Here we term the A coefficients as aberration coefficients, where the order of the coefficients is given by the sum of the powers of the variables in that term of the expansion. The order of A_n is then

$j + k + l$. Note that this definition is slightly different from the aberration coefficients defined in optical and electron microscopy in which a magnification parameter is included.

For easy identification, the A coefficients can be written in the form

$$\langle x|\theta^2\phi\rangle \text{ or } \langle y|v_1^j v_2^k v_3^l\rangle$$

Table 1.2 The dominant aberrations of quadrupole probe-forming systems.

Order and Type	Coefficient	Remarks
1st Intrinsic	$\langle x\|x\rangle$	Magnifications: not necessarily equal
	$\langle y\|y\rangle$	
	$\langle y\|\phi\rangle$	Astigmatism terms: zero for a correctly focused image
	$\langle y\|\phi\rangle$	
1st Parasitic	$\langle x\|U_n\rangle$	Translation (These terms simply cause a displacement of the image)
	$\langle y\|V_n\rangle$	
	$\langle x\|\alpha_n\rangle$	Tilt (These terms simply cause a displacement of the image)
	$\langle y\|\beta_n\rangle$	
2nd Intrinsic	$\langle x\|\theta\delta\rangle$	Chromatic aberration
	$\langle y\|\phi\delta\rangle$	
	$\langle x\|x\delta\rangle$	These terms have little effect, due to the small object size
	$\langle y\|y\delta\rangle$	
2nd Parasitic	$\langle x\|\phi\rho_n\rangle$	Rotation terms: note the coupling between x- and y-planes
	$\langle y\|\theta\rho_n\rangle$	
	$\langle x\|\theta\varepsilon_n\rangle$	Aberration due to excitation changes (e.g. power supply ripple)
	$\langle y\|\phi\varepsilon_n\rangle$	
3rd Intrinsic	$\langle x\|\theta^3\rangle$	Spherical or aperture aberration
	$\langle x\|\theta\phi^2\rangle$	
	$\langle y\|\theta^2\phi\rangle$	
	$\langle y\|\phi^3\rangle$	
5th Intrinsic	$\langle x\|\theta^5\rangle$	Negligible in current systems.
	$\langle y\|\theta^5\rangle$	These remain after spherical aberration has been corrected
	etc	using octupoles.

where

U_n = x-translation of quadrupole n
V_n = y-translation of quadrupole n
α_n = tilt of quadrupole n in xz-plane
β_n = tilt of quadrupole n in yz-plane
ρ_n = rotation of quadrupole n about optical axis
ε_n = percentage change in excitation of quadrupole n.

where the variable on the LHS is the dependent variable and the expression on the RHS is the combination of independent variables to which the coefficient relates.

Coefficients which are independent of the misalignment variables, P, are real or intrinsic aberrations, and relate to a perfect quadrupole system. Those containing P are the result of imperfect construction or alignment and are called parasitic aberrations.

In the above expansion the number of possible aberration coefficients which could affect the spatial resolution of the probe are numerous. Table 1.2 shows some of the dominant aberration coefficients which have been found to affect the performance of current energetic-ion microprobe systems. This table does not include aberration effects due to harmonic contamination of the quadrupole field due to poor lens construction or the effects of fringing fields. These effects are discussed later in chapter 6.

For a perfect probe-forming system, the image is a demagnified version of the object:

$$x_i = x_o/D_x$$
$$y_i = y_o/D_y$$

for all values of θ_o and ϕ_o.

Here, because $\langle x|x\rangle$ and $\langle y|y\rangle$ are usually $\ll 1$ in microprobe systems we use the demagnifications $D_x = 1/\langle x|x\rangle$ and $D_y = 1/\langle y|y\rangle$. Additional terms in the above expression degrade the image and therefore for the best performance their effect must be minimised.

1.11 Review of present microprobe systems using magnetic quadrupole focusing

This section is intended to indicate, as far as possible, the present state of quadrupole probe-forming systems in operation by several experimental groups around the world. For a more complete and thorough review the reader is referred to an excellent and comprehensive review article by Cookson [CO79]. The first international conference specifically held for the latest developments in ion microprobe research [DE82] should also indicate to the reader the rapid expansion taking place in this area of research. The emphasis here will be placed on the beam optical aspects of the lens systems underlying the operation of these probes, and for the description of microprobe hardware, data accumulation and operation etc, it is left to the reader to consult the respective references.

Table 1.3 shows the quadrupole lens configurations currently used by groups around the world. At the time of writing, the groups which have consistently achieved spot sizes of under 10 μm with sufficient beam intensity for elemental analysis are those at Harwell, Heidelberg, Karlsruhe, Bochum, Melbourne, Oxford, Sandia and Rossendorf. It seems probable that the common failure of probe systems in not achieving spot sizes of under 10 μm is the underestimation of the so-called parasitic aberrations. These effects, such as bad quadrupole design, misalignments etc, are discussed at length in chapter 6.

The first focused-ion microprobe, built at Harwell in 1969 [CO70, CO72], is based on the antisymmetric quadruplet CDCD where the first and last quadrupoles, and the inner two quadrupoles, are excited equally. Such a configuration displays features similar to that of a cylindrical lens, that of equal demagnifications in the x- and y-planes, and is frequently termed the ‘Russian Quadruplet’ because its first-order properties were

Table 1.3 Major quadrupole lens configurations used so far.

Configuration	Notes	Laboratory	Reference
C D C D	Quadruplet arranged in the Russian configuration. The inner two quads and the outer two quads are electrically coupled. C = converging quadrupole D = diverging quadrupole.	Harwell Lower Hutt Studsvik Melbourne Manchester Surrey Namur	CO70, CO72 CO78 BR77 LE78
C D C D	Quadruplet arranged as two separated doublets with an intermediate focus	ETH Zurich UCLA	SU74 CH74, SI76
C D C D	Unsymmetric quadruplet	Bochum	WI78
C D C	Spaced triplet	Harwell	CO76
C D C	Triplet	Bochum	BI82
C D C	Coupled triplet	Oxford	WA81, WA82
C D C	Coupled triplet	Rossendorf	GE82
C D	Doublet	Karlsruhe Heidelberg Sandia	HE76a, HE78a NO77, BO78b, BO77 DO82

first investigated by Dymnikov *et al* [DY65a] in 1965. Figure 1.15(a) shows a schematic diagram of the Harwell Russian Quadruplet system, together with some relevant first-order beam-optical properties. For a working distance of 21 cm (the distance of the image to the last quadrupole) the nominal demagnifications are 5.6 in each plane. Early calculations [HA66a] have shown that for a beam with angular divergence in the x- and y-planes of 0.89 and 0.49 mrad respectively, the intrinsic aberrations would be expected to add less than 1μm to the size of the spot. The Harwell group were also sufficiently aware that parasitic aberrations would play a major role in limiting the spatial resolution of the focused beam and subsequently designed a system of accurately constructed lenses which could be easily rotated with respect to each other. The success of the Harwell quadruplet system in focusing down a 3 MeV proton beam to spot sizes of 4 μm in diameter with a typical beam density of about 30 pA μm^{-2} has stimulated many other groups to adopt a similar configuration (see table 1.3).

The use of a quadruplet of quadrupoles with a large spacing between the first and second doublet enables large demagnifications to be achieved in both planes. Figure 1.15(b) shows a schematic diagram of the ETH Zurich microprobe [BO78a, SU74] with its associated relevant first-order properties, which is similar in some respects to the probe constructed at UCLA [CH74, SI76]. The calculated demagnifications in both the x- and y-planes of the Zurich system are of the order of 100, thereby allowing the object dimensions to be relatively large and limiting any halo around the focused spot caused by scattering from the slit edges. In spite of the high demagnification, however, spot sizes of only 11 μm by 26 μm have been achieved for beam densities of about 2 pA μm^{-2} [BO78a]. At UCLA [SI76], the best resolution of 8 μm × 8 μm with a current density of less than 1 pA μm^{-2} has been produced. As will be indicated later, it is likely that in both cases the resolutions are limited by the high degree of both intrinsic and parasitic aberrations present in such high demagnification systems, and the relatively low current densities obtained are a consequence of having to limit the beam solid angle by collimation in order to reduce the effects of these aberrations to acceptable levels.

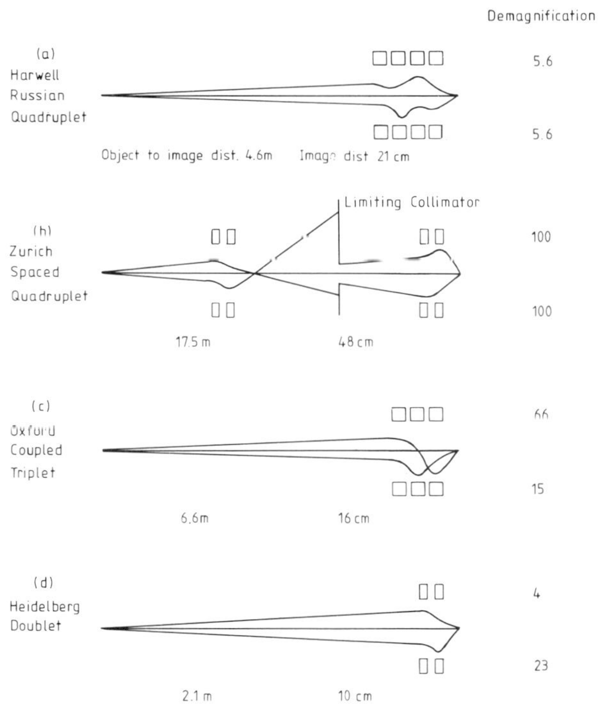

Figure 1.15 A schematic diagram showing trajectories and some first-order beam properties of four microprobe configurations.

The unsymmetric quadruplet configuration of the microprobe at Bochum [WI78] which has lens excitations in the ratio 1:6.44:10.21:6.96 achieved a spot size of 40 μm × 60 μm, limited by a combination of a poor object to image distance ratio and large aberrations. A more recent arrangement, however, in which the lens configuration has been changed to that of a 'modified triplet' has achieved spot sizes of about 2 μm with a beam density of 20 pA μm^{-2} [BI82, WI81].

The Oxford coupled triplet [WA81, WA82], shown schematically in figure 1.15(c) has the first and second lenses excited equally and has x- and y-demagnifications of 66 and 15 respectively for a working distance of 16 cm. By careful consideration and minimisation of the intrinsic and parasitic aberrations pertaining to this configuration, spot sizes of 1 μm × 1 μm are routinely achieved with current densities of 80 pA μm^{-2}. A coupled triplet configuration at Rossendorf (GDR), with the first and last quadrupoles coupled in excitation, has achieved spot sizes of below 4 μm for a total beam current of 200 pA[GE82]. The Harwell three lens system [CO76] is based on the geometry of the original quadruplet configuration with the second lens left inactive. The x- and y-plane demagnifications have values 8.8 and 18.9 and resolutions of 2 μm have been reached in one direction. This system has been used at Harwell using nuclear reactions to detect light elements and current densities are typically five times greater than for the Russian Quadruplet, i.e. 150 pA μm^{-2}.

The Heidelberg doublet [NO77, BO78b, BO77], shown in figure 1.15(d), achieves widely disparate demagnifications in the x- and y-planes of 4 and 23 at the smaller image distance of 10 cm and beam sizes of 2 μm × 2 μm have been obtained at current densities of 20 pA μm^{-2}. In general, for such low-demagnification systems, the beam-optical aberrations are low, and in this particular case, parasitic aberrations have been minimised by extremely accurate machining and pre-alignment of the quadrupoles. The small spot size has been achieved by using small object dimensions, and halo due to slit scattering has been reduced by an ingenious object-slit design [NO75]. The Sandia doublet [DO82] is optically similar to the Heidelberg doublet and achieves almost identical performance figures. A similar system developed at Karlsruhe [HE76a, HE78a], uses a larger spacing between the quadrupoles to give an increased spot brightness of typically 120 pA μm^{-2}, although necessarily at the expense of one of the demagnification parameters, the values of which are 2.4 and 39 in the two respective planes.

From this discussion of some of the better known microprobe systems, it has been seen that focusing of high-energy ion beams down to small spot sizes can be achieved with several quadrupole configurations, not all with the same degree of success. In order to try to simplify the situation at this stage, we can in general categorise probe-forming systems into three types:

(1) For systems which operate under conditions of low demagnification, the beam optical aberrations do not present much of a problem, and although parasitic effects such as rotational misalignment, etc, still play a prominent part in the achievement of high resolution, the spot size seems to be determined by the dimensions and quality of the object aperture.

(2) Systems which have intrinsically high demagnification figures suffer from correspondingly high beam-optical and parasitic aberration coefficients and hence the object dimensions play only a minor part in the achievement of small spot sizes.

(3) For efficient use of quadrupole lenses in the achievement of high spatial resolution it seems that the quadrupole configuration chosen should have sufficiently high demagnification, thereby avoiding problems in object-slit design, but not so large that the associated aberrations become unmanageable. In order to achieve this aim it is necessary to investigate as thoroughly as possible the many configurations of quadrupoles which lie in this category.

1.12 Brief review of theoretical work carried out relevant to the beam optics of quadrupole probe-forming systems

As shown in the last few sections, two or more quadrupole lenses are required for energetic ion-probe formation. It seems likely at this stage that the use of quadrupole quadruplets and to a lesser extent quadrupole triplets will require some degree of coupling between several of the lenses for ease of operation. For instance, the Harwell Russian Quadruplet, with the first and last lenses and the inner lenses coupled, and the Oxford triplet with the first two lenses coupled, require only two current supplies and hence the experimental optimisation of such systems for probe formation is much simplified. Even with this practical limitation in mind, the magnitude of a systematic evaluation of possible probe-forming systems, together with the associated parameters relevant to high spatial resolution presents problems. Investigations into the first-, second- and third-order beam-optical properties of the doublet, triplet, the three combinations of coupled triplet and nine possible combinations of coupled quadruplet would provide a basis for such a survey into quadrupole probe-forming systems. In addition, there are many possible geometries, i.e. lens spacing, lens length, etc, which would need to be considered. Finally, the importance of parasitic aberrations must not be underestimated, and can be a large factor when high spatial resolutions are required.

Previous work along the above lines has been fragmented and mostly incomplete. An excellent starting point, however, is the two books by P W Hawkes [HA70, HA66b] which comprehensively cover the work carried out in this field up to 1971.

Most of the work carried out so far involves analytical techniques to solve the path of a charged particle through combinations of quadrupoles, and this approach is considered in chapter 3. As will be shown, the first- and second-order beam-optical properties of a quadrupole or quadrupole combination can be derived adequately by simple matrix techniques using a fast computer. It is only when higher-order properties (e.g. the spherical or aperture aberrations) are considered that the procedure becomes unwieldy and requires approximations to be made.

Formulae for the third-order aberration coefficients for a single long quadrupole based on the rectangular model (see chapter 2) have been derived for the case of a stigmatic object [DY65b, c] and also for the astigmatic situation [OV67, FI68a]. Coefficients derived from these formulae have been calculated and tabulated by Hawkes [HA70]. The dependence of the spherical aberration coefficients on object position [DY66] and lens excitation [OV68] has been calculated and illustrated graphically. A more recent article justifying the use of the rectangular model in spherical aberration calculations is to be found in Durand [DU77].

Third-order calculations based on the bell-shaped model of the short quadrupole (see chapter 2) have been undertaken [HA67a,b,c,d] and aberration coefficients listed in the book by Hawkes [HA66a]. Formulae for the third-order aberration coefficients have been derived by Dymnikov *et al* [DY65d,e] and Szilagyi *et al* [SZ72]. Aberrations expressed in terms of position of the object have been investigated for the case of high magnification [SZ74, BL74] and also calculated as functions of lens excitation [SZ73]. In both cases the results are presented graphically. Comparisons have also been made between the aberration coefficients extracted using a bell-shaped profile and the rectangular model [SZ74].

Formulae for the chromatic and aperture aberrations using the modified rectangular model (the rectangular model with half-bell termination) and graphs illustrating their behaviour are shown in Kanaya *et al* [KA68a, KA69, KA68b].

The quadrupole doublet represents the simplest system for probe formation and extensive work has been carried out on its beam-optical properties [HA70, HA66b and references therein] using the rectangular and bell-shaped field models. The third-order aberrations of a range of stigmatic doublets have been calculated by Fishkova *et al* [FI67] using the thin lens approximation and by the same group using the rectangular model [FI68a]. Results of calculations of the third-order properties of a range of doublets using both the symmetric and asymmetric bell-shaped models are listed by Hawkes [HA70].

The first-order properties of a range of quadrupole triplets have been studied for several geometries by Ueda [UE73, UE75] using the bell-shaped model. Most of the earlier published information on triplets is in graphical form, some of which is reproduced by Hawkes [HA70]. The third-order properties of a triplet of quadrupoles producing unit magnification have been calculated [FO68a] and are tabulated in [HA70]. The chromatic aberrations of the single quadrupole, doublet and triplet have been studied by Regenstreif [RE63] using the thin lens approximation.

Although several quadruplets of quadrupoles have been studied, only those which display some simplifying symmetry and coupling of excitations have been considered. The best known probe-forming quadruplet to date is based on the antisymmetric quadruplet CDCD, with the first and last and inner excitations coupled. The properties of this quadruplet were first studied using the rectangular model by Dymnikov *et al* [DY65f] in Leningrad, which has subsequently led to this system being termed the 'Russian Quadruplet'. This configuration was considered important because of its orthomorphic properties; the magnifications are identical in each plane and therefore this system has a similarity with the axially-symmetric or cylindrical lens. The third-order aberrations of this configuration have also been considered for a particular geometry and the correction of these aberrations using octupoles has also been described [FI68b, YA68]. Dhuicq [DH68] has also studied this configuration for unequal quadrupole lengths (while still maintaining antisymmetry) and has calculated octupole strengths required in the third and last quadrupoles for elimination of the spherical aberration of a particular geometry.

A second antisymmetric quadruplet (CCDD with the same coupling as above) has been studied by Vosburgh [VO67] using the rectangular model, and the first-order properties and chromatic aberration effects are summarised in Kawakatsu [KA68c] and briefly reviewed in Hawkes [HA70].

Using the bell-shaped model, Hawkes has investigated both these antisymmetric configurations and results of first-order properties and aberration coefficients under conditions of high magnification are tabulated in [HA70].

Dhuicq [DH61] has derived equations for the first-order properties of a symmetric quadruplet, CDDC with the coupling as above, and the results of his work are summarised in Hawkes [HA66b].

Combinations of quadrupoles have been studied by Moses with a view to minimising both the aperture aberrations [MO68, MO70] and the chromatic aberrations [MO71]. Comparisons between results of calculations for round lenses and quadrupole lenses led him to a generalised comment that when both the field strength and linear dimensions are comparable for the two types of lenses, then the spherical aberrations of quadrupoles appear to be about fourteen times larger than those of cylindrical lenses and the chromatic aberration about four times as great. Lyubchik *et al* [LY74] have developed generalised expressions to fifth order for systems consisting of an arbitrary number of quadrupoles. Simultaneous correction of the third-order aberrations in various microprobe devices using octupoles has been achieved using fast computer techniques.

The problems of parasitic aberrations, although important in high-resolution probe-forming quadrupole systems, are difficult to categorise, since their relative magnitude depends on the configuration considered and the design and construction of the lenses. In general, however, it is possible to divide this class of aberration into two types: those which occur due to harmonic contamination of the quadrupole field due to imperfect design or poor construction of the quadrupole lenses, and the effects external to each lens element such as lens misalignment, external magnetic fields, etc.

Calculations of the effects on the quadrupole field of pole misalignment have been undertaken by Cobb and Cole [CO65]. Meyer [ME61], Deltrap

[DE64] and Hardy [HA67e] have calculated the effects of such internal misalignments on the magnetostatic potential and have derived formulae for the resulting aberrations.

Several papers have been concerned about the second class of parasitic aberrations, i.e. investigations into the effects of displacement of the quadrupole magnetic axis from the beam axis, tilt and rotational misalignment of perfectly constructed quadrupole lenses. Formulae have been devised by Kawakatsu *et al* [KA68c], Meads [ME63, ME67] and Lobb [LO72]. Third-order beam transport matrices for such aberrations have been constructed by Heck [HE77], and coefficients relating to some current microprobe configurations have been calculated by Grime *et al* [GR82].

2

Description of the quadrupole field

2.1 The quadrupole magnetic field (cross section)

A brief description of the quadrupole field in its pure form has already been given in §1.9. The pure quadrupole field can only be derived from a quadrupole lens showing exact fourfold symmetry, with the poles hyperbolic in shape and of infinite extent. The field (referring back to the coordinate system shown in figure 1.13) is then

$$B_x = Gy$$
$$B_y = Gx$$

i.e. the magnetic field components are such that the x-component is proportional to the distance from the central axis in the y-direction and the y-component is proportional to the distance from the axis in the x-direction.

G, the magnetic field gradient, is defined as

$$G = B_p/r_o \simeq \frac{2\mu_o nI}{r_o^2} \qquad (2.1)$$

where r_o is the quadrupole aperture radius (m), B_p is the magnetic field at the pole tip (T) and nI is the total ampere-turns per pole.

The idealised quadrupole lens, however, is difficult to construct since the poles need to be truncated at some point. Additionally, due to the imperfections of mechanical alignment and assembly of the lens, there will necessarily be some departure from pure fourfold symmetry. The following discussion and the treatment of the practical quadrupole field follows in part the excellent paper by Cobb and Cole [CO65].

The magnetic scalar potential derived from the two-dimensional Laplace equation for a quadrupole with perfect fourfold symmetry and no variation of field with length takes the form

$$\phi(\mathrm{r},\theta) = \sum_{n=2,6,10\,\ldots} k_{2n}\,(\sin n\theta)r^n \qquad (2.2)$$

where k_{22} is the coefficient for the pure quadrupole field, with k_{26} and k_{210} the dodecapole and 20-pole harmonic coefficients. These higher-order terms satisfy the symmetry conditions imposed on a quadrupole in the solution of the Laplace equation and are non-vanishing if the pole shape is not a perfect hyperbola.

If the fourfold symmetry is not maintained in the quadrupole, then other higher-order terms enter into the expansion series for ϕ. These are the terms k_{33} (sextupole), k_{44} (octupole), k_{55} (decapole), etc. For the general coefficient k_{mn}, n is the multipole index such that $n=2$ represents a quadrupole, $n=3$ (sextupole), $n=4$ (octupole), etc, and m refers to a boundary condition $\phi(r,\theta)=-\phi(r,\theta+\pi/m)$ in the solution to Laplace's equation.

The general expression for the scalar potential of a practical quadrupole may then be written as

$$\phi(r,\theta) = \sum_{n=2}^{\infty} \sum_{m=2}^{\infty} v_{mn} k_{mn} r^{n} \sin(n\theta - \alpha_{mn}) \qquad (2.3)$$

where

$$v_{mn} = \begin{cases} 1 & \text{for } n - m, 3m, 5m, \ldots \\ 0 & \text{for all other } n. \end{cases}$$

α_{mn} has been included since higher-order terms will not necessarily have the same phase as the quadrupole term.

From $\boldsymbol{B}=-\triangle\phi$ can be obtained the general expression for the magnetic field, i.e.

$$B_{r} = \sum_{n=2}^{\infty} \sum_{m=2}^{\infty} n v_{mn} k_{mn} r^{n-1} \sin(n\theta - \alpha_{mn})$$
$$B_{\theta} = \sum_{n=2}^{\infty} \sum_{m=2}^{\infty} n v_{mn} k_{mn} r^{n} \cos(n\theta - \alpha_{mn}) . \qquad (2.4)$$

As can be observed from equation (2.2), dodecapole, 20-pole and higher-order harmonic components arise in a perfectly symmetrical quadrupole due to practical considerations in the design and construction of the pole shape. An excellent and concise report on how the pole can be shaped to eliminate the dodecapole component has been given by Parzen [PA76] and some of the conclusions are summarised here.

For a quadrupole with a hyperbolic pole shape it is usual to cut the hyperbola at some angle θ, where θ is measured with respect to the pole axis at the aperture centre (figure 2.1). The manner in which the pole is shaped after the cut-off angle is not a sensitive factor and is shown in figure 2.1 to be continued parallel to the pole axis. Parzen indicates that for

$$\theta = \theta_1 = 27°$$

and

$$r_1/r_0 = 1.122$$

the first undesirable harmonic, the dodecapole contaminant, vanishes. The good field aperture

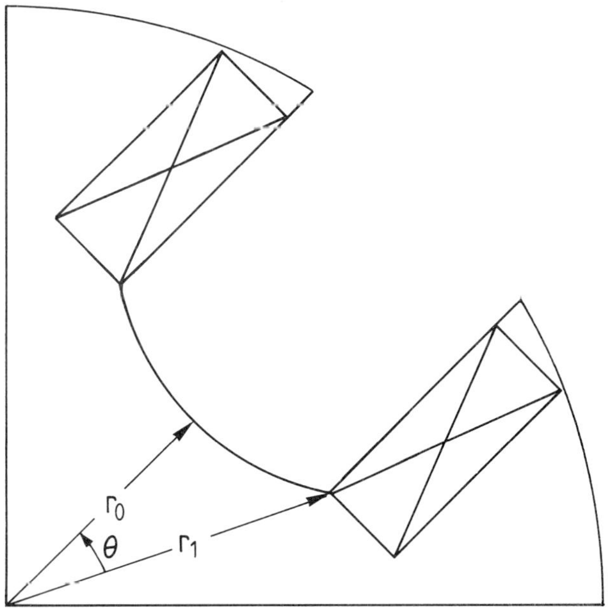

Figure 2.1 Quadrupole with a truncated hyperbolic pole.

$\Delta r_{0.01}$, expressed as a fraction of the aperture radius r_o and defined as the region in which the field gradient does not change by more than 1% in any direction, is found to be about 0.80 at $\theta_1 = 27°$. The magnitude of $\Delta r_{0.01}$ at $\theta_1 = 27°$ is then limited by the 20-pole harmonic contaminant. Assumed in the above deduction is that the field excitation coil is sufficiently hidden by the pole so that the field quality inside the quadrupole aperture is not significantly affected. A reasonable criterion for this condition is that the coil is not closer than r_1, the radial position of the hyperbola at the cut-off angle θ_1. An advantage of hyperbolic pole-shaping is that $\Delta r_{0.01}$ is not sensitive to the pole-shape angle θ and the 80% good field aperture is maintained over a wide range of θ. A practical disadvantage of the hyperbolic pole shape is that the pole-shape profiling is complex.

An alternative and more simple way of producing a relatively pure quadrupole field is to use circular pole-shaping (figure 2.2). If the circular part of a pole of radius R is cut off at an angle θ, then the first undesirable harmonic, the dodecapole contaminant, vanishes when

$$\theta = \theta_1 = 31.5°$$
$$r_1/r_0 = 1.785$$

and

$$R/r_0 = 1.094.$$

The magnitude of the good field aperture $\Delta r_{0.01}$ at $\theta = \theta_1 = 31.5°$ and $R = 1.094r_0$ is about 0.73 (compared to 0.80 for the hyperbolic case) and is limited by the 20-pole contaminant. $\Delta r_{0.01}$ is sensitive to the geometry of the pole and falls to 0.5 for $30° < \theta < 33°$. As in the case for the hyperbolic pole, the coil must be sufficiently hidden from the pole aperture in order that the field quality in this region is not affected, i.e. the coil must not be closer than r_1, the radial position of the pole at the cut-off angle θ_1. Since $r_1 = 1.785r_0$ for the geometry at which the dodecapole component vanishes in the circular case, compared with $r_1 = 1.122r_0$ in the hyperbolic case, then the quadrupole must be constructed correspondingly larger. Studies have been carried out using several different circular pole shapes in order to reduce the size of the quadrupole lens while maintaining good $\Delta r_{0.01}$ figures. Those readers interested further should consult the report by Parzen [PA76].

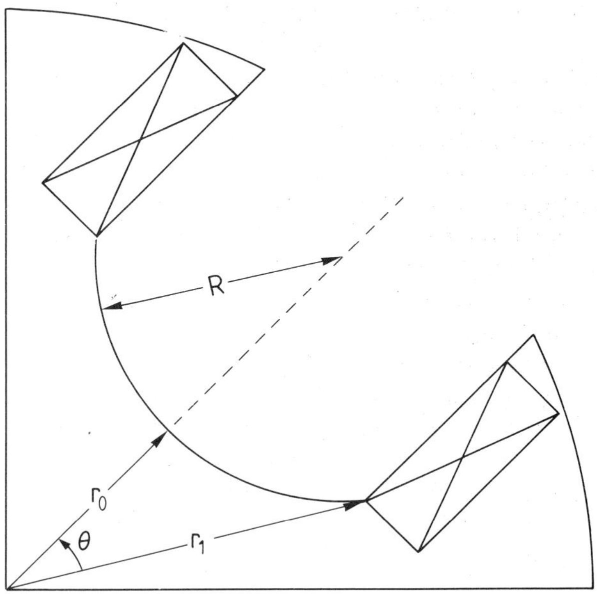

Figure 2.2 Quadrupole with a truncated circular pole.

A second alternative to the terminated hyperbolic pole shape is the even more simple straight-line pole (figure 2.3). In this case, the dodecapole component vanishes when

$$\theta = \theta_1 = 20.11°$$
$$r_1/r_0 = 1.064.$$

However, the magnitude of the good field aperture $\Delta r_{0.01}$ at $\theta_1 = 20.11°$ is in this case 0.53, considerably less than the figures obtained for the circular or hyperbolic poles.

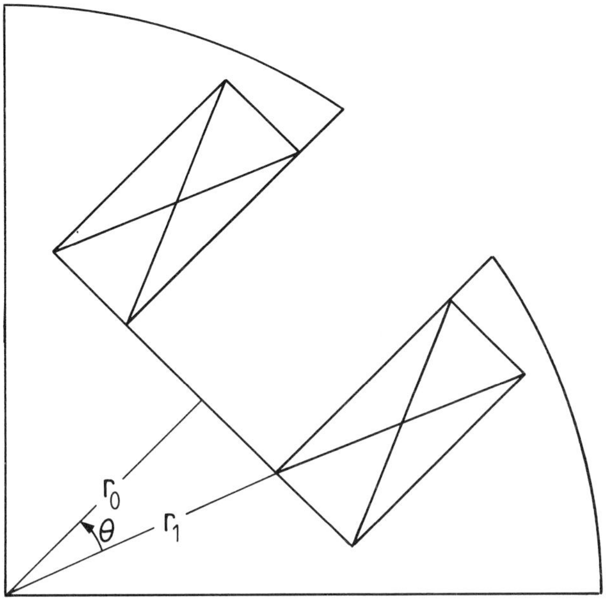

Figure 2.3 Quadrupole with a straight-line pole.

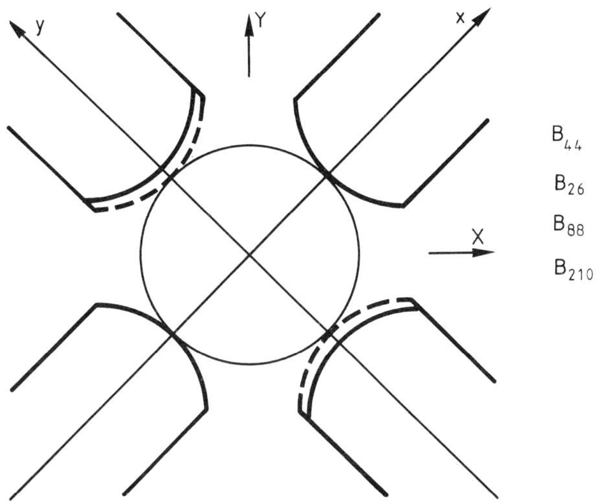

Figure 2.4 Harmonic components generated by moving two poles radially away from the aperture centre along the *y*-axis.

Many other quadrupoles have been designed with increasingly complex pole shapes such that both the dodecapole and 20-pole contaminant harmonics disappear simultaneously. Again, for further information the reader should consult [PA76] and references therein.

Quadrupoles constructed with defects in symmetry have higher-order harmonic contaminants such as sextupole, octupole, decapole, etc, components of magnetic field mixed in with the basic quadrupole field. These higher-order terms have been analysed by Cobb and Cole [CO65] and Halbach [HA69]. Using a rotating coil and a harmonic wave analyser, Cobb and Cole were able to correlate possible symmetry defects present in a quadrupole lens with the corresponding higher multipole components. The results presented in figures 2.4–2.11 are a summary of some of the conclusions from [CO65] and indicate which multipole components are present for some common

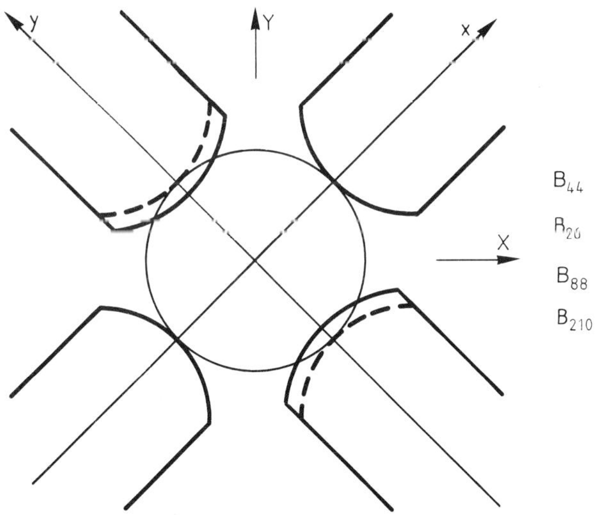

Figure 2.5 Harmonic components generated by moving two poles radially towards the aperture centre along the *y*-axis.

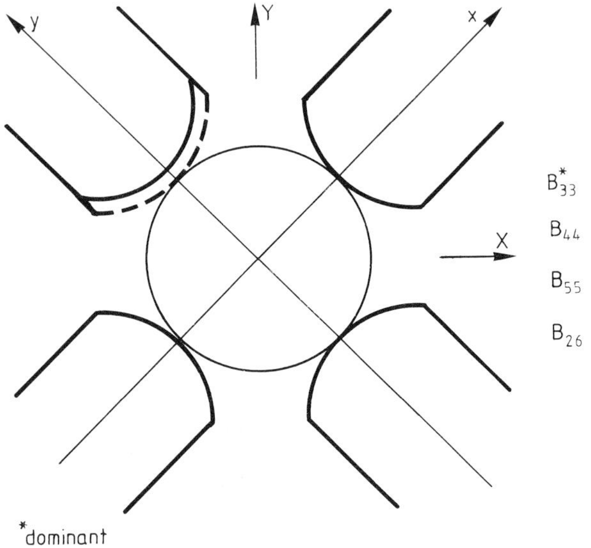

Figure 2.6 Harmonic components generated by moving the pole at positive *y* radially away from the aperture centre.

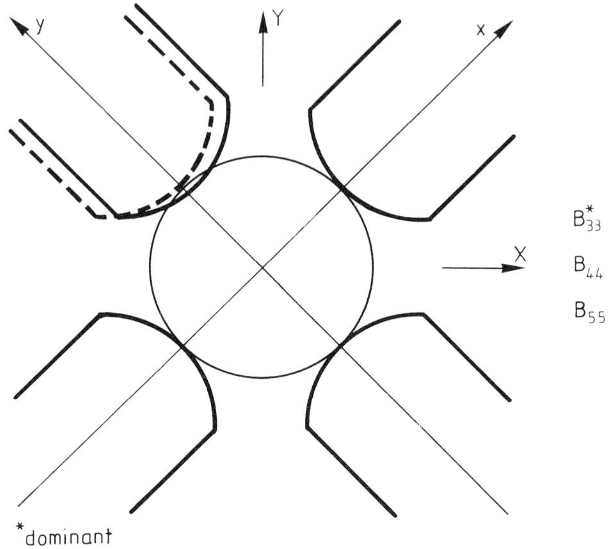

Figure 2.8 Harmonic components generated by moving the pole at positive *y* in the positive *X* direction.

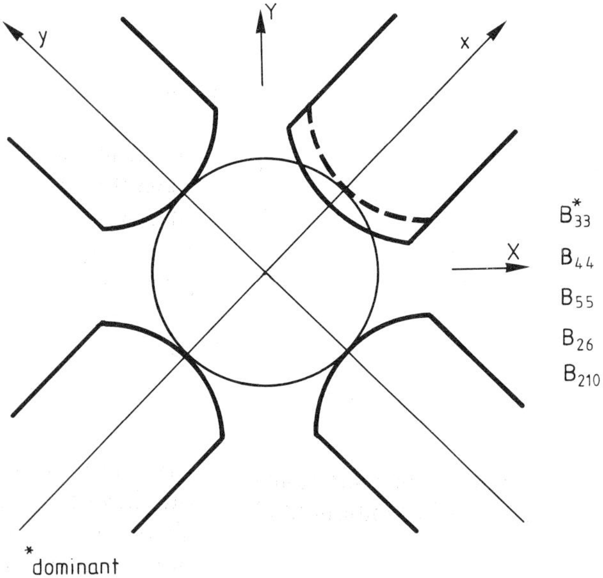

Figure 2.7 Harmonic components generated by moving the pole at positive *x* radially towards the aperture centre.

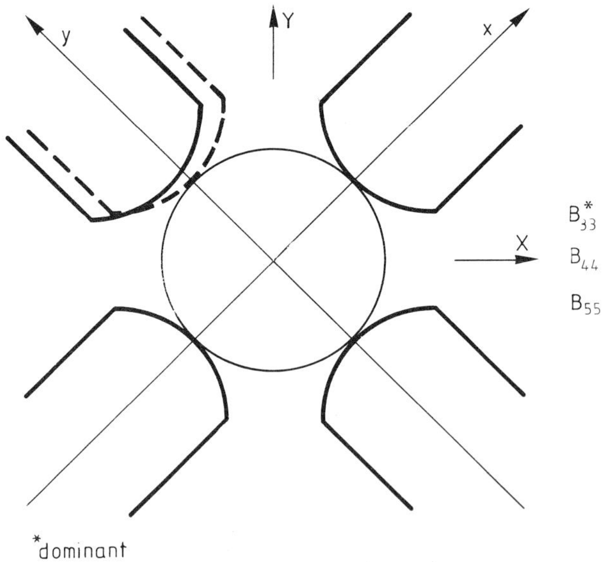

Figure 2.9 Harmonic components generated by moving the pole at positive *y* in the negative *X* direction.

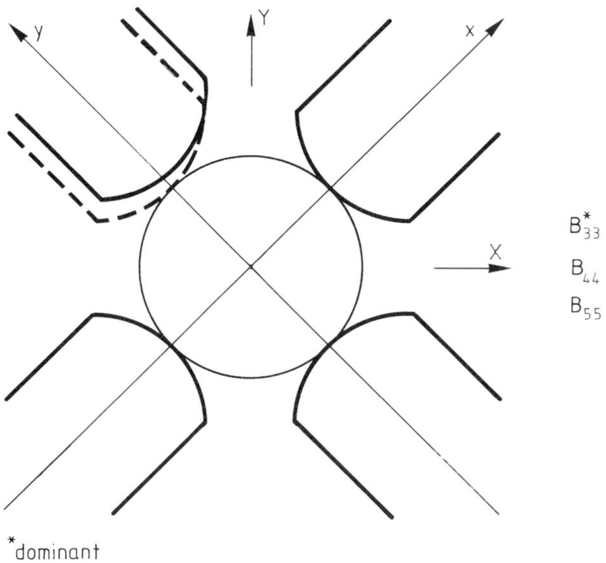

Figure 2.10 Harmonic components generated by moving the pole at positive y in the positive Y direction.

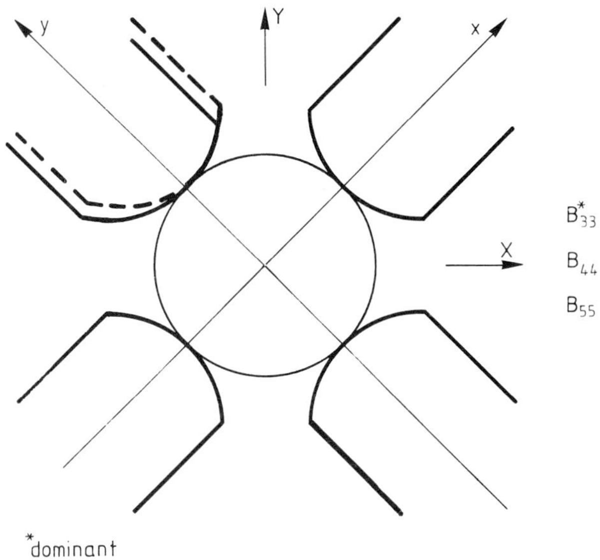

Figure 2.11 Harmonic components generated by moving the pole at positive y in the negative Y direction.

quadrupole lens asymmetries. From these results, it can be seen that the sextupole harmonic is most easily generated by a displacement of a single pole whereas the octupole component requires a radial movement of two poles along an axis.

Typical of quadrupole magnets used in the transport of nuclear particles is a dominant harmonic component in the range 0.1–1% of the basic quadrupole field near full aperture [LO72]. From equations (2.4), it is seen that a harmonic of order $2n$ varies with the radius as r^{n-1}. Thus the particle trajectories that are furthest from the quadrupole axis are those that are influenced most by the contaminant harmonic fields, and it would be expected that, of these harmonic fields, the components with low n, e.g. sextupole and octupole, are the harmonics which would affect the particle trajectories to the greatest extent.

2.2 The quadrupole magnetic field (longitudinal profile)

The focusing properties of a quadrupole lens are determined to a significant extent by the dependence of the field on the longitudinal coordinate z. In general, the longitudinal field profile will be different for each design of quadrupole lens. We define a normalised shape function $f(z)$ which represents the field profile in the longitudinal direction such that

$$L = \int_{-\infty}^{\infty} f(z)\mathrm{d}z \qquad (2.5)$$

where L is the effective length of the quadrupole lens and $f(z)$ has a maximum value of unity.

The shape functions that are amenable to theoretical treatment and permit the equations of motion of particles passing through the quadrupole to be solved analytically, and also enable the aberrations to be evaluated in terms of tabulated functions are limited.

For long quadrupoles ($L/d >> 1$, where d is the pole gap) the rectangular model is convenient and has been used extensively in theoretical studies of lenses, especially in the determination of the first-order focusing properties. The shape function (shown in figure 2.12(a)) has the form

$$f(z) = 1 \text{ for } |z| < L/2 \,. \qquad f(z) = 0 \text{ for } |z| > L/2 \,. \tag{2.6}$$

For short quadrupoles, with L/d close to unity, the bell-shaped model has been used in analytical investigations. The shape function (figure 2.12(b)) has the form

$$f(z) = [1 + (z/d)^2]^{-2}. \tag{2.7}$$

For quadrupoles of intermediate length the composite bell-shaped model has been used. The shape function takes the form of a rectangular profile terminated at each end with half bells (figure 2.12(c)) such that

$$f(z) = 1 \text{ for } |z| \leqslant L/2 \qquad f(z) = \left(1 + \frac{(|z| - L/2)^2}{d^2}\right)^{-2} \text{ for } |z| > L/2. \tag{2.8}$$

Other less-used shape functions have been used in analytical studies. For further details of these models, and of the analytical procedures involved in the calculations, the reader should consult the two books by P W Hawkes [HA66b] and [HA70] and the references therein.

For greater accuracy in calculating the trajectories through quadrupoles, for example in the determination of higher-order aberrations in multi-lens systems, the models described above may not be adequate. With the advent of fast computers, functions which fit the experimentally determined field profiles more accurately can be used, and numerical techniques can be used to evaluate these more complex functions.

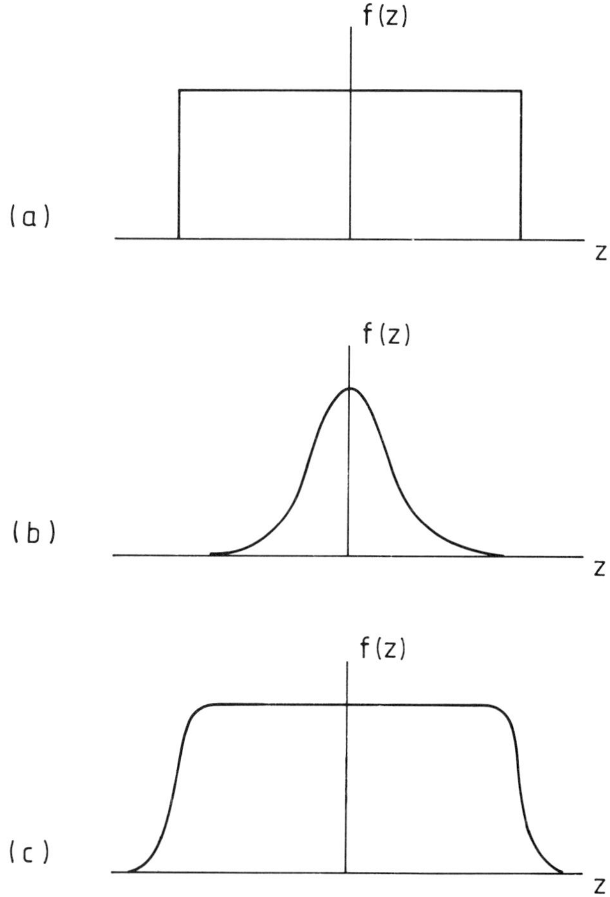

Figure 2.12 Longitudinal field profiles used in the analytical treatment of quadrupole lenses: (a) rectangular model, (b) bell-shaped model, (c) composite bell-shaped model.

Thomson [TH67] has investigated very short lenses ($L/d < 1$) and has used the shape functions

$$f(z) = A\, e^{-z^2/b^2}$$

and

$$f(z) = A\,(1 + z^2/b^2)^{-n},\ n = 2, 4, 6, \ldots, 12. \quad (2.9)$$

A and B were selected so that $f(z)$ matched the experimentally measured profiles.

Vosburgh [VO67] has used the shape function

$$f(z) = \operatorname{sech}\,[a(z^2 + bz)] \quad (2.10)$$

for magnetic quadrupoles of unity pole gap to length ratio. For magnetic quadrupoles of pole gap and length dimensions of 3cm, he found that, when $a = 0.238$ and $b = 0.5$, the function matches the experimentally measured field profiles very closely.

For a longer lens of dimensions $d = 8$cm and $L = 15$cm, P A Moller has suggested the composite model consisting of a rectangular profile terminated by the functions

$$f(z) = \left[1 + \left(\frac{|z| - 5}{3.8}\right)^{3.7}\right]^{-1} \quad (2.11)$$

This model was subsequently used by Dhuicq [DH68] in numerical calculations of the beam-optical properties of quadrupole multiplets.

In work determining the influence of the end-fields on particle trajectories in a quadrupole lens system, Ovsyannikova and Yavor [OV71] have used a similar composite bell-shaped profile of the form

$$\begin{aligned} f(z) &= 1 && \text{for } |z| < L/2 \\ f(z) &= 1 + \left(\frac{(|z| - L/2)^2}{d^2}\right)^{-v} && \text{for } |z| \geqslant L/2 \end{aligned} \quad (2.12)$$

where d, L and v have been adjusted to fit the measured profile.

At Oxford, G W Grime [GR80] has used the function

$$f(s) = \frac{1 + e^{c_0}}{1 + e^{P(s)}}$$

where (2.13)

$$P(s) = c_0 + c_1 s + c_2 s^2 + c_3 s^3 + \ldots\ .$$

The variable s is the distance in the z-direction in units of d, the pole gap dimension. This model is based on the function used by H A Enge [EN67] in the evaluation of the fringing fields of dipole magnets.

For the experimentally measured longitudinal field profile of the quadrupole type used in the Oxford microprobe [WA82], a good fit occurs for $c_0 = -6.0$, $c_1 = 3.5$ and c_2 to $c_5 = 0$.

In practice, the quadrupole effective length L is found experimentally to be

$$L = l + Ca$$

where l is the pole length, a is the bore radius and the coefficient C lies between 0.9 and 1.1 [GR60, BA66].

It is of interest to note that insertion of the values of the coefficients c_0 and c_1 as given above into the shape function as described in equation (2.13) gives quadrupole effective lengths for the Heidelberg, Oxford and Harwell type quadrupoles of $(l + 1.1a)$, $(l + 0.9a)$ and $(l + 1.1a)$ respectively (see table 6.1).

2.3 The quadrupole field (three-dimensional)

The general solution in cylindrical coordinates of Laplace's equation in three dimensions can be expressed as [LO72]

$$\phi = \sum_{n=1}^{\infty} \cos(n\theta + \gamma_n) \sum_{i=0}^{\infty} a_{n+2i}(z)\, r^{n+2i}. \tag{2.14}$$

Since $\boldsymbol{B} = -\nabla\phi$, then

$$B_r = \sum_{n=1}^{\infty} \cos(n\theta + \gamma_n) \sum_{i=0}^{\infty} (n + 2i)\, a_{n+2i}(z)\, r^{n+2i-1}$$

$$B_\theta = -\sum_{n=1}^{\infty} \sin(n\theta + \gamma_n) \sum_{i=0}^{\infty} n\, a_{n+2i}(z)\, r^{n+2i-1} \tag{2.15}$$

$$B_z = \sum_{n=1}^{\infty} \cos(n\theta + \gamma_n) \sum_{i=0}^{\infty} a'_{n+2i}(z)\, r^{n+2i}$$

where a_{n+2i} is given by the recursion formula

$$a_{n+2i}(z) = \frac{(-1)^i a_n^{(2i)} n!}{4^i i!\, (n+i)!} \qquad a_n^{(m)} = \frac{\mathrm{d}^m}{\mathrm{d}z^m}(a_n(z))$$

and is calculated from the shape function $a_n(z)$. The quadrupole field expansions may also be expressed in cartesian coordinates [ST65] as

$$B_x = y[G - \tfrac{1}{12}(3x^2 + y^2)\, G''] + \text{higher orders}$$

$$B_y = x[G - \tfrac{1}{12}(3y^2 + x^2)\, G''] + \text{higher orders} \tag{2.16}$$

$$B_z = xy[G' - \tfrac{1}{12}(x^2 + y^2)\, G'''] + \text{higher orders}$$

where primes denote differentiation with respect to z and

$$G = \left.\frac{\partial B_x}{\partial y}\right|_{x=y=0} = \left.\frac{\partial B_y}{\partial x}\right|_{x=y=0}$$

is the transverse field gradient at the quadrupole axis. In the central region of a quadrupole of sufficient length the field gradient is essentially independent of z and consequently $B_z = 0$. In the end-fields, however, this is not the case, and in this region a z-component of magnetic field is introduced together with higher harmonic contamination of the x and y field components. Lobb [LO72] has shown that these contaminant harmonic fields lead to a variation of effective length with distance from the quadrupole axis. This variation of effective length with radius can be a major source of geometric aberration, and as a consequence of quadrupole symmetry the first geometric aberration is of third order [BA66]. Attempts have been made to minimise this departure from the quadrupole field at the quadrupole ends. For further information the reader should consult [GR60], [ST65] and [PA76] and references therein.

We shall return to the calculation of the three-dimensional quadrupole field in §4.3.

In summary, all real quadrupole lenses will have contaminant harmonic fields included in the field distribution as a consequence of (a) the pole construction being necessarily different from the hyperbolic shape required for a pure quadrupole field, (b) deviation from absolute fourfold symmetry due to physical construction of the lens and (c) fringe fields introduced by end effects.

2.4 Determination of the field profile for practical lenses

In order to use equations (2.15) to calculate the magnetic fields, the longitudinal field profile for the element must either be measured or calculated. The measurement of quadrupole fields using a rotating coil has been described by Fyvie and Lobb [FY74], who show that not only the longitudinal profile, but also the harmonic contamination may be measured by this method.

Alternatively, the magnetic field may be calculated from the shapes of the poles and coils by solving the Laplace equation numerically for the appropriate boundary conditions. The usual method of solution is to subdivide the region of interest into a regular mesh of elementary areas or volumes (depending on whether the field is to be calculated in two or three dimensions). Then by a series of iterations over all the elements, the fields at the nodes of the mesh are adjusted until the boundary conditions across the interfaces of adjacent elements are satisfied. Several programs exist to carry out these calculations and the reader is referred to a recent review article by Bonjour [BO80b].

2.5 Electrostatic quadrupoles

The fields in the uniform field region of an electrostatic quadrupole were given in §1.9 as

$$\begin{aligned} E_x &= -G_e x \\ E_y &= G_e y \end{aligned} \qquad (2.17)$$

where G_e is the electrostatic field gradient ($\mathrm{Vm^{-2}}$). By integrating the field along the $y=0$ axis from the axis to the pole tip, the relation between the field gradient and the applied pole voltages may be obtained, i.e.

$$G_e = 2V_p/a_0^2 \qquad (2.18)$$

where a_0 is the bore radius of the lens (m) and V_p is the pole voltage (V).

In the absence of space charge, the electrostatic potential is derived from the same equation $\nabla^2\phi=0$ as the magnetic scalar potential. Thus the expressions for the three-dimensional electrostatic quadrupole field will be similar to those already derived for the magnetic case. This subject is discussed in more detail in §4.3.

3

Calculation of the imaging properties of quadrupole systems: matrix methods

3.1 Introduction

We have seen in §1.9 that in order to produce a focused beam of charged particles using quadrupole lenses at least two elements of opposite polarity must be used, and that in practice, combinations of two, three or four lenses have been used in microprobe systems. It is obvious that there are many possible combinations of quadrupoles, and it is desirable at this stage to develop the theoretical methods of assessing the suitability of quadrupole systems for microprobe applications, so that a systematic survey of all possible combinations may be undertaken.

In addition, as discussed in §1.11, practical experience with microprobe systems has shown that a system may not behave as predicted by simple theory because of unexpected higher-order or parasitic aberrations which may become significant when the image diameter is very small. Thus the methods of calculation must be developed with two requirements in mind: (a) rapid calculation of the quadrupole excitations required to form an image in a specific location and initial assessment of the suitability of the configuration as a probe-forming system, and (b) detailed analysis of the image so that the effects of higher-order aberrations and construction tolerances may be assessed.

The starting point for any beam-optical calculation is the differential equation of motion of a charged particle in electromagnetic fields. This equation is difficult to solve analytically and direct numerical solution would be time consuming if used to carry out the many repetitive calculations required to locate an image. Thus in order to satisfy the first requirement, an approximate solution of the equation of motion for a quadrupole of arbitrary length and excitation is embodied in a matrix for the element which transfroms the ray coordinates at the entrance of the quadrupole to those at the exit. This matrix is known as the transfer matrix and the transfer matrix of a complete system may then be found by a series of matrix multiplications.

The detailed analysis of the image to sufficient accuracy to allow all the aberrations to be investigated can be carried out only by direct numerical solution of the equation of motion. This will be discussed in the next chapter.

In this chapter, the use of matrix methods in beam optics is outlined and the expressions for the matrix elements of a quadrupole lens are derived. It should be stressed that this is not a complete treatment of matrix methods, but is intended as a guide to writing matrix-based beam optics programs.

3.2 Matrix methods in beam optics

The coordinates of a particle emerging from an optical system may be expressed in terms of the entrance coordinates as a polynomial series whose coefficients are the aberration coefficients, so that

$$\begin{aligned} x_f &= m_{11}x_i + m_{12}y_i + m_{13}\theta_i + m_{14}\phi_i + m_{15}\,\delta + \dots \\ y_f &= m_{21}x_i + m_{22}y_i + m_{23}\theta_i + m_{24}\phi_i + m_{25}\,\delta + \dots \\ \theta_f &= m_{31}x_i + m_{32}y_i + m_{33}\theta_i + m_{34}\phi_i + m_{35}\,\delta + \dots \\ \phi_f &= m_{41}x_i + m_{42}y_i + m_{43}\theta_i + m_{44}\phi_i + m_{45}\,\delta + \dots \\ \delta &= \delta \qquad\qquad + \text{higher order terms} \end{aligned} \tag{3.1}$$

where the subscript i refers to the entrance coordinates and the subscript f refers to the exit coordinates and, for example, $m_{11} = \langle x|x\rangle$ represents the first-order magnification.

These equations may be written as a matrix equation:

$$\begin{bmatrix} x_f \\ y_f \\ \theta_f \\ \phi_f \\ \delta \end{bmatrix} = \begin{bmatrix} m_{11} & m_{12} & m_{13} & m_{14} & m_{15} & \dots & \dots \\ m_{21} & m_{22} & m_{23} & m_{24} & m_{25} & \dots & \dots \\ m_{31} & m_{32} & \dots & \dots & \dots & \dots & \dots \\ m_{41} & \dots & \dots & \dots & \dots & \dots & \dots \\ 0 & 0 & 0 & 0 & 1 & 0 & \dots \end{bmatrix} \begin{bmatrix} x_i \\ y_i \\ \theta_i \\ \phi_i \\ \delta \\ \cdot \\ \cdot \\ H \end{bmatrix} \tag{3.2}$$

where H represents higher-order functions of the entrance coordinates. Alternatively, we may write:

$$\mathbf{X}_f = \mathbf{M}\cdot\mathbf{E}_i\ ,$$

where $\mathbf{X}_f$ is the matrix of the exit coordinates, $\mathbf{M}$ is the element transfer matrix, whose components must be derived from the equation of motion, and $\mathbf{E}_i$ is the matrix of the entrance coordinates expanded to the required polynomial order.

The overall effect of an optical system may be calculated by a series of transformations, namely

$$\begin{aligned} &\text{Element 1: } \mathbf{X}_1 = \mathbf{M}_1\cdot\mathbf{E}_0 && \text{Expand } \mathbf{X}_1 \text{ to give } \mathbf{E}_1 \\ &\text{Element 2: } \mathbf{X}_2 = \mathbf{M}_2\cdot\mathbf{E}_1 && \text{Expand } \mathbf{X}_2 \text{ to give } \mathbf{E}_2 \\ &\text{Element 3: } \mathbf{X}_3 = \mathbf{M}_3\cdot\mathbf{E}_2 \\ &\qquad\vdots \\ &\text{Element } n\text{: } \mathbf{X}_n = \mathbf{M}_n\cdot\mathbf{E}_{n-1} && \text{Expand } \mathbf{X}_{n-1} \text{ to give } \mathbf{E}_{n-1} \end{aligned} \tag{3.3}$$

where $\mathbf{X}_n$ is the particle coordinate matrix after the nth element.

This method allows the exit coordinates of a single particle to be calculated, but is of limited use, since it does not calculate the overall transfer matrix (i.e. the matrix of the aberration coefficients) directly. This may be done by deriving a matrix $\mathbf{P}$ for each element which will transform the expanded coordinate matrix directly, i.e.

$$\mathbf{E}_n = \mathbf{P}_n \cdot \mathbf{E}_{n-1} \tag{3.4}$$

so that we may write

$$\begin{aligned}
\mathbf{E}_n &= \mathbf{P}_n \cdot \mathbf{E}_{n-1} \\
\mathbf{E}_{n-1} &= \mathbf{P}_{n-1} \cdot \mathbf{E}_{n-2} \\
&\;\;\vdots \\
\mathbf{E}_2 &= \mathbf{P}_2 \cdot \mathbf{E}_1 \\
\mathbf{E}_1 &= \mathbf{P}_1 \cdot \mathbf{E}_0
\end{aligned} \tag{3.5}$$

or

$$\mathbf{E}_n = \mathbf{P}_n \cdot \mathbf{P}_{n-1} \cdot \; . \; . \; . \; \cdot \mathbf{P}_2 \cdot \mathbf{P}_1 \cdot \mathbf{E}_0 \tag{3.6}$$

giving the overall transfer matrix of the system as

$$\mathbf{P} = \mathbf{P}_n \cdot \mathbf{P}_{n-1} \cdot \; . \; . \; . \; \cdot \mathbf{P}_2 \cdot \mathbf{P}_1 . \tag{3.7}$$

The $\mathbf{P}$ matrix is of the form

$$\begin{bmatrix} x_{\mathrm{f}} \\ y_{\mathrm{f}} \\ \theta_{\mathrm{f}} \\ \phi_{\mathrm{f}} \\ \delta \\ x_{\mathrm{f}}^2 \\ x_{\mathrm{f}}y_{\mathrm{f}} \\ x_{\mathrm{f}}\theta_{\mathrm{f}} \\ \vdots \\ H \end{bmatrix} = \begin{bmatrix} p_{11} & p_{12} & p_{13} & p_{14} & \cdots & \\ p_{21} & p_{22} & p_{23} & p_{24} & \cdots & \\ & & \vdots & & & \\ & & & & p_{ij} & \\ & & & & & \\ & & & & & p_{NN} \end{bmatrix} \begin{bmatrix} x_i \\ y_i \\ \theta_i \\ \phi_i \\ \delta \\ x_{\mathrm{f}}^2 \\ x_i y_i \\ x_i\theta_i \\ \vdots \\ H \end{bmatrix} \tag{3.8}$$

It will be seen by comparison with equation (3.2) that the first five rows of the $\mathbf{P}$ matrix are identical with the $\mathbf{M}$ transfer matrix. The remaining terms in the $\mathbf{P}$ matrix may be derived from the first-order $\mathbf{M}$ matrix elements by simple algebra. For example, to derive the terms in the x_{f}^2 row we have

$$x_{\mathrm{f}}^2 = (m_{11}x_{\mathrm{i}} + m_{12}y_{\mathrm{i}} + m_{13}\theta_{\mathrm{i}} + m_{14}\phi_{\mathrm{i}} + m_{15}\delta)^2$$

or

$$\begin{aligned} x_{\mathrm{f}}^2 = m_{11}^2\, x_{\mathrm{i}}^2 + m_{12}^2 y_{\mathrm{i}}^2 + m_{13}^2\, \theta_{\mathrm{i}}^2 + \ldots \\ + 2m_{11}m_{12}x_{\mathrm{i}}y_{\mathrm{i}} + 2m_{11}m_{13}x_{\mathrm{i}}\theta_{\mathrm{i}} + \ldots . \end{aligned} \tag{3.9}$$

Thus the $\langle x^2|x^2\rangle$ term, $p_{66} = m_{11}^2$, the $\langle x^2|xy\rangle$ term, $p_{67} = 2m_{11}m_{12}$, etc.

If we assume that the particle coordinates are the four position coordinates and the fractional momentum-spread present in all particle beams, δ, then the full expressions for the $\mathbf{E}$ matrices up to third order are ('T' indicates 'Transpose')

$$\mathbf{E}^{(1)} = (x\ y\ \theta\ \phi\ \delta)^{\mathrm{T}} \qquad \text{(5 terms)}$$

$$\mathbf{E}^{(2)} = (\mathbf{E}^{(1)}\ x^2\ xy\ x\theta\ x\phi\ x\delta\ y^2\ y\theta\ y\phi\ y\delta\ \theta^2\ \theta\phi\ \phi^2\ \phi\delta\ \delta^2)^{\mathrm{T}} \qquad \text{(20 terms)}$$

$$\begin{aligned}\mathbf{E}^{(3)} = (&\mathbf{E}^{(2)}\ x^3\ x^2y\ x^2\theta\ x^2\phi\ x^2\delta\ xy^2\ xy\theta\ xy\phi\ xy\delta\ x\theta^2 \\ &x\theta\phi\ x\theta\delta\ x\phi^2\ x\phi\delta\ x\delta^2\ y^3\ y^2\theta\ y^2\phi\ y^2\delta\ y\theta^2 \\ &y\theta\phi\ y\theta\delta\ y\phi^2\ y\phi\delta\ y\delta^2\ \theta^3\ \theta^2\phi\ \theta^2\delta\ \theta\phi^2\ \theta\phi\delta \\ &\theta\delta^2\ \phi^3\ \phi^2\delta\ \phi\delta^2\ \delta^3)^{\mathrm{T}} \qquad \text{(55 terms)}\end{aligned} \tag{3.10}$$

The forms taken by the $\mathbf{P}$ matrices are

$$\mathbf{P}^{(1)} = \begin{bmatrix} \mathbf{M}^{(1)} \end{bmatrix} \qquad (5 \times 5)$$

$$\mathbf{P}^{(2)} = \left[\begin{array}{c|c} \multicolumn{2}{l}{\mathbf{M}^{(2)}\ (5 \times 20)} \\ \hline \mathbf{O} & \mathbf{D}^{(2)} \\ (15 \times 5) & (15 \times 15) \end{array}\right] \qquad (20 \times 20) \tag{3.11}$$

$$\mathbf{P}^{(3)} = \left[\begin{array}{c|c|c} \multicolumn{3}{c}{\mathbf{M}^{(3)}\ \ (5 \times 55)} \\ \hline \mathbf{O} & \mathbf{D}^{(2)} & \mathbf{O} \\ (15 \times 5) & (15 \times 15) & (15 \times 35) \\ \hline \multicolumn{2}{c|}{\mathbf{O}} & \mathbf{D}^{(3)} \\ \multicolumn{2}{c|}{(35 \times 20)} & (35 \times 35) \end{array}\right] \qquad (55 \times 55)$$

where $\mathbf{M}^{(n)}$ is the nth-order $\mathbf{M}$ matrix, $\mathbf{O}$ is the null matrix and $\mathbf{D}^{(n)}$ is the matrix of nth-order derived elements.

It will be seen that the higher-order matrices contain large areas of zeroes and this fact may be exploited, when writing computer programs, to reduce the length of time for matrix multiplication.

It is convenient at this point to derive the transfer matrix of a drift space, both as an example of a transfer matrix and also since the result is required in the next section. In a field-free drift space of length L the path of the particle is a straight line and so

$$\begin{aligned} x_\mathrm{f} &= x_\mathrm{i} + \theta_\mathrm{i} L \\ y_\mathrm{f} &= y_\mathrm{i} + \phi_\mathrm{i} L \\ \theta_\mathrm{f} &= \theta_\mathrm{i} \\ \phi_\mathrm{f} &= \phi_\mathrm{i} \end{aligned} \tag{3.12}$$

and the $\mathbf{M}$ matrix becomes

$$\mathbf{M}^{(1)} = \begin{bmatrix} 1 & 0 & L & 0 & 0 \\ 0 & 1 & 0 & L & 0 \\ 0 & 0 & 1 & 0 & 0 \\ 0 & 0 & 0 & 1 & 0 \\ 0 & 0 & 0 & 0 & 1 \end{bmatrix} \tag{3.13}$$

The remaining columns of $\mathbf{M}^{(2)}$ and $\mathbf{M}^{(3)}$ contain zeroes.

To summarise this section, it has been shown that, given expressions for the matrix elements of the components of an optical system, the aberration coefficients of the system may be obtained by a series of matrix multiplications. In many cases, however, we need to calculate the system parameters required to produce an image in a specific location, and to do this we must examine how the imaging properties may be derived from the matrix elements of the system.

3.3 Paraxial imaging properties derived from the first-order matrix elements

Figure 3.1 shows an optical system AB which transforms the beam defined by an aperture at O into the plane of observation I. In order to obtain the optimum performance as a probe-forming system, all rays from a point object at O must pass through a single point in I, regardless of their initial divergence, i.e. I must be a geometrical image of O in both the xz- and yz-planes. Any separation between the x and y image planes will be referred to as the astigmatism and a correctly focused system will be described as stigmatic. In order to achieve stigmatic focusing, it is necessary to calculate the image distance v from the matrix elements of the system. In addition, in order to make an initial assessment of the probe-forming properties of the system, it is desirable to know the system demagnifications when correctly focused.

Let us assume that the first-order transfer matrix from O to B has been calculated to be

$$\mathbf{M}_\mathrm{OB} = \begin{bmatrix} b_{11} & 0 & b_{13} & 0 \\ 0 & b_{22} & 0 & b_{24} \\ b_{31} & 0 & b_{33} & 0 \\ 0 & b_{42} & 0 & b_{44} \end{bmatrix} \tag{3.14}$$

where, because in most simple optical systems the paraxial motions in the xz- and yz-planes are independent, the mixed x-plane and y plane terms are zero. Even in cases where the motions in the two planes are not independent, the transfer matrix can

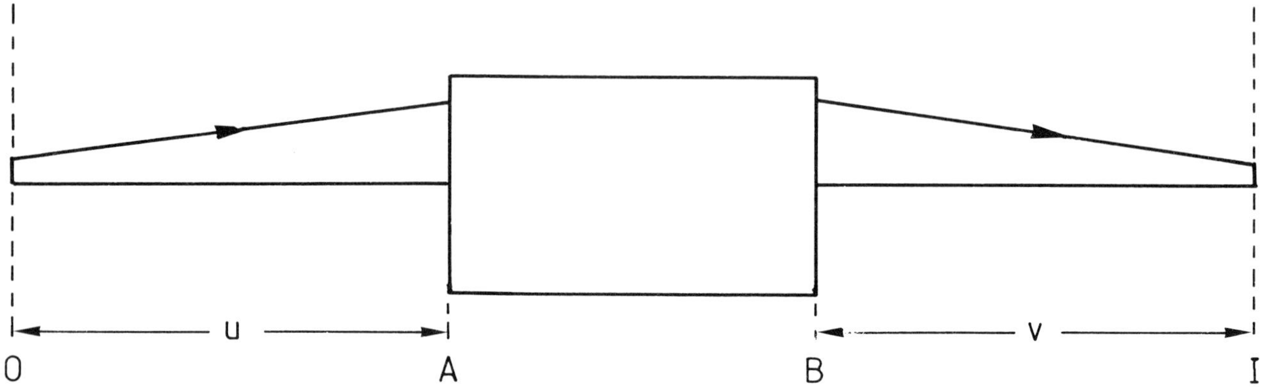

Figure 3.1 Schematic diagram of an imaging system.

often be decomposed into two matrices, one of which describes the focusing action without cross terms. For example, the magnetic solenoid may be treated as a focusing system followed by a rotation.

By pre-multiplying by the matrix for the image distance, (equation (3.13)), using different values in the x- and y-planes to allow for the intrinsic astigmatism of quadrupole systems) we obtain

$$\mathbf{M}_{\mathrm{OI}} = \begin{bmatrix} 1 & 0 & v_x & 0 \\ 0 & 1 & 0 & v_y \\ 0 & 0 & 1 & 0 \\ 0 & 0 & 0 & 1 \end{bmatrix} \begin{bmatrix} b_{11} & 0 & b_{13} & 0 \\ 0 & b_{22} & 0 & b_{24} \\ b_{31} & 0 & b_{33} & 0 \\ 0 & b_{42} & 0 & b_{44} \end{bmatrix} \tag{3.15}$$

or

$$\mathbf{M}_{\mathrm{OI}} = \begin{bmatrix} b_{11} + v_x b_{31} & 0 & b_{13} + v_x b_{33} & 0 \\ 0 & b_{22} + v_y b_{42} & 0 & b_{24} + v_y b_{44} \\ b_{31} & 0 & b_{33} & 0 \\ 0 & b_{42} & 0 & b_{44} \end{bmatrix} \tag{3.16}$$

When correctly focused, the x- and y-values are independent of the divergences θ and ϕ, so that $<x|\theta>$ and $<y|\phi> = 0$ and thus

$$\begin{aligned} b_{13} + v_x b_{33} &= 0 \qquad x\text{-plane focusing} \\ b_{24} + v_y b_{44} &= 0 \qquad y\text{-plane focusing} \end{aligned} \tag{3.17}$$

These equations allow the image distances to be calculated from the system parameters. In order to obtain the demagnification of the system, we may either use the expressions

$$\begin{aligned} D_x &= \frac{1}{<x|x>} = \frac{1}{b_{11} + v_x b_{31}} \\ D_y &= \frac{1}{<y|y>} = \frac{1}{b_{22} + v_y b_{42}} \end{aligned} \tag{3.18}$$

or, by using a consequence of Liouville's theorem that the determinant of the transfer matrix of a non-accelerating lens is unity [BA66], when $<x|\theta> = <y|\phi> = 0$ we may write

$$\begin{aligned} <\theta|\theta> &= \frac{1}{<x|x>} = D_x \\ <\phi|\phi> &= \frac{1}{<y|y>} = D_y \end{aligned} \tag{3.19}$$

and so

$$\begin{aligned} D_x &= b_{33} \\ D_y &= b_{44}. \end{aligned} \tag{3.20}$$

Thus in order to optimise the excitations of the lenses to provide a stigmatic image at a specified distance v_0 from the end of the lens element, it is necessary to select initial values for the excitations and, using these to calculate $\mathbf{M}_{OB}$, obtain v_x and v_y from equation (3.17). An optimisation routine may then be used to vary the excitations so as to minimise a function of the distance of the focal planes from the target (e.g. minimise $(v_0 - v_x)^2 + (v_0 - v_y)^2$).

3.4 First-order matrix elements of a quadrupole lens

In the preceding sections we have discussed the use of matrix methods in general terms. We now need to examine in detail the form taken by the matrix element of the quadrupole lens.

The motion of a charged particle in electrostatic and magnetic fields is governed by the well-known equation of motion

$$\frac{\mathrm{d}}{\mathrm{d}t}(m\boldsymbol{v}) = e\,(\boldsymbol{E} + \boldsymbol{v} \times \boldsymbol{B}) \qquad (3.21)$$

where $m\boldsymbol{v}$ is the relativistic momentum of the particle given as

$$m\boldsymbol{v} = m_0\boldsymbol{v}\,(1 - |\boldsymbol{v}|^2/c^2)^{-1/2}$$

with m_0 the rest mass of the particle (kg), $\boldsymbol{v}$ the velocity (ms^{-1}), c the velocity of light (ms^{-1}), e the charge of the particle (C), $\boldsymbol{B}$ the magnetic field (T) and $\boldsymbol{E}$ the electrostatic field (Vm^{-1}). Apart from a small term representing radiative damping, equation (3.21) is relativistically correct. The accurate solution for the position of the particle as a function of time is not straightforward and will be discussed in the next chapter. In order to derive a more simple first-order solution for the matrix elements, it is possible to make two assumptions:

(1) The mass of the particle remains constant. This approximation is only significant for electrostatic fields since the relativistic mass is not changed by magnetic fields.

(2) The motion is paraxial, i.e. it is assumed that all divergences and displacements from the axis are small. This implies that the axial velocity remains constant.

Under these assumptions, the equation of motion may be expanded into the following cartesian components:

$$\frac{\mathrm{d}v_x}{\mathrm{d}t} = \frac{e}{m}(-v_z B_y + E_x)$$
$$\frac{\mathrm{d}v_y}{\mathrm{d}t} = \frac{e}{m}(v_z B_x + E_y) \qquad (3.22)$$
$$\frac{\mathrm{d}v_z}{\mathrm{d}t} = 0.$$

Using the expressions for the fields in a uniform quadrupole lens:

(1) Magnetic

$$B_x = G_m y \qquad B_y = G_m x \qquad (3.23)$$

(2) Electrostatic

$$E_x = -G_e x \qquad E_y = G_e y \qquad (3.24)$$

where G_m and G_e are the axial field gradients, we obtain

$$\frac{dv_x}{dt} = -\frac{e}{m}x(G_m v_z + G_e)$$
$$\frac{dv_y}{dt} = \frac{e}{m}y(G_m v_z + G_e). \tag{3.25}$$

It is more convenient to work with the axial position z as the independent variable so that the particle trajectory (x and y as a function of z) may be calculated. The equations become

$$\frac{d^2x}{dz^2} = -k^2x \tag{3.26}$$

$$\frac{d^2y}{dz^2} = k^2y \tag{3.27}$$

where

$$k^2 = \frac{e}{mv_z^2}(G_m v_z + G_e). \tag{3.28}$$

Equation (3.26) represents simple harmonic motion in the xz-plane (converging) and has the solutions

$$x = x_i \cos kz + \theta_i (\sin kz)/k$$
$$\theta = -x_i k\sin kz + \theta_i \cos kz \tag{3.29}$$

where x_i and θ_i are the coordinates at $z = 0$ (assumed to be the start of the quadrupole field). Equation (3.27) represents motion in the diverging plane and has the solution

$$y = y_i\cosh kz + \phi_i(\sinh kz)/k$$
$$\phi = y_i k\sinh kz + \phi_i\cosh kz \tag{3.30}$$

Thus for a uniform quadrupole field of length l we obtain the following transfer matrix:

$$\mathbf{M}^{(1)} = \begin{bmatrix} \cos kl & 0 & (1/k)\sin kl & 0 \\ 0 & \cosh kl & 0 & (1/k)\sinh kl \\ -k\sin kl & 0 & \cos kl & 0 \\ 0 & k\sinh kl & 0 & \cosh kl \end{bmatrix} \tag{3.31}$$

This matrix is correct for a quadrupole field with sharp field boundaries. In practice, the fringing fields of a real lens will cause the effective length to be greater than the mechanical length. A reasonably accurate expression for this increase has been found empirically to be

$$l_e = l_0 + 1.1\, a_0 \tag{3.32}$$

where l_0 is the mechanical length and a_0 the bore radius of the lens (see §2.2).

The action of the lens is determined by the product kl, which is sometimes referred to as the angular length of the lens, since in the converging plane the motion of the particle is oscillatory, as described by the circular functions in the matrix elements. When kl is a multiple of 2π, the particle emerges from the lens with exit coordinates in the converging plane equal to the entrance coordinates. In the diverging plane, the defocusing action increases monotonically with kl due to the presence of the hyperbolic functions. The strength parameter k is a quantity (with unit m^{-1}) which relates the field strengths required for a given focusing performance to the mass and energy of the particle. The expressions for k in the magnetic and electrostatic cases are

$$k_m = \left(\frac{e\,G_m}{m\,v_z}\right)^{1/2}$$
$$k_e = \left(\frac{e\,G_e}{m\,v_z^2}\right)^{1/2} \tag{3.33}$$

It is more convenient to relate the k-values to the acceleration voltage of the particle and the excitations applied to the poles of the lens. In chapter 2, the gradients of magnetic and electrostatic quadrupoles were given as

$$G_m = \frac{B_p}{a_0} \simeq \frac{2\mu_0 nI}{a_0^2}$$
$$G_e = \frac{2V_p}{a_0} \qquad (3.34)$$

where a_0 is the quadrupole aperture radius, B_p is the magnetic pole-tip field (T), nI is the total ampere-turns per pole and V_p is the pole voltage (V).

Thus, using the non-relativistic expression for the kinetic energy of the particle,

$$eV = \frac{1}{2} mv_z^2$$

where V is the accelerating potential, we obtain

$$k_m = \left(\frac{B_p}{a_0}\right)^{1/2} \left(\frac{e}{2mV}\right)^{1/4}$$
$$k_e = \frac{1}{a_0} \left(\frac{V_p}{V}\right)^{1/2}. \qquad (3.35)$$

In practical units, the excitations required to obtain a specific k are

$$B_p = 1.414 \times 10^{-3}\, k_m^2\, a_0(A\ V_m/Q)^{1/2} \qquad \text{(Tesla)}$$

and (3.36)

$$V_p = K_e^2\, a_0 V_m \times 10^4 \qquad \text{(Volts)}$$

where a_0 is the aperture radius (cm), V_m is the accelerating voltage (MV), A is the mass number of the particle and Q is the charge number (in units of proton charge).

Table 3.1 shows values of the excitations required to obtain k-values from 0 to 10 for 1MeV protons passing through a 1 cm radius quadrupole. Electrostatic quadrupoles are limited to smaller k-values because of problems associated with breakdown for voltages greater than approximately 30kV (corresponding to $k=17$ for a 1 cm quadrupole). In contrast, the upper limit for magnetic quadrupoles is approximately $k=80$, and is set by saturation in the material of the poles, which by careful design and choice of material can be greater than 1T.

Table 3.1 Excitations required to give specified k values for 1 MeV protons in 1 cm radius magnetic and electrostatic quadrupoles.

k (m^{-1})	Magnetic pole-tip field (T)	Pole voltage (kV)
1.0	1.41×10^{-3}	0.10
2.0	5.65×10^{-3}	0.40
3.0	1.27×10^{-2}	0.90
4.0	2.26×10^{-2}	1.60
5.0	3.53×10^{-2}	2.50
6.0	5.09×10^{-2}	3.60
7.0	6.93×10^{-2}	4.90
8.0	9.05×10^{-2}	6.40
9.0	1.14×10^{-1}	8.10
10.0	1.41×10^{-1}	10.0

To convert these values for use with a particle of mass A and charge Q accelerated through V_m MV, multiply these excitations by $a_0(AV_m/Q)^{1/2}$ for the magnetic case and a_0V_m for the electrostatic case.

3.5 The chromatic aberration of a quadrupole lens

The second-order chromatic aberration coefficients of a quadrupole may be derived from the first-order matrix elements by varying the value of k to incorporate the spread of particle energy within the beam. The effects of different particle energies on the performance of an optical system are usually expressed in terms of the fractional shift in momentum from the mean

$$\delta = \frac{\Delta(mv)}{mv}. \tag{3.37}$$

The effect of a momentum shift of δ on the k-values may be shown to be, for $\delta << 1$:

$$\begin{aligned} k' &= k_m (1 - \delta/2) \text{ (magnetic)} \\ k' &= k_e (1 - \delta) \text{ (electrostatic)} \end{aligned} \tag{3.38}$$

where k' is the effective k-value.

If these expressions are used in the matrix elements from equation (3.31) (again assuming that $\delta << 1$), the second-order aberration coefficients may be extracted. These are

$$\begin{aligned} \langle x|x\delta\rangle &= fkl \sin kl \\ \langle x|\theta\delta\rangle &= f[(\sin kl)/k - l \cos kl] \\ \langle y|y\delta\rangle &= -fkl \sinh kl \\ \langle y|\phi\delta\rangle &= f[(\sinh kl)/k - l \cosh kl)] \\ \langle\theta|x\delta\rangle &= f(kl \cos kl + \sin kl) \\ \langle\theta|\theta\delta\rangle &= fkl \sin kl \\ \langle\phi|y\delta\rangle &= -fk(kl \cosh kl + \sinh kl) \\ \langle\phi|\phi\delta\rangle &= -fkl \sinh kl \end{aligned} \tag{3.39}$$

where $f = 0.5$ for magnetic quadrupoles and $f = 1$ for electrostatic quadrupoles.

The difference between the chromatic aberration coefficients for magnetic and electrostatic quadrupoles may be exploited to construct an achromatic lens by combining a convergent magnetic quadrupole with a divergent electrostatic lens. For certain combinations of the excitations the chromatic aberrations may be reduced to zero. This is discussed further in §7.5.

3.6 Higher-order aberrations in quadrupoles

The potential distribution in a quadrupole lens is symmetric about the centre of the lens (i.e. $\phi(x,y) = -\phi(-x,-y)$) and so the potential must be an even function of radius. In this case, since the force on the particle is proportional to $\nabla\phi$, the force must be an odd function of radius. Thus the next imaging aberration to appear after the first-order focusing is third-order. The most important third-order aberration for microprobe systems is spherical aberration, which shows as a variation of the image position with the entrance angle of the ray, the coefficients being $\langle x|\theta^3\rangle$, $\langle x|\theta\phi^2\rangle$, $\langle y|\theta^2\phi\rangle$ and $\langle y|\phi^3\rangle$. In order to calculate these coefficients from the equation of motion, the paraxial approximation must be abandoned, thus allowing the change in axial velocity as the particle goes off axis to be incorporated. In addition, the full three-dimensional fringing field of the quadrupoles must be included, since this region includes higher-order harmonics which may affect the third-order aberrations (see §6.1). Under these circumstances the equations of motion take the form

$$\ddot{x} + k^2 g(z)x = -k^2[Ax^3 + Bxy^2 + C\dot{x}\,(x^2 - y^2) + Dxy\dot{y} + Gy\dot{x}\dot{y}] + Hx(x^2 - 3y^2) \qquad (3.40)$$

where $g(z)$ is the axial potential distribution and A to H are functions of the quadrupole excitation and $g(z)$ [HA70].

The solution of these equations and the extraction of the third-order coefficients is discussed by several authors, e.g. [FU75] and [LE70], and the reader is referred to these sources for further details.

The calculation of higher-order aberration coefficients by analytical solution becomes a complex task which requires a knowledge of the detailed field profile for full accuracy. For this reason, numerical solution may be preferred, since changes to the field distribution may be incorporated without additional algebra.

3.7 Limitations of matrix methods

Although matrix methods have the advantage of simplicity and speed of calculation, their accuracy and usefulness, particularly for the stringent requirements of microprobe design, are limited by several factors.

(1) The approximations made during the solution of the equation of motion limit the accuracy of the results to a specific order. Thus if higher-order aberration coefficients (e.g. fifth-order) are required, the equation of motion must be solved again to the desired accuracy and the relevant coefficients extracted. This is not usually a problem, but can be significant when any higher-order aberration dominates the image.

(2) The matrix elements derived above are valid only for quadrupole fields. If the transfer matrices for other field harmonics are required, the solutions must be obtained individually. This has been carried out to third order for magnetic dipoles [MA71], electrostatic sectors [MA72, MA75b] and octupoles [HE76b] and to second order for sextupoles [BR67].

(3) The effects of element misalignment must be included explicitly. This is achieved by inserting a transformation matrix before each element to express the ray coordinates in terms of the reference frame of the misaligned element and the inverse transformation at the end of the element to return to the laboratory frame.

Thus the matrix of the misaligned element becomes

$$\mathbf{M}' = \mathbf{T}^{-1}\,\mathbf{M}\,\mathbf{T}$$

where **M** is the matrix of the correctly aligned element and **T** is the misaligned matrix. The detailed form of **T** is discussed in [HE77] and [LO72].

Subject to these limitations, matrix programs are widely used. Most accelerator laboratories have access to a first-order matrix program used for the design of beam transport systems, for which the lower-order properties are adequate. For more stringent applications, higher-order programs have been published. These include TRANSPORT [BR67, BR70, BR80], which is a comprehensive second-order program with many misalignment and optimisation features, and IONBEAM [HE76a], which is a third-order program with similar features. IONBEAM also treats octupoles.

Programs of this type are suitable for most stages of microprobe design, but for a detailed examination of the image, especially when the geometrical image is very small and aberrations dominate, low order matrix programs should be used with care.

4

Calculation of the imaging properties of quadrupole systems: numerical raytracing

In the last chapter it was shown that the imaging properties of quadrupole systems may be calculated to a reasonable degree of accuracy using transfer matrices based on an approximate paraxial solution of the equation of motion. In order to increase the accuracy of the solution and to incorporate deviations from perfect field distributions such as misalignments, fringing fields and contaminant harmonics of the quadrupole field, it becomes more advantageous to solve the equation of motion numerically for a representative set of particles in the beam and to derive the properties of the image from the trajectories of the particles leaving the system. This approach has only become possible with the availability of fast digital computers and although it may be argued that a reliance on numerical methods tends to obscure the underlying theory, the ability to predict in detail the effect of any perturbation on the performance of an imaging system outweighs this disadvantage for all practical purposes.

Since most computer installations will have available library subroutines for the efficient solution of differential equations, a detailed description of the numerical techniques used will not be presented here. Instead, following an outline of the numerical methods, this chapter will concentrate on the way in which a practical raytracing package may be constructed around a suitable differential equation solver. The discussion will be based on the program OXRAY developed at Oxford for the analysis of microbeam systems. In addition to the raytracing routines, the sections relating to the calculation of the fields, data handling, the extraction of aberration coefficients and the generation of image intensity profiles will be described.

4.1 The numerical solution of ordinary differential equations

A system of N simultaneous first-order differential equations may be written as

$$\begin{aligned} y_1' &= f_1(z, y_1, y_2, \ldots, y_N) \\ y_2' &= f_2(z, y_1, y_2, \ldots, y_N) \\ &\cdot \\ &\cdot \\ &\cdot \\ y_N' &= f_N(z, y_1, y_2, \ldots, y_N) \end{aligned} \tag{4.1}$$

where z is the independent variable and $y_i' = \mathrm{d}y_i/\mathrm{d}z$.

Since most systems of equations containing higher derivatives (including the second-order equation of motion) may be reduced to first-order form by the introduction of new variables, there is no loss of generality incurred by considering only first-order equations (most library routines treat first-order equations only).

These equations must be solved numerically to obtain the N values of y_i (the solutions) at the desired value z_I of the independent variable. In beam-optics terminology, the y_i will be the components of the position and velocity vectors of the particle and z the axial distance through the system; z_I will be in a field-free region in image space. Once the trajectory of the particle at one point in image space has been obtained, the subsequent motion of the particle is rectilinear and the displacement of the particle at any point in image space may be found by simple geometry.

In order to specify the solution fully, N initial conditions must be specified. These may take the form of the values of y_i at $z=0$ (initial-value problems) or the values of y_i at some intermediate value of z (boundary-value problems). In beam optics, the path of the particle at the entrance to the system in usually specified and so initial-value methods are appropriate. Rarely, it may be necessary to specify the maximum extent of the beam (for example at an intermediate slit) in which case a boundary-value method must be used.

Most strategies for the solution of equations (4.1) operate by stepping in discrete steps in z using a Taylor series expansion to extrapolate the next value of y_i from the preceding values:

$$y_n = y_{n-1} + hy_{n-1}' + (h^2/2!)y_{n-1}'' + \ldots \tag{4.2}$$

where y_n represents the value of one of the components of y_i after the nth step in z, and h is the size of the increment in z.

The values of the derivatives in this equation must be calculated at z_{n-1} using previously calculated values of y and so the equation may be rewritten using backward differences:

$$y_n = y_{n-1} + hy_{n-1}'(1 + c_1(y_{n-1} - y_{n-2}) + c_2(y_{n-1} - y_{n-3}) + \ldots) \tag{4.3}$$

where the c_n are numerical constants whose value depends on the method in use.

The disadvantage of this simple extrapolation method is that the derivatives are always evaluated retrospectively using previously calculated values of y_i. The accuracy and speed of response of the system to rapid changes in y_i may be improved by calculating an approximate value of y_n using equation (4.3) and then using this value to improve the accuracy of the derivatives in equation (4.2), which may then be used to obtain a revised value of y_n. This 'predictor–corrector' method is the basis of many of the initial-value routines available, although the details of the process, such as the error control, the selection of order and step size and the generation of the first few points in the iteration vary considerably.

Errors in the solution of the equation arise from two sources. The fact that the series in equation (4.2) is terminated prematurely, and that the derivatives are calculated using discrete values leads to a truncation error, while the limited precision of digital arithmetic causes rounding errors. The truncation error depends on the order of the solution and the size of the step in z, and should be controlled by the routine so that it is within a specific limit at each step. Rounding errors will accumulate randomly, but their effect may be minimised by using high-precision arithmetic. It should be noted that although the local single-step error is controlled by the routine, any errors which do occur will propagate through the system and cannot be corrected at a later stage. Consequently, the overall accuracy of the solution may be limited by the sensitivity of the system to small perturbations. For the design of sub-micron microprobe systems, the accuracy required in the solution is high, since in the image plane a resolution of better than 10^{-3} μm is required, whereas at other points in the system the beam may deviate by several mm from the axis. This implies a fractional accuracy on the displacement of 10^{-7} or better. On most computers, the accuracy of basic precision arithmetic is of the order of eight or nine decimal digits, and hence in order to reduce the effects of rounding errors, additional or double precision should be used if possible.

A more detailed discussion of the numerical techniques for the solution of differential equations may be found in any of the standard texts (for example, Fox and Mayers [FO68a] or Ralston and Wilf [RA67]). In a more recent article, Hawkes [HA80] discusses the use of numerical techniques in solving the equations of motion in particle optics.

Other points to be considered when selecting a differential equation solving routine are the efficiency (i.e. the number of times the derivatives must be calculated to achieve a solution with sufficient accuracy), the smallest usable step size (this parameter will determine how the routine responds to rapid field changes) and the degree of flexibility in the specification of the local step error. Some routines also provide facilities to interrupt the solution at specified intermediate points which would, for example, allow the trajectory of a particle within the system to be calculated.

The routine selected for the OXRAY program was the NAG library routine D02QAF [NA82] although it should be emphasised that an exhaustive study of the routines available was not undertaken and it is possible that the other routines could be equally suitable. The double precision (64-bit word length) version of D02QAF has sufficient resolution to trace rays across step field boundaries, which avoids the errors introduced by treating each element in isolation. In order to achieve this accuracy, the routine performs around 800 field calculations to trace a single ray through a typical quadrupole microprobe system. This takes between 1 and 2 s of CPU time on a VAX-11 computer.

4.2 Solving the differential equations of motion

The equation of motion of a particle moving in magnetic and electrostatic fields has already been introduced in chapter 3 (equation (3.21)) as

$$\frac{\mathrm{d}}{\mathrm{d}t}(mv) = e(\boldsymbol{E} + \boldsymbol{v} \times \boldsymbol{B}), \qquad (4.4)$$

where

$$m = m_0\,(1 - |\boldsymbol{v}|^2/c^2)^{-1/2}.$$

The coordinate system used in the treatment of probe-forming systems was defined in § 1.10 and is shown in figure 4.1 together with some other parameters used in the solution.

In order to transform the equations to first order, we change the independent variable from time t to axial distance z by means of the substitution

$$\frac{\mathrm{d}}{\mathrm{d}t} = \frac{\mathrm{d}}{\mathrm{d}z}\left(\frac{\mathrm{d}z}{\mathrm{d}t}\right) = v_z\,\frac{\mathrm{d}}{\mathrm{d}z},$$

where v_z is the z component of velocity. By expanding equation (4.4) into its cartesian components and introducing the variables x and y we then obtain

$$\begin{aligned}
\frac{\mathrm{d}v_x}{\mathrm{d}z} &= \frac{e}{mv_z}(v_y B_z - v_z B_y + E_x)\\
\frac{\mathrm{d}v_y}{\mathrm{d}z} &= \frac{e}{mv_z}(v_z B_x - v_x B_z + E_y)\\
\frac{\mathrm{d}v_z}{\mathrm{d}z} &= \frac{e}{mv_z}(v_x B_y - v_y B_x + E_z) \qquad (4.5)\\
\frac{\mathrm{d}x}{\mathrm{d}z} &= \frac{v_x}{v_z}\\
\frac{\mathrm{d}y}{\mathrm{d}z} &= \frac{v_y}{v_z}
\end{aligned}$$

The numerical solution of these equations requires a routine to calculate the right-hand sides of equations (4.5) using values of x, y, z, v_x, v_y and v_z supplied by the solving routine.

It should be noted that using MKS units for the quantities in the above equations will result in values for the components with greatly different orders of magnitude (e.g. x, $y \sim 10^{-4}$ m; v_x, $v_y \sim 10^5$ m s^{-1}; $v_z \sim 10^7$ m s^{-1}). The effect of rounding errors in the solution will be greatly reduced if equations (4.5) are scaled to give an order of magnitude of about unity.

The solving routine will require initial values of v_x, v_y, v_z, x and y before the solution can proceed. The initial values of x and y are simply the displacement of the particle at the object, but the initial values of the velocities must be derived from the energy of the particle and the divergences of the ray at the object.

The velocity of a particle of rest mass m_0, charge e, accelerated through a potential difference of V volts is given by

$$|\boldsymbol{v}| = c\left(1 - \frac{1}{(1 + eV/m_0c^2)^2}\right)^{1/2} \qquad (4.6)$$

and since

$$\theta = v_x/v_z$$

and (4.7)

$$\phi = v_y/v_z$$

we have

$$|\boldsymbol{v}|^2 = v_z^2\,(1 + \theta^2 + \phi^2). \qquad (4.8)$$

These equations allow the velocity components to be calculated.

Using these initial conditions, the ray may be traced from the object at $z = 0$ to a plane in the image field-free region z_{E} (see figure 4.1), where the particle displacements x_{E} and y_{E} may be obtained directly and the divergences θ_{E} and ϕ_{E} obtained from equations (4.7). The displacement of the particle at any subsequent plane z_{I} may now be found by simple geometry:

$$\begin{aligned}
x_{\mathrm{I}} &= x_{\mathrm{E}} + \theta_{\mathrm{E}}(z_{\mathrm{I}} - z_{\mathrm{E}})\\
y_{\mathrm{I}} &= y_{\mathrm{E}} + \phi_{\mathrm{E}}(z_{\mathrm{I}} - z_{\mathrm{E}}).
\end{aligned} \qquad (4.9)$$

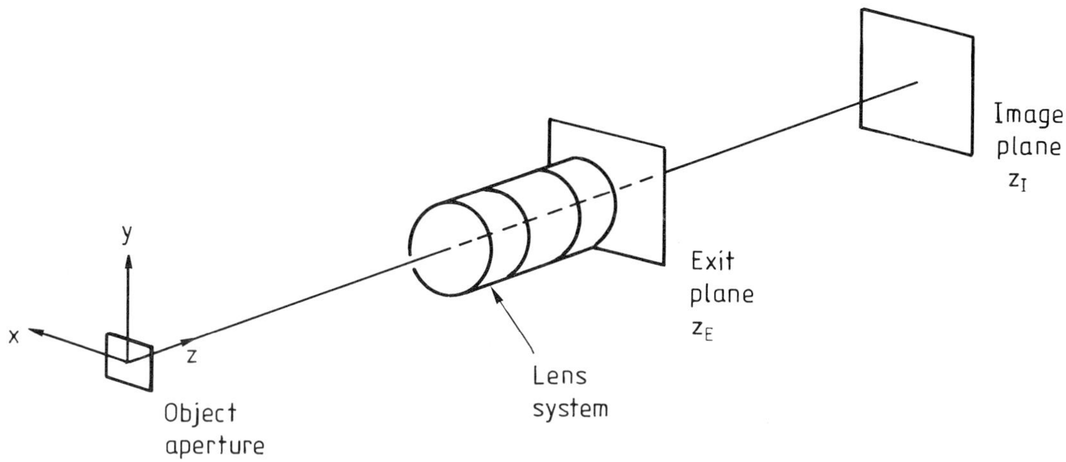

Figure 4.1 The coordinate system used in the analysis of probe-forming systems.

4.3 Calculating the electrostatic and magnetic fields

In order to evaluate the right-hand sides of equations (4.5) we need to calculate, with sufficient accuracy, the magnetic or electrostatic fields due to the elements at any point (x, y, z) in the system. To exploit fully the versatility of the raytracing method, the field calculations must incorporate arbitrary field harmonics, fringing fields and fields due to misaligned elements.

4.3.1 Field distributions of cylindrical harmonics
Using a cylindrical polar coordinate system with the origin at the centre of the element and the z-axis given by the optical axis of the element, the expressions for the magnetic fields of a cylindrical harmonic of order n (equations (2.15)) are

$$\begin{aligned} B_r &= C_n \cos(n\theta + \gamma_n) \sum_{i=0}^{\infty} (n + 2i)\, a_{n+2i}(z) \times r^{n+2i-1} \\ B_\theta &= -C_n \sin(n\theta + \gamma_n) \sum_{i=0}^{\infty} n\, a_{n+2i}(z)\, r^{n+2i-1} \\ B_z &= C_n \cos(n\theta + \gamma_n) \sum_{i=0}^{\infty} a'_{n+2i}(z)\, r^{n+2i} \end{aligned} \tag{4.10}$$

where the coefficients a_n are derived from the axial field gradient distribution $f_n(z)$ by

$$a_{n+2i} = \frac{(-1)^i a_n^{(i)}\, n!}{4^i i!(n+i)!} \tag{4.11}$$

with

$$a_n^{(m)} = \frac{\mathrm{d}^m}{\mathrm{d}z^m} (f_n(z))$$

and

$$a_n{}'_{+2i} = \frac{\mathrm{d}}{\mathrm{d}z} (a_{n+2i}).$$

It will be seen from these equations that the field for each harmonic consists of a sine or cosine term which describes the variation of the field with θ (γ_n is the phase of the harmonic relative to the reference axis of the element) multiplied by an infinite series of terms which describes the variation with radius. The axial or z-dependence of the field is contained in the coefficients of this series.

The order of the derivative of $f_n(z)$ increases by 2 for successive terms in the series, so that the B_θ and B_r components require $f_n(z)$ and the even derivatives while for B_z we require the odd derivatives. Since higher-order derivatives of the shape function may be difficult to calculate, it is permissible to terminate the series after the first two terms. The error introduced by this truncation is small, due to the strong dependence of the denominator of equation (4.11) on i.

It is necessary, therefore, to select an expression for the shape function whose first three derivatives may be calculated and which can be made to fit the observed fringing-field profiles. As was mentioned in §2.2, the function is based on that used by Enge [EN67] for the treatment of dipole fringing fields, with the numerator modified to match the uniform field region at $s = 0$, i.e.

$$f_n(z) = \frac{1 + \mathrm{e}^{c_0}}{1 + \mathrm{e}^{P_n(s)}} \tag{4.12}$$

where

$$P_n(s) = c_0 + c_1 s + c_2 s^2 + c_3 s^3 + c_4 s^4 + c_5 s^5$$

and

$$s = |z - z_1|/a_0$$

with a_0 the bore radius of the element and z_1 the lower limit of the fringing-field region.

The form of this expression is shown in figure 4.2.

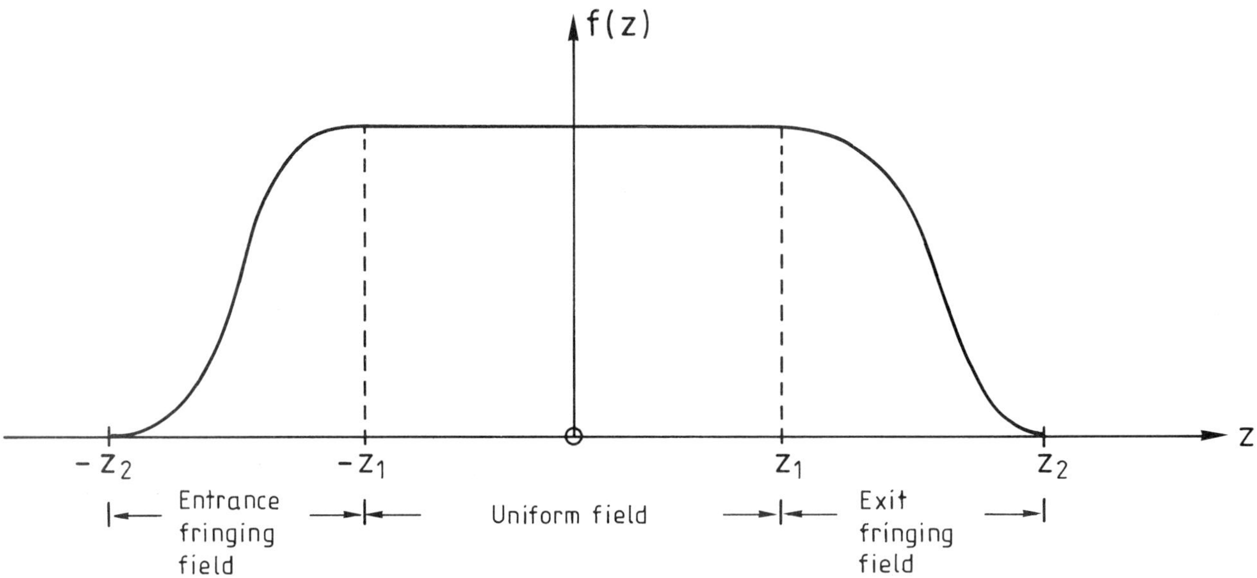

Figure 4.2 The form of the fringing-field function showing the various regions of the element.

By a suitable choice of the coefficients c_0 to c_5 and the boundary z_1, the expression has been made to fit the observed field profiles of typical quadrupoles (see §6.1). In figure 4.2, z_1 and z_2 are the limits of the fringing-field region. For $z<z_1$, the field is uniform ($f_n(z)$ is unity and all higher derivatives are zero). For $z>z_2$, the field is zero. Where the fringing fields of two elements overlap, the algebraic sum of the fields is taken. This approximation does not take account of any mutual interaction between the lenses which may modify the resulting field distribution.

The effective length of an element is found by integrating the transverse field along the z-axis of the element (where x and y are selected to give a non-zero value of B_r), i.e.

$$B_0 l_e = \int_{-z_2}^{z_2} B_r(z)\,\mathrm{d}z \tag{4.13}$$

where B_0 is the value of the field at $z=0$ and l_e is the effective length.

4.3.2 *Misalignments*

In order to calculate the fields using the expressions in equation (4.10) it is necessary to transform the particle coordinates from the cartesian frame of the system into the cylindrical polar coordinates of the element (r, θ, z'). For an element whose optical axis coincides with the z-axis of the system and whose $\theta=0$ plane coincides with the xz-plane, this may be carried out simply using the transformations

$$\begin{aligned} r &= (x^2+y^2)^{1/2} \\ \theta &= \tan^{-1}(y/x) \\ z' &= z - z_c \end{aligned} \tag{4.14}$$

where z_c is the distance of the centre of the element from the object.

Similarly, when the polar components of the fields have been calculated, they must be transformed back into cartesian components for the differential equations using

$$\begin{aligned} B_x &= B_r\cos\theta - B_\theta\sin\theta \\ B_y &= B_r\sin\theta + B_\theta\cos\theta \\ B_z &= B_z'. \end{aligned} \tag{4.15}$$

When the element is misaligned, however, the coordinates of the particle in the frame of the system must first be transformed into cartesian coordinates in the frame of the misaligned element before the transformations in equations (4.14) are carried out. The possible misalignments are translations (displacements of the centre of the element parallel to the three system axes), rotation of the element about its optical axis and tilts of the element about its x- and y-axes.

The translations are calculated simply as

$$\begin{aligned} x' &= x - u \\ y' &= y - v \\ z' &= z - z_c \end{aligned} \tag{4.16}$$

where the primed coordinates refer to the element axes, unprimed coordinates refer to the laboratory frame, u is the translation parallel to the x-axis and v is the translation parallel to the y-axis.

The rotational misalignments are then treated using a transformation of the form

$$\begin{bmatrix} x' \\ y' \\ z' \end{bmatrix} = \mathbf{R} \begin{bmatrix} x \\ y \\ z \end{bmatrix} \tag{4.17}$$

where $\mathbf{R}$ is the rotation matrix given by

$$\mathbf{R} = \begin{bmatrix} \cos\beta\cos\rho & \sin\alpha\sin\beta\sin\rho + \cos\beta\sin\rho & \sin\alpha\sin\rho - \cos\alpha\sin\beta\cos\rho \\ -\cos\beta\sin\rho & \cos\alpha\cos\rho - \sin\alpha\sin\beta\sin\rho & \sin\alpha\cos\rho + \cos\alpha\sin\beta\sin\rho \\ \sin\beta & -\sin\alpha\cos\rho & \cos\alpha\cos\rho \end{bmatrix} \tag{4.18}$$

and the rotations have been carried out in the following order:

(1) rotation about the x'axis by β
(2) rotation about the new y' axis by α
(3) rotation about the new z' axis by ρ

When these transformations have been carried out, the cartesian coordinates of the particle in the frame of the element may be converted to polar coordinates (using the first two of equations (4.14)) and the components of the field calculated as before.

Finally, if there has been any rotational misalignment, the cartesian field components must be transformed back into the frame of the system using the inverse of equation (4.17), i.e.

$$\begin{bmatrix} B_x \\ B_y \\ B_z \end{bmatrix} = \mathbf{R}^{\mathrm{T}} \begin{bmatrix} B_x' \\ B_y' \\ B_z' \end{bmatrix} \qquad (4.19)$$

where, since **R** is orthogonal, we have used the fact that the inverse of **R** is equal to its transpose, $\mathbf{R}^{\mathrm{T}}$.

4.3.3 Numerical values

In order to give numerical values to the fields calculated by the above process, it is necessary to relate the coefficients c_n in equation (4.10) to an easily measured excitation parameter of the element. This is discussed in § 2.5.

In this respect, the current through the coils is not satisfactory since the calculation of the fields from the current requires a detailed knowledge of the magnetic circuit of the element. As an alternative, the fields are calculated in terms of the field at the tip of the pole, since this may be measured directly using a Hall probe or similar device.

For a given pole-tip field B_{T}, the coefficients C_n are given by

$$C_n = B_{\mathrm{T}} k_n / n\, a_0{}^{n-1} \qquad (4.20)$$

where a_0 is the bore radius (m) and k_n is the relative strength of the harmonic of order n, which may be obtained either by calculation [BO80b, BL83] or by analysis of rotating coil measurements [FY74, CO65].

4.3.4 Electrostatic fields

In the absence of space charge, the electrostatic potential satisfies the Laplace equation $\nabla^2\phi=0$ in exactly same way as the magnetic scalar potential. Hence the electrostatic fields may also be calculated using equation (4.10). In this case, however, the fields are expressed in terms of the voltages applied to the electrodes (in volts) and the coefficients are given by

$$C_n = V_{\mathrm{p}} k_n / a_0^n \qquad (4.21)$$

where k_n is the relative strength of the harmonic, as in the magnetic case.

When incorporating electrostatic quadrupoles into focusing systems, it should be remembered that, owing to the difference in origin between the magnetic and electrostatic forces, the focusing action will take place in planes at 45° to the focusing planes of the magnetic quadrupoles when the two fields are parallel.

4.4 Practical aspects of numerical raytracing

Using the method described in the preceding sections it is possible to calculate the trajectory of a particle in the image space of a system given the position and direction of the particle at the object, together with details of the magnetic and electrostatic fields in the lenses. In this section, we discuss the practical details of how this basic routine may be incorporated in a beam-optics package.

The discussion relates specifically to the program OXRAY but the general principles may equally be applied to other programs.

4.4.1 Data handling

The handling of the large amounts of data required to fully specify a lens system may be simplified by categorising the data under three headings:

(a) Element data

These comprise the properties of each element, which are determined when the element is constructed, such as the length, radius and fringing-field profile. In order to avoid specifying these data many times, it may be stored in a separate file for each type of element, which is read by the program whenever that element is used. The parameters in this category are the bore radius a_0 (m) and then, for each harmonic in the field distribution,

(1) field type (magnetic or electrostatic)
(2) the harmonic order and relative strength (n, k_n)
(3) the upper and lower fringe-field boundaries z_1, z_2 (m)
(4) the fringing field coefficients c_0 to c_5.

(b) System data

This heading includes data which relate to the way in which the elements are positioned and excited to form a focusing system. Because of the uncertainty in locating the effective field boundaries of the elements, it is more accurate to specify the position of an element by the distance of its centre from the object rather than by the lengths of drift spaces between the elements as is done in matrix programs.

In this case, the system data comprise the mass, charge and energy of the particle for which the system is designed, the number of elements (N), and then for each element;

(1) the position of the element z_c (m)
(2) the type of element (i.e. the name of the file where the element data are stored)
(3) the pole-tip field or pole voltage B_T or V_T (T or kV)
(4) any misalignments of the element relative to the optical axis of the system.

(c) Object data

This category contains the initial coordinates of the particle at the object, i.e. the displacement from the axis (x, y), the divergence from the axis (θ, ϕ) and the fractional momentum-shift from the mean (δ). These parameters are normally supplied to the routine by the controlling program according to the type of calculation being carried out.

The element and system data are stored in separate disk files as formated text so that they may be created using a text editor, although OXRAY contains routines to create and modify data files directly. When the program is run, the selected system and element data are read from disk into data arrays in memory.

4.4.2 Locating the image planes

The first consideration in analysing the performance of a focusing system is to locate the image plane. Using matrix methods this is straightforward (see §3.3), but when raytracing, the solution requires more effort.

The usual method of locating the image planes of a focusing system using raytracing is to trace through the system an axial ray ($x = y = \theta = \phi = 0$) and a paraxial ray ($x = y = 0$; $\theta = \phi = \alpha$; $\alpha \ll 1$) and to determine the locations of the intersections of the projections of these rays onto the xz- and yz-planes in image space. This is shown for the xz-plane in figure 4.3. If the exit coordinates of the axial ray in the xz-plane are x_0 and θ_0, and for the paraxial ray x_1 and θ_1, then the image distance is given by

$$v_x = -(x_1 - x_0)/(\theta_1 - \theta_0) \ , \qquad (4.22)$$

measured positive in the beam direction from z_E.

A similar expression may be used to obtain the y-image distance.

The choice of the paraxial angle α is important. If α is too large, the imaging will be dominated by higher-order off-axis aberrations, while if it is too small, truncation and rounding errors in the solution become significant. This is shown in figure 4.4 in which the x-image distance v_x and the astigmatism $(v_x - v_y)$ calculated by this method are plotted as a function of α for the Oxford coupled triplet system. The diagram shows that the result is essentially independent of α over a range from 10^{-5} mrad to 10^{-2} mrad. The value chosen for α should be as large as possible within the above range, since the effect of truncation errors will be reduced, and since the fractional tolerance requirements are less stringent, fewer calculations will be required to trace the paraxial ray. (In the OXRAY program $\alpha = 10^{-2}$ mrad.)

Used in conjunction with a minimisation routine, as described in §3.3, this method of locating the image planes may be used to determine accurately the excitations required to form an image at a specific location. The amount of time spent raytracing may be reduced by locating the image approximately using matrices and only using raytracing to achieve the precise optimisation of the excitations.

4.4.3 *Calculating the aberration coefficients*

The raytracing method allows great versatility in the calculation of aberration coefficients, since coefficients of any order in any of the system variables may be calculated.

The principle of the method is to trace sufficient rays with different values of the desired system variables to allow a polynomial of the desired order to be fitted to the observed image-plane displacements. As an example, consider the extraction of aberration coefficients in three variables v_1, v_2, and v_3 up to order n_1, n_2 and n_3 respectively. The variables v may be any of the intrinsic variables (x, y, θ, ϕ, δ) or the parasitic variables (misalignments etc) and the highest-order coefficients calculated will be of the form $\langle x|v_1^{n_1} v_2^{n_2} v_3^{n_3}\rangle$. The full polynomial relating the x-coordinate of the ray to the three variables takes the form

$$x = c_{000} + c_{100}v_1 + c_{010}v_2 + c_{001}v_3 + c_{200}\, v_1^2 + \ldots + c_{ijk}\, v_1^i v_2^j v_3^k + \ldots \qquad (4.23)$$

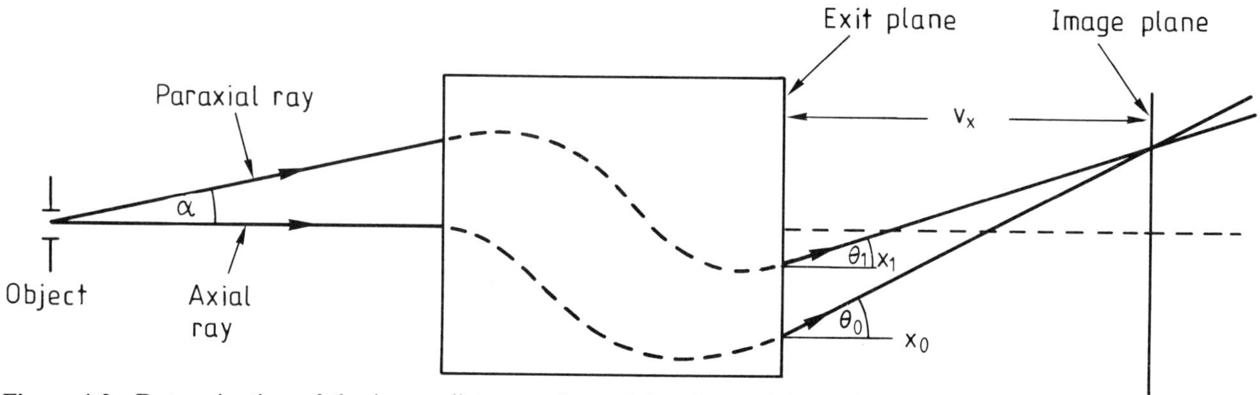

Figure 4.3 Determination of the image distance using axial and paraxial rays (xz-plane).

where

$$0 \leqslant i \leqslant n_1$$

$$0 \leqslant j \leqslant n_2$$

$$0 \leqslant k \leqslant n_3$$

and c_{ijk} is the coefficient $<x|v_1^i v_2^j v_3^k>$.

Similar expressions hold for y, θ and ϕ.

The total number of terms in equation (4.23) (and hence the number of unknown aberration coefficients) is

$$N = (n_1 + 1)\,(n_2 + 1)\,(n_3 + 1)$$

and so in order to solve for the coefficients we require N equations similar to equation (4.23) using different values of the variables v. These equations are obtained by stepping each of the parameters v_i through $n_i + 1$ different values to obtain a regular mesh of values and tracing a ray at each step. This process yields N values of the image-plane displacement x_m, and if the value of v_i corresponding to the mth ray is ${}_mv_i$ we may write the N equations in matrix form

$$\boldsymbol{X} = \mathbf{V} \cdot \boldsymbol{C}_x \qquad (4.24a)$$

where $\boldsymbol{X}$ is the matrix of the ray coordinates

$$\boldsymbol{X} = (x_1\; x_2\; x_3\; .\; .\; .\; x_m\; .\; .\; .\; x_N)^{\mathrm{T}}$$

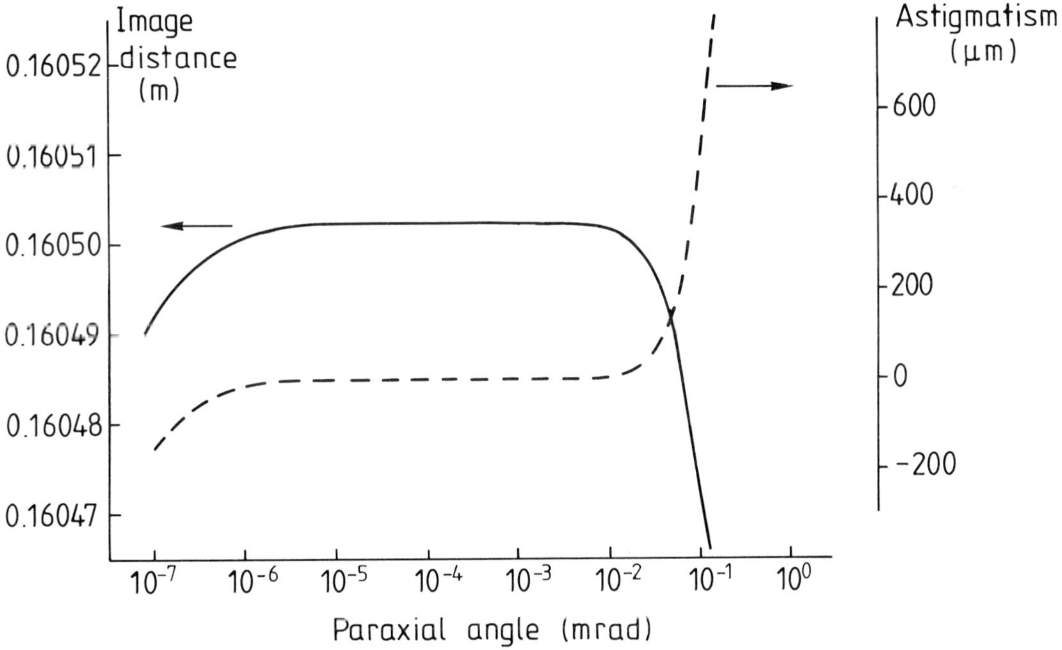

Figure 4.4 Variation of image distance and astigmatism with paraxial angle (Oxford coupled triplet system).

V is the matrix containing the polynomial functions of the independent variables

$$\mathbf{V} = \begin{bmatrix} 1 \; {}_1v_1 \; {}_1v_2 \; {}_1v_3 \; \ldots \\ 1 \; {}_2v_1 \; {}_2v_2 \; {}_2v_3 \; \ldots \\ \vdots \\ 1 \; {}_mv_1 \; \ldots \; {}_mv_1^i \, {}_mv_2^j \, v_3^k \; \ldots \\ \vdots \\ 1 \; {}_Nv_i \; \ldots \end{bmatrix}$$

and C_x is the matrix of the aberration coefficients:

$$C_x = (c_{000} \; c_{100} \; c_{010} \; \ldots \; c_{ijk} \; \ldots)^{\mathrm{T}}.$$

Similar equations can also be obtained for y, θ and ϕ

$$Y = \mathbf{V} \cdot C_y \tag{4.24b}$$

$$\Theta = \mathbf{V} \cdot C_\theta \tag{4.24c}$$

$$\Phi = \mathbf{V} \cdot C_\theta \tag{4.24d}$$

where in each case the C matrix contains the aberration coefficients relating the image coordinate to the system variables v_1, v_2, v_3. **V** is the same for each equation.

The solution of equations (4.24) for C is obtained by multiplying both sides of the equation by the inverse of **V**. This gives

$$\mathbf{V}^{-1}\mathbf{V}C_x = \mathbf{V}^{-1} \cdot X$$

or

$$C_x = \mathbf{V}^{-1} \cdot X \,. \tag{4.25}$$

Inverting a square matrix is a standard procedure which again should be available as a library routine at most computer installations (see, for example, [NA82]), and since **V** is common to all equations, the inversion need only be carried out once. Since matrix inversion requires large amounts of memory, the maximum size of **V**, and hence the total number of coefficients which may be extracted simultaneously, may be limited. In OXRAY the system variables may be selected from the following:

x, y, θ, ϕ	object coordinates
δ, m	fractional particle momentum- and mass-shift
u_n, v_n	x- and y-displacement of element n
$\rho_n, \alpha_n, \beta_n$	rotation and tilts of element n
ε_n	fractional excitation change of element n

The selection of the range of variation of each variable is again important. If the variables are shifted far from their mean value, the solution may be affected by aberrations of higher order than the equations, while if the shifts are too small, rounding errors may influence the accuracy. In addition, matrix inversion is very sensitive to rounding errors, and so in order to improve the accuracy it is desirable to scale the values of v_i to be of the order of unity before the matrix **V** is constructed. This prevents the numerical values of the elements of **V** becoming either very large or very small as the order increases. When the inversion has been carried out, the elements of **V** may be scaled back to their natural values before obtaining the aberration coefficients.

To show how the values of the aberration coefficients depend on the range of the values to generate **V**, the third-order aberration coefficients $\langle x|\theta^3\rangle$, $\langle x|\theta\phi^2\rangle$, $\langle y|\theta^2\phi\rangle$ and $\langle y|\phi^3\rangle$ were calculated for the Oxford triplet with maximum values of θ and ϕ ranging from 10^{-2} – 10 mrad.

The results obtained using a maximum order of 3 (i.e. four distinct values of θ and ϕ; 16 rays) are shown in figure 4.5. It will be seen that, apart from at the extremes of the range, the values of the coefficients are constant over a wide range of maximum angle. The deviation at large angles is due to the influence of fifth-order aberrations, as shown by figure 4.6, in which the values of the third-order coefficients obtained using a fifth-order fit (six distinct values of θ; 36 rays) are plotted as a function of maximum angle. It is seen that the third-order coefficients remain constant over a greater range when the fifth-order terms are included in the calculation.

4.4.4 Generating image intensity maps

The intensity distribution of the beam in the image plane may be calculated in two ways. The direct method is to generate a large number of rays with random object coordinates (defined by the object and collimator slit settings) and to trace each ray

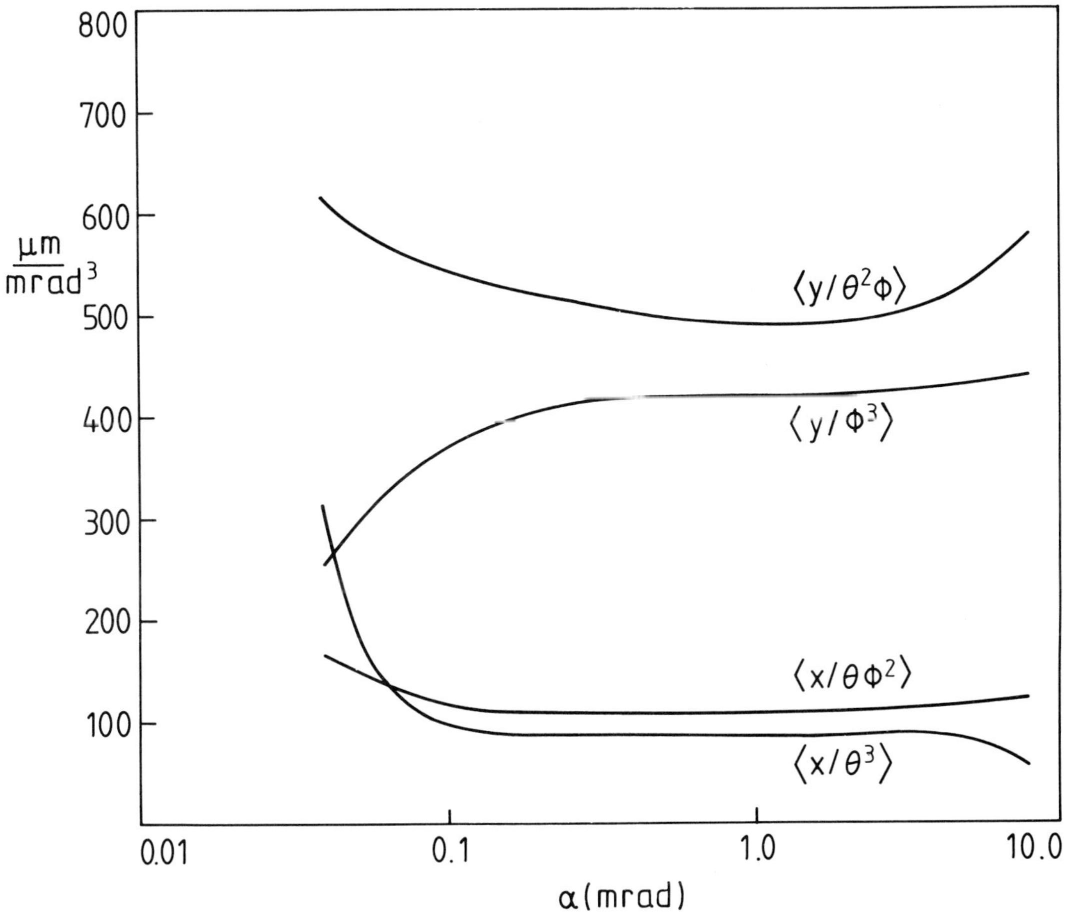

Figure 4.5 Variation of third-order coefficients with maximum object divergence: third-order fit.

through the system. The image coordinates of each ray may be stored and used subsequently to generate a two-dimensional frequency distribution of the points in the image plane, which may then be plotted either as a scatter plot or a contour map.

The advantage of this method is that if the generation of the random object coordinates (and other variables such as energy spread or vibration of the elements) is representative, the resulting intensity distribution is an accurate reflection of the real

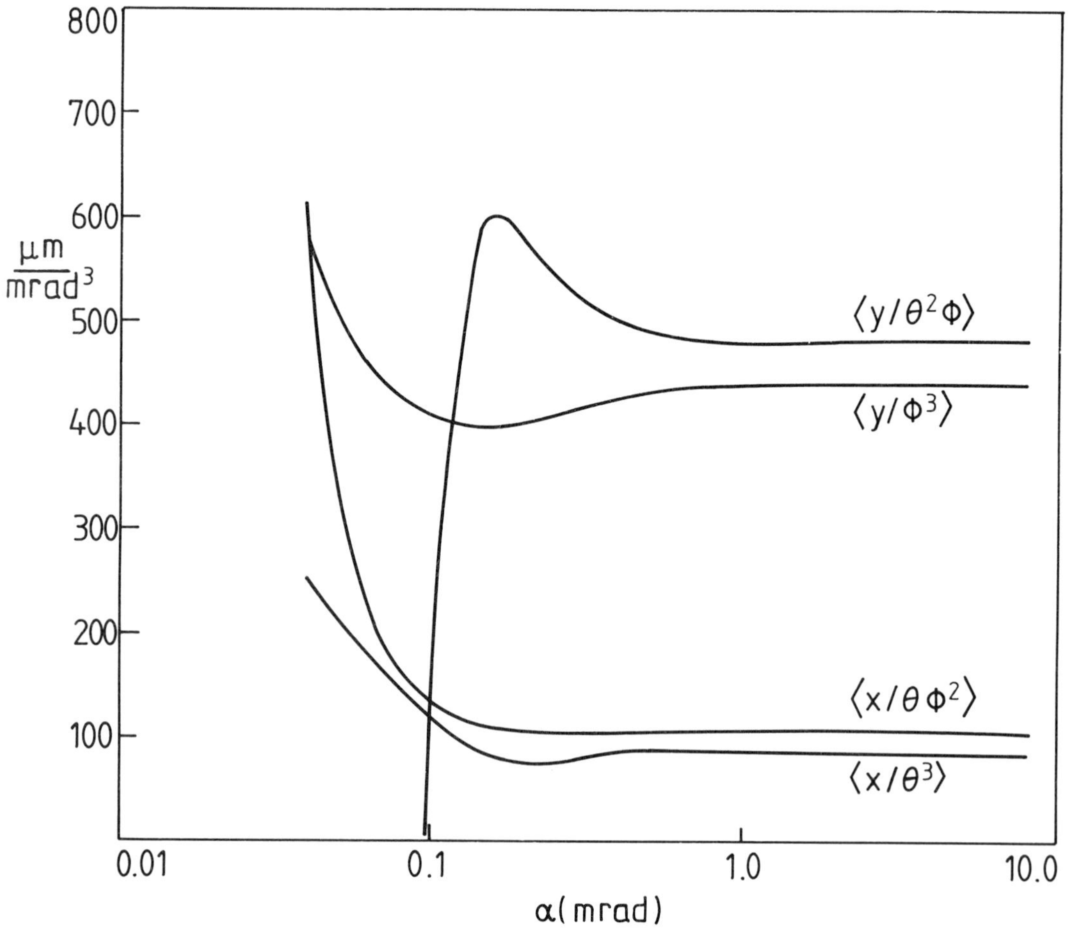

Figure 4.6 Variation of third-order coefficients with maximum object divergence: fifth-order fit.

Figure 4.7 Image intensity distributions obtained for a point object using the Oxford triplet with 10% 12-pole harmonic contamination in the final element: (a) calculated directly, (b) using first-order aberration coefficients, (c) using up to second-order coefficients, (d) using up to third-order coefficients, (e) using up to fourth-order coefficients, (f) using up to fifth-order coefficients. 3000 points used in each distribution. Each plot represents an area of 20 × 20 μm² and the entrance divergence is 0.1 mrad.

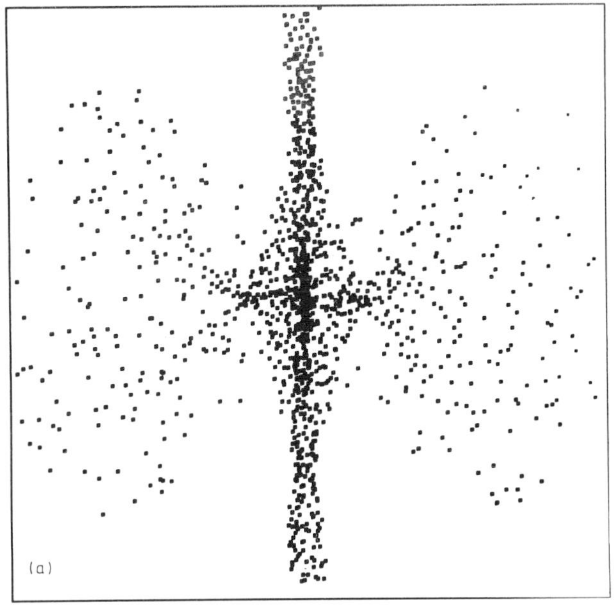

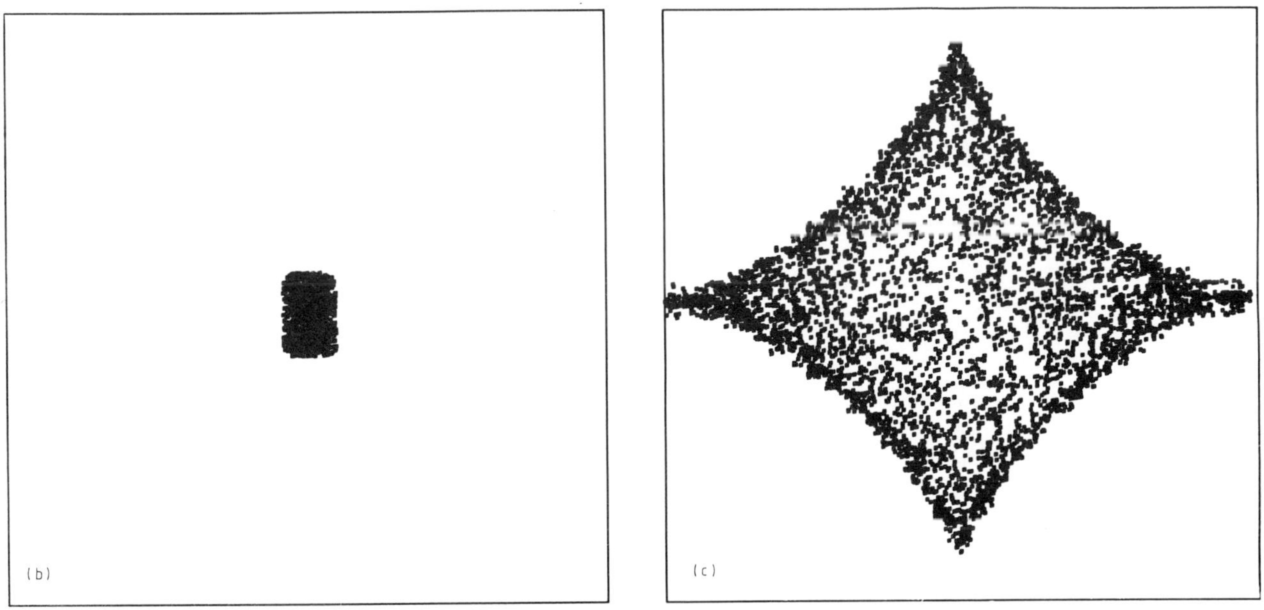

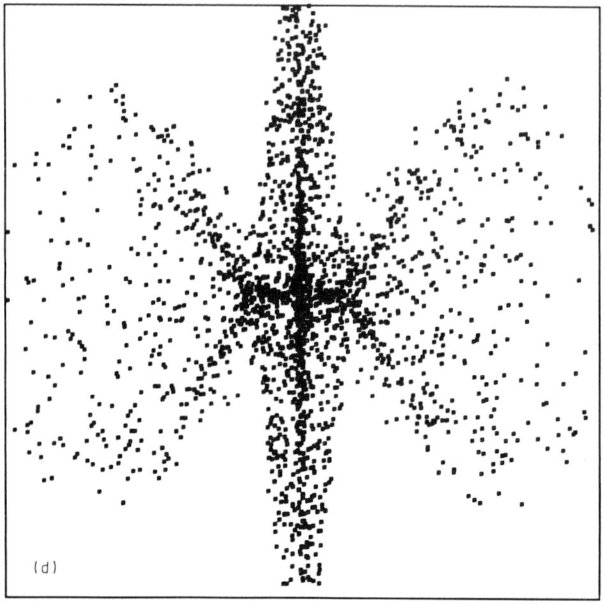

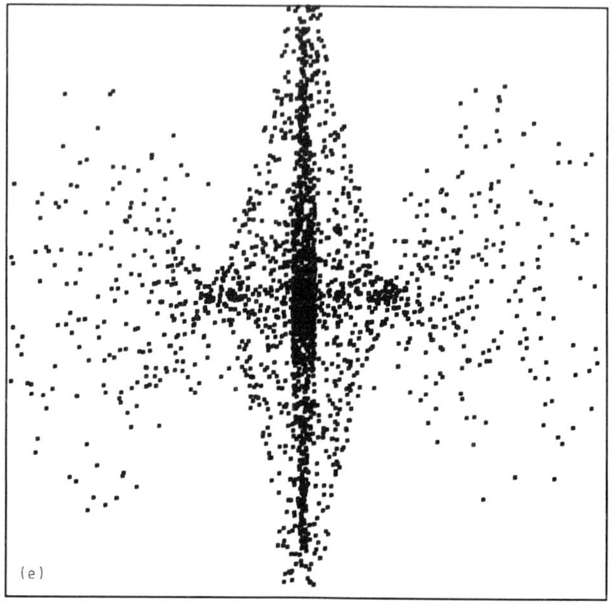

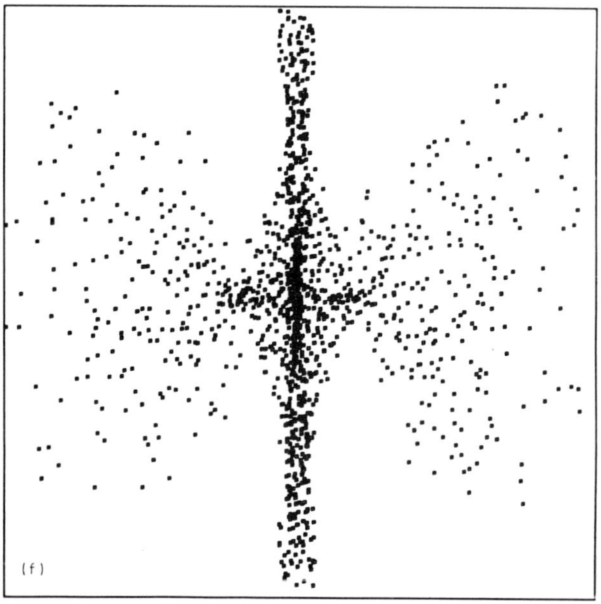

case, i.e. no arbitrary restrictions of order or accuracy have been imposed. The major drawback is that, in order to obtain a smooth distribution, of the order of 2000 rays must be traced. This is very time consuming and may be prohibitively expensive.

The alternative approach is to produce a representative set of aberration coefficients (as described in the preceding section) and then by using random values for the object variables obtain the image coordinates from equation (4.23). In this way, the many points required to obtain the image distribution may be calculated very quickly, but the detail shown depends on the aberration coefficients which were selected to represent the imaging. In order to demonstrate this, an artificially exaggerated dodecapole harmonic component was introduced into the final element of the Oxford Triplet [WA82] to produce an image with complex structure. The aberration pattern of a point object was then calculated first by direct raytracing and then by using aberration coefficients of maximum order 1, 2, 3, 4 and 5. The resulting intensity distributions are plotted in figure 4.7. The diagrams show the increase in detail with increasing order and further show that in this case a representative image is only obtained for orders greater than third.

4.5 Assessment of accuracy

The accuracy of the OXRAY raytracing package is difficult to assess. It is possible to make an internal estimate of the size of random errors (e.g. truncation and arithmetic errors), but systematic errors, such as the use of insufficiently precise expressions for the magnetic fields, may only be estimated by comparison with other calculations or measurements, of which few are available to the necessary accuracy. An indication of the internal accuracy is obtained from figure 4.4, in which the image distance, which is calculated from the ratio of very small displacements and divergences at the exit of the system, is seen to be constant over a range of four orders of magnitude of the entrance divergence.

A further check of the internal accuracy has been carried out by Blower [BL83]. As a demonstration of his result that the sum of the rotation aberration coefficients in one plane of a microprobe system is zero (see §6.4), he has calculated this sum using raytracing for many different quadrupole configurations. In all cases the result was zero, to an extremely high accuracy. This indicates that not only the raytracing but also the expressions for the misaligned magnetic fields are internally consistent.

The comparison of the raytracing program with other beam-optics calculations is less conclusive. This is due to the fact that the only programs available for comparison have been matrix programs, with their inherent limitations.

Comparison with experiment is even more difficult, since the errors arising in the measurement of aberration coefficients of microprobe systems tend to be large. The comparison was attempted for the Oxford coupled triplet between measured coefficients and calculated values obtained using OXRAY and two matrix programs: TRANSPORT [BR67, BR70, BR80] and IONBEAM [HE76a]. The results are presented in table 4.1. This table is reproduced from [GR82], which also gives details of the way in which the aberration coefficients were measured. Note that there are two columns of calculations using OXRAY. The calculations labelled SCOFF were carried out using a sharp cut-off fringing field in order to correspond to the conditions used in the matrix programs. The other column labelled EFF uses an extended fringing field to correspond more closely to experimental conditions. The results in table 4.1 show that OXRAY agrees well with matrix programs, especially for the first- and second-order properties, and although the errors in the measurements make a detailed comparison difficult, the agreement is good enough to suggest that the program is functioning correctly.

Table 4.1 Measured and calculated coefficients for the Oxford triplet.

Coefficient	Units	TRANSPORT	IONBEAM	OXRAY (SCOFF)	OXRAY (EFF)	Measured
$Q_1 = -Q_2$	T	0.145665	0.145880	0.144499	0.156034	0.146
Q_3		0.174776	0.177580	0.173848	0.189018	0.176
D_x		60.55	62.89	60.59	67.80	n
D_y		−14.13	−14.33	−14.13	−15.21	n
$\langle x\|U_1\rangle$	μm/μm	−0.5125	0.53*	−0.5121	−0.48	−0.51
$\langle x\|U_2\rangle$		−0.4239	0.44*	−0.4235	−0.43	−0.42
$\langle x\|U_3\rangle$		1.9200	2.02*	1.9191	1.90	1.88
$\langle y\|V_1\rangle$		−3.7124	3.63*	−3.7106	−3.74	n
$\langle y\|V_2\rangle$		7.9961	7.82*	7.9920	8.07	n
$\langle y\|V_3\rangle$		−3.2131	3.15*	−3.2114	−3.27	n
$\langle x\|x\delta\rangle$	μm/μm%	−0.0488	−0.0484	−0.0476	−0.0508	n
$\langle y\|y\delta\rangle$		0.1363	0.1364	0.1307	0.0812	n
$\langle x\|\theta\delta\rangle$	μm/mrad%	−258.5	−256.6	−254.6	−246.31	191.3*
$\langle y\|\phi\delta\rangle$		774.4	775.6	760.8	782.97	854.9*
$\langle x\|\phi\rho_1\rangle$	μm/mrad^2	6.972*	7.270*	6.962	6.57	11.0*
$\langle x\|\phi\rho_2\rangle$		12.974*	12.762*	12.953	13.52	14.7*
$\langle x\|\phi\rho_3\rangle$		19.941*	19.552*	−19.925	−20.09	13.3*
$\langle y\|\theta\rho_1\rangle$		29.853*	29.939*	−29.818	−30.02	20.2*
$\langle y\|\theta\rho_2\rangle$		55.575*	56.007*	−55.56	−59.90	46.6*
$\langle y\|\theta\rho_3\rangle$		85.432*	85.618*	85.40	89.61	73.2*
$\langle x\|\theta\varepsilon_1\rangle$	μm/mrad%	23.19*	22.74*	23.31	21.71	8.2*
$\langle x\|\theta\varepsilon_2\rangle$		21.57*	21.05*	−21.44	−23.37	14.0*
$\langle x\|\theta\varepsilon_3\rangle$		256.15*	248.58*	256.48	261.77	205.1*
$\langle y\|\phi\varepsilon_1\rangle$		279.55*	286.29*	278.96	268.51	295.8*
$\langle y\|\phi\varepsilon_2\rangle$		1242.8*	1212.68*	−1243.9	−1281.3	1283.5*
$\langle y\|\phi\varepsilon_3\rangle$		190.91*	195.14*	190.68	200.41	203.4*
$\langle x\|\theta^3\rangle$	μm/mrad^3	—	26.73	35.76	87.92	290
$\langle x\|\theta\phi^2\rangle$		—	15.23	14.83	111.1	n
$\langle y\|\phi^3\rangle$		—	−1065	−90.65	−437.9	−512
$\langle y\|\theta^2\phi\rangle$		—	66.69	−76.52	−427.6	n

Q_n is the pole-tip field of quadrupole n
D_x and D_y are the demagnifications
* sign not determined

n not measured
— not calculated (second-order program)
Other symbols are as defined in table 1.2.

5

The imaging properties of quadrupole probe-forming systems

5.1 Introduction

The discussion in §1.12 indicated that it would be desirable to undertake a systematic evaluation of all the combinations of quadrupole lenses so that the most promising systems for microprobe applications may be selected. The effort involved in calculating and presenting data for all possible combinations of quadrupoles would be prohibitive, and so in this book we have selected a representative set of systems for which we have calculated the imaging properties of interest. The results of this survey, which form a large part of this book, are presented in Appendix 1, and the remainder of this chapter shows how they may be applied.

Since there seems to be no obvious reason at this stage for considering any system using more than four quadrupole lenses, only doublets, triplets and quadruplets have been considered. In addition, in order to simplify the calculation and presentation of the data, it is assumed that the excitations of the lenses are coupled so that there are only two independent excitation parameters. Thus in the case of the triplets, two of the lenses have equal excitations, while for the quadruplets the lenses are excited in pairs. It should be noted that only the magnitude of the excitations are equal; coupled lenses may still be of opposite polarity.

Coupling of the lenses also leads to a degree of convenience in the construction of practical microprobe systems since not only is it possible to operate the system with only two power supplies but also having only two power supplies to adjust makes the task of focusing the system much easier.

The systems selected for investigation, therefore, are the doublet, the three alternating coupled triplet combinations and the nine coupled quadruplet configurations, i.e.

CD		
CDC	CDC	CDC
CDCD	CDCD	CDCD
CDDC	CDDC	CDDC
CCDD	CCDD	CCDD

Each of these configurations has been investigated for several interlens spacings. As an example, to clarify the notation, CDCD represents a system of four quadrupoles arranged converging, diverging, converging, diverging (in the order in which the beam passes through the system) with the first two lenses coupled and the third and fourth lenses coupled. The results of calculations of the first-, second- and third-order properties of these configurations are presented in Appendix 1. The first- and second-order properties were calculated using the matrix techniques described in chapter 3 and are shown graphically. The third-order aberration coefficients were calculated using the raytracing program OXRAY described in chapter 4 and are presented in tabular form.

It will be seen that several of the configurations listed above are absent from Appendix 1. These are CCDD and CCDD which feature no probe-forming properties over the range of excitations considered, and CCDD, which is equivalent to a doublet.

Parasitic aberrations have not been considered in this chapter since it is not feasible to attempt a systematic survey of the many parasitic aberrations which affect probe-forming systems. Parasitic aberrations are considered in more general terms in the next chapter.

5.2 The choice of system parameters

In order to make the results of Appendix 1 independent of the quadrupole lengths, all distances and displacements are expressed in units of the quadrupole effective length (QL) (equation (4.13)). It is assumed that all the quadrupoles in a system are of equal length.

The object distance (i.e. the distance between the object aperture and the effective field boundary of the first lens) is taken throughout to be 20 QL. The effects of varying the object distance are discussed in §5.5.

Each configuration is considered for interlens spacings of 0,1,2 and 5 quadrupole lengths. As will be seen from the curves for the doublets, the introduction of lens spacings for this configuration results in a degradation of the first-order properties, specifically the x-demagnification. The lens separations for the triplet therefore are chosen to be between the first and second lenses, thereby maintaining a close quadrupole pair as the last two elements. For similar reasons, the lens spacings for the quadruplets are chosen to be between the second and third lenses.

The first- and second-order properties were calculated using a rectangular field profile for the quadrupole lens, while for the calculations of the third-order properties using raytracing, a longitudinal profile based on the Oxford lens [WA82] has been used (§2.2). The third-order properties are therefore derived using realistic fringing fields. The effect of the fringing fields on the third-order properties are discussed in the next chapter.

5.3 Interpretation of the results in Appendix 1

For each configuration, eight graphs and a table of spherical aberration coefficients are presented. The type of configuration is included diagrammatically for the first- and second-order properties and in abbreviated form for the third-order tables.

The first graph of each group shows the values of the two excitation parameters (K QL)† (defined by equations (3.33)) for which a stigmatic focus is obtained. This plot, commonly called a 'load curve', represents the complete imaging behaviour of the lens system over a range of K QL values from 0 to π with no restriction imposed on the image distance or image parameters.

The plot of excitation versus image distance enables the user to select, in conjunction with the load curve, the excitations of the lenses which will give a focus at a particular image distance (measured in units of quadrupole length from the effective field boundary of the last lens in the system). The image distance is limited, for practical reasons, to between 0 and 5QL. The remaining imaging parameters are presented as a function of image distance so that the effects of varying the image distance, which in the design of a probe-forming system must be determined at an early stage, may be readily seen. Although the graphs are self-explanatory, it should be noted that the chromatic aberration curves for $<x|\theta\delta>$ *and* $<y|\phi\delta>$ are normalised to the quadrupole length. The quantities plotted are therefore $<x|\theta\delta>/\mathrm{QL}$ and $<y|\phi\delta>/\mathrm{QL}$. In these plots, δ is the fractional momentum shift $\delta p/p$ of the particle and the angles θ and ϕ are measured in radians.

The values plotted on all the graphs, apart from the load curve, have been restricted to 'reasonable' values, so that the detail of potentially useful systems is not masked by inappropriate scaling. These restrictions are 'global', so that if a parameter exceeds the bounds on one graph, the point is omitted from all the other graphs apart from the load curve. This is the reason for the apparent discontinuities in some of the plots which enable the reader to see instantly that a parameter has gone 'out of bounds'.

The restrictions used in plotting the graphs are, for the demagnification, $|D_x|,|D_y| <500$ and for the chromatic aberration $|<x|x\delta>|$, $|<y|y\delta>| <1000$ and $|<x|\theta\delta/\mathrm{QL}|$, $|<y|\phi\delta>/\mathrm{QL}\,| <1000$.

It will be seen that, for many of the configurations, a focus may be achieved at the same image distance corresponding to up to four different values of the quadrupole excitations. These different 'branches' may be related to the path of the beam as it traverses the lens system. If the demagnification in a particular plane is negative, then using the definition $D_x = <\theta|\theta>$ introduced in chapter 3, it follows that a ray must emerge from the lens system with the opposite slope to that with which it entered the system.

For a positive image distance, this can only take place if the ray does not cross the axis or if the ray

† Note that K is used here in preference to k, since the computer-generated plots in Appendix 1 also use K.

crosses the axis an even number of times. Similarly, if the demagnification is positive, the rays must cross the axis an odd number of times. For all the systems considered here, the number of crossovers is never more than one, so it may be stated that if $D<0$ there is no crossover and if $D>0$ there is a crossover. Thus an insight into the trajectory of the beam within the system may be obtained from the plots of D_x and D_y. In addition we may identify four possible situations which correspond to the various branches of the load curve for multifocusing systems:

A	$D_x<0$, $D_y<0$	no crossover before focus
B	$D_x>0$, $D_y<0$	crossover in the x-plane
C	$D_x<0$, $D_y>0$	crossover in the y-plane
D	$D_x>0$, $D_y>0$	crossover in both planes

In cases where there may be possible misinterpretation, the branches have been labelled A, B, C and D, corresponding to the above cases.

It is worth commenting at this stage that, in general, crossovers are associated with higher demagnification figures and correspondingly higher aberration coefficients.

The spherical aberration coefficients are presented in tabular form and again are normalised to the quadrupole length QL. In the tabulation, each lens configuration is labelled by a schematic representation. As an example, 20$\overline{\mathrm{C1D}}$0C represents a triplet system with polarities C D C in which the first two lenses are coupled. The normalised object distance is 20 quadrupole lengths with a one quadrupole length spacing between the first and second lens and zero spacing between the second and third lens.

Where possible, the aberration coefficients are calculated for every branch investigated. The different branches are not labelled in the tables. Since the aberration coefficients are tabulated together with quadrupole excitation values and image distance, direct correlation with the associated branch in the first- and second-order plots is straightforward.

The coefficients are given to three significant figures where possible, and are presented in scientific notation with the exponent in brackets (for example, 2.61 (2) is equivalent to 2.61×10^2).

5.4 Calculation of the image size using Appendix 1: Worked example

Consider the design of a probe-forming system utilising magnetic quadrupoles with a 10 cm effective length and a radius of 1.5 cm in the following configuration:

Particle	4 MeV proton
Arrangement of lenses	Doublet (CD—no space)
Object distance	200 cm (20 QL)
Required image distance	15 cm (1.5 QL)
Object aperture dimensions	$20 \times 20\ \mu m^2$ ($x=y=\pm 10\ \mu m$)
Entrance angle divergence	1×1 mrad^2 ($\theta=\phi=\pm 0.5$ mrad)
Momentum spread	0.1% ($\delta=\pm 0.0005$)

This system (CD) is the first configuration encountered in Appendix 1 and the relevant optical parameters can be extracted as follows:

(1) From the image distance curves we find that for a normalised image distance of 1.5, the excitation of the second quadrupole is KL = 1.1 (k_1 = 11m^{-1}).
(2) Using the load curve, the excitation of the first quadrupole is KL = 0.9 (k_2 = 9m^{-1}).
(3) Using table 3.1 or equation (3.36), the pole-tip fields can be estimated to be B_1= 0.34 T and B_2=0.51 T.
(4) Using the demagnification curves, the values of the demagnifications for the image distance of 1.5QL are $D_x = -3.6$ and $D_y = -20$.
(5) From the second-order plots, the chromatic aberration coefficients at the required image distance are

$$<x|x\delta> = 2.55$$

$$<y|y\delta> = 2.08$$

$$<x|\theta\delta>/QL = 49 \quad \therefore \quad <x|\theta\delta> = 4.9$$

$$<y|\phi\delta>/QL = 44 \quad \therefore \quad <y|\phi\delta> = 4.4$$

(6) From the third-order aberration table, the relevant spherical aberration coefficients are

$$<x|\theta^3>/QL = -7.80\times 10^3 \qquad \therefore \quad <x|\theta^3> = -7.80\times 10^2$$

$$<x|\theta\phi^2>/QL = -7.2\times 10^4 \qquad \therefore \quad <x|\theta\phi^2> = -7.20\times 10^3$$

$$<y|\theta^2\phi>/QL = -1.4\times 10^4 \qquad \therefore \quad <y|\theta^2\phi> = -1.4\times 10^3$$

$$<y|\phi^3>/QL = -3.2\times 10^4 \qquad \therefore \quad <y|\phi^3> = -3.2\times 10^3$$

(units : rad^{-3}) (units : mrad^{-3})

(7) Having obtained the relevant coefficients, the first-order spot size and the contributions due to aberrations can be calculated. Using the expression for the ray displacements given in §1.10, the first-order spot has dimensions

$$20\ \mu m/|D_x| = 5.6\ \mu m\ (x\text{-direction})$$
$$20\ \mu m/|D_y| = 1.0\ \mu m\ (y\text{-direction}).$$

The chromatic aberration increases the spot size equally in the positive and negative x-directions by an amount

$$\Delta x_{ch} = |<x|x\delta>\, x_{max}\, \delta_{max}| + |<x|\theta\delta>\, \theta_{max}\, \delta_{max}|.$$

The total broadening in the x-direction is then $2\Delta x_{ch}$, i.e.

$$\begin{aligned}2\Delta x_{ch} &= 2\,(2.55\times 10\times 10^{-6}\times 0.0005 + 4.9\times 0.5\times 10^{-3}\times 0.0005)\ m\\ &= 2.47\times 10^{-6}\ m.\end{aligned}$$

Hence the total broadening, $2\Delta x_{ch}$ = 2.47 μm.
Similarly, in the y-direction, the chromatic broadening is

$$\Delta y_{ch} = |<y|y\delta>\, y_{max}\, \delta_{max}| + |<y|\phi\delta>\, \phi_{max}\, \delta_{max}|.$$

Therefore:

$$\begin{aligned}2\Delta y_{ch} &= 2\,(2.08\times 10\times 10^{-6}\times 0.0005 + 4.4\times 0.5\times 10^{-3}\times 0.0005)\ m\\ &= 2.22\times 10^{-6}\ m,\end{aligned}$$

i.e. the total broadening $2\Delta y_{ch}$=2.22 μm. Adding the chromatic effects to the first-order image dimensions gives an aberrated spot with maximum dimensions 8.1 × 3.22 μm^2.
The effect of spherical aberration is to draw out the image of a point object into four cusps (see for example figure 6.6a). An examination of the tables of spherical aberration coefficients shows that the two coefficients relating to each plane of the system are always of the same sign. Thus the maximum

extent of the spherical aberration pattern, and hence the contribution to the broadening of the spot, is given for the positive x-direction as

$$\Delta x_{\text{sph}} = \left| \langle x|\theta^3\rangle\, \theta^3_{\max} + \langle x|\theta\phi^2\rangle\, \theta_{\max}\, \phi^2_{\max} \right|$$

and the total contribution to the broadening is $2\Delta x_{\text{sph}}$. Similarly, the total broadening in the y-direction is $2\Delta y_{\text{sph}}$ where

$$\Delta y_{\text{sph}} = \left| \langle y|\phi^3\rangle\, \phi^3_{\max} + \langle y|\theta^2\phi\rangle\, \theta^2_{\max}\, \phi_{\max} \right|.$$

Using the numerical values of the spherical aberration coefficients extracted from the tables we obtain

$$\begin{aligned} 2\Delta x_{\text{sph}} &= 2\,[7.8 \times 10^2 \times (0.5 \times 10^{-3})^3 + 7.2 \times 10^3 \\ &\quad \times (0.5 \times 10^{-3})^3] \\ &= 2.0 \times 10^{-6}\ \text{m} = 2.0\ \mu\text{m} \\ 2\Delta y_{\text{sph}} &= 2\,[3.2 \times 10^3 \times (0.5 \times 10^{-3})^3 + 1.4 \times 10^3 \\ &\quad (0.5 \times 10^{-3})^3] \\ &= 2.3 \times 10^{-6}\ \text{m} = 2.3\ \mu\text{m}. \end{aligned}$$

The final maximum dimensions of the spot are then $10.1 \times 5.5\ \mu\text{m}^2$.

It should be emphasised that these represent the maximum limits of the beam spot. The distribution of intensity within these limits will depend on the distribution of tne incident beam within the entrance phase space. This is demonstrated in figure 5.1, in which the actual intensity distribution calculated directly by raytracing using OXRAY (see §4.4.4) is plotted. The broken lines indicate the extremes of the image calculated above. The diagram shows that the values calculated from Appendix 1 provide a good estimate of the extremes of the image, although the majority of the beam flux is contained within a smaller area.

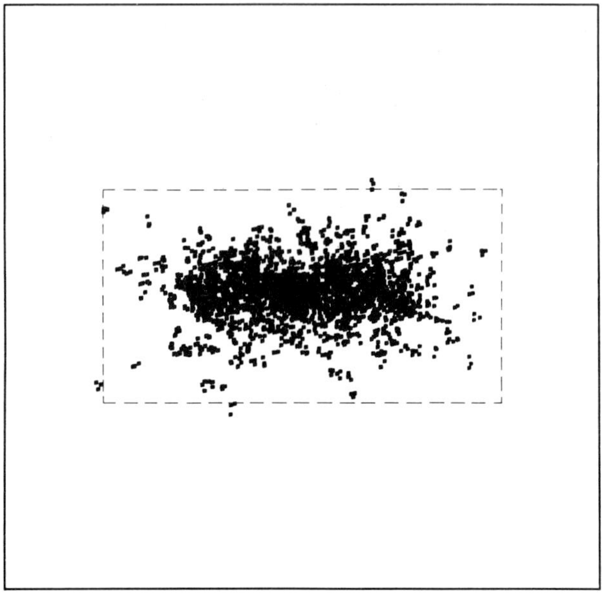

Figure 5.1 The image intensity distribution of the 20COD system described in §5.4 as calculated directly by OXRAY. The dotted lines indicate the image size predicted from Appendix 1. The distribution is built up from 1000 rays and represents an area of $15 \times 15\ \mu\text{m}^2$.

5.5 The dependence of imaging properties on object distance

The results presented in Appendix 1 were calculated for a fixed object distance of 20 quadrupole lengths. This restriction was introduced in order to reduce the amount of information to be tabulated to reasonable proportions. However, it is possible to make the results more general using simple relationships between the aberration coefficients and the object distance. It is found that over a restricted range of object distances, for a fixed image distance the following quantities remain approximately constant as the normalised object distance (U_N) is varied:

Parameter	Invariant quantity
Excitation	k_n
Demagnification	D_x/U_N, D_y/U_N
Object chromatic aberration	$\langle x\|x\delta\rangle$, $\langle y\|y\delta\rangle$
Angle chromatic aberration	$\langle x\|\theta\delta\rangle/U_N$, $\langle y\|\phi\delta\rangle/U_N$
Spherical abberation	$\langle x\|\theta^3\rangle U_N^3$, $\langle x\|\theta\phi^2\rangle/U_N^3$
	$\langle y\|\phi^3\rangle/U_N^3$, $\langle y\|\theta^2\phi\rangle/U_N^3$

This invariance is demonstrated in figures 5.2–5.4, which show respectively the variation in excitations, normalised demagnifications and normalised chromatic aberrations for a C2D0C triplet configuration as a function of normalised object distance.

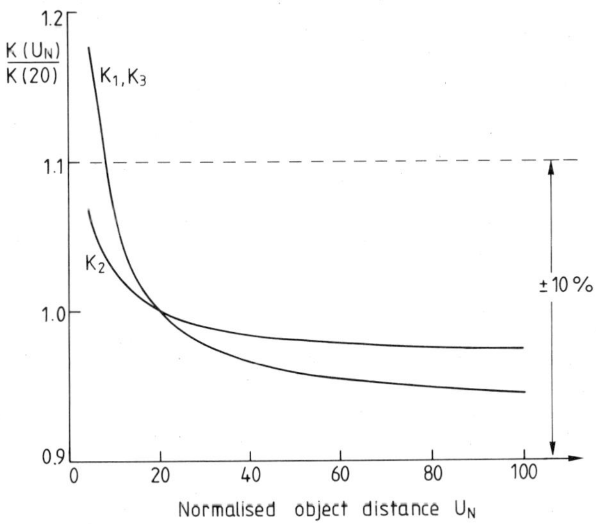

Figure 5.2 Excitations as a function of object distance with constant image distance for a C2D0C system.

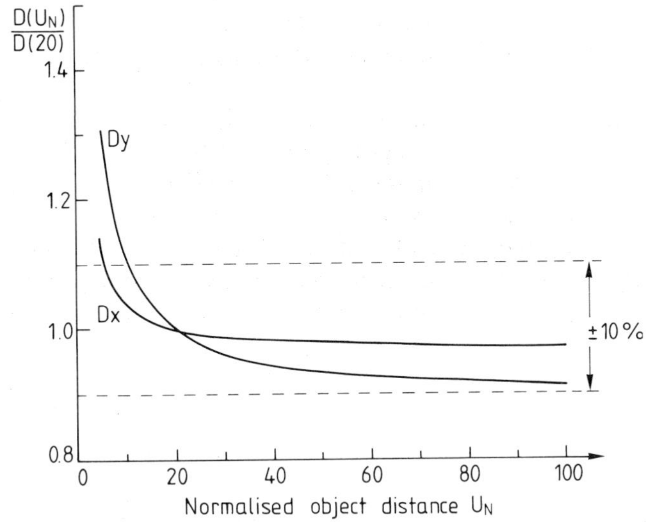

Figure 5.3 Normalised demagnification as a function of object distance at a constant image distance for a C2D0C system.

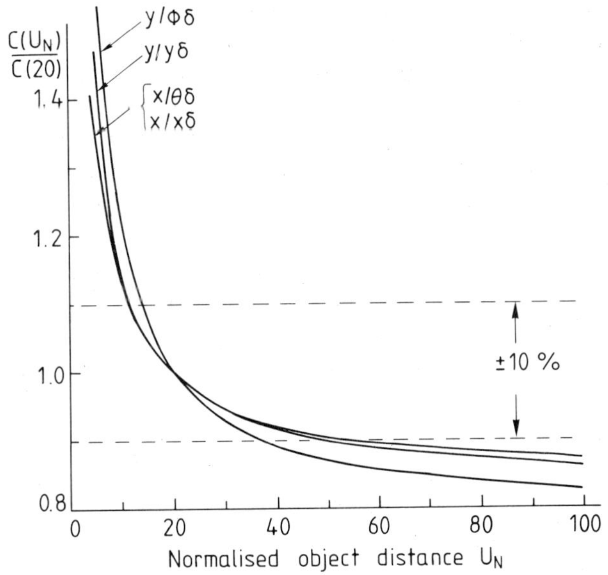

Figure 5.4 Normalised chromatic aberration as a function of object distance at a constant image distance for a C2D0C system.

The image distance in each case is constant and has a value of 0.75QL.

The quantities are plotted as a ratio of their value at U_N to that at $U_N = 20$. The diagrams show that over a range of U between approximately 10 and 50 the imaging properties for this configuration remain within 10% of their tabulated values at $U_N = 20$. Calculations have demonstrated this behaviour for other systems, although it should be emphasised that the range of object distances over which the approximation is valid may be different.

The above relationships may now be used to calculate the imaging parameters at the new object distance from the parameters determined for $U_N = 20$. To continue the example introduced in the previous section, let us now calculate the optical parameters for an object distance of $U_N = S$, where $10<S<50$. We can approximate as follows:

(1) For a stigmatic image at the normalised image distance of 1.5QL, the quadrupole excitations remain as before.

(2) Since the values D_x/U_N and D_y/U_N are approximately constant, the demagnifications change to $D_s = S/20\, D_{20}$ where $D_s = D_x$ or D_y at $U_N = S$.

(3) The object chromatic aberration coefficients $<x|x\delta>$ and $<y|y\delta>$ remain unchanged.

(4) The angle chromatic aberration coefficients $<x|\theta\delta>$ and $<y|\phi\delta>$ then become

$$<x|\theta\,\delta>_s = (S/20)<x|\theta\,\delta>_{20}$$

$$<y|\phi\,\delta>_s = (S/20)<y|\phi\,\delta>_s.$$

(5) The spherical aberration coefficients become

$$<x|\theta^3>_s = (S/20)^3 <x|\theta^3>_{20}$$

$$<x|\theta\,\phi^2>_s = (S/20)^3 <x|\theta\,\phi^2>_{20}$$

$$<y|\theta^2\phi>_s = (S/20)^3 <y|\theta^2\phi>_{20}$$

$$<y|\phi^3>_s = (S/20)^3 <y|\phi^3>_{20}.$$

Subscript $_S$ denotes parameters calculated for a normalised object distance of S.

In conclusion, we have shown in this chapter how, using the graphs and tables in Appendix 1, together with some simple normalising relationships, it is possible to predict the imaging properties of a broad spectrum of quadrupole probe-forming systems. The accuracy with which the parameters can be extracted is necessarily limited by the errors in reading the graphs and the approximations involved, but it is hoped that this section will help guide the designer toward the most suitable system for his requirements, and once having selected a system, he is recommended to carry out more detailed calculations using one of the beam-optics programs available.

6

Parasitic aberrations in quadrupole probe-forming systems

The aim of this chapter is to try to deal with the practical aspects of achieving maximum spatial resolution (i.e. the smallest beam-spot size) of a microprobe system, with the emphasis placed upon the understanding of the parasitic beam-optical aberrations. It is beyond the scope of this book to carry out a systematic survey of the effects of parasitic aberrations in the same manner as the investigations already carried out into the intrinsic aberrations of probe-forming systems. Nevertheless, it is hoped that an investigation into the effects of parasitic aberrations on several specific microprobe systems will indicate which of the parasitic aberrations are important.

It is assumed in this chapter that a typical microprobe system is represented as shown in figure 6.1. A demagnified image of an object aperture is produced at the target position by means of magnetic quadrupole elements. Adjustable collimator slits before the focusing system serve to limit the angular divergence of the beam in both the x- and y-planes.

The effects of parasitic aberrations are calculated using OXRAY and the results are depicted either by aberration coefficients (see §1.10) or by scatter plots (see §4.4.4). Where the results are presented as scatter plots, it is taken throughout that the object dimensions are zero. The scatter plots therefore depict the shape and increase in the beam spot due to a particular parasitic aberration in relation to the intrinsic spherical aberration of the system.

In several sections in this chapter the effects of parasitic aberrations in three existing microprobe systems are considered. The chosen systems are the Heidelberg doublet [NO77], the Oxford triplet [WA82] and the Harwell quadruplet [CO72], representing a reasonably broad base of con-

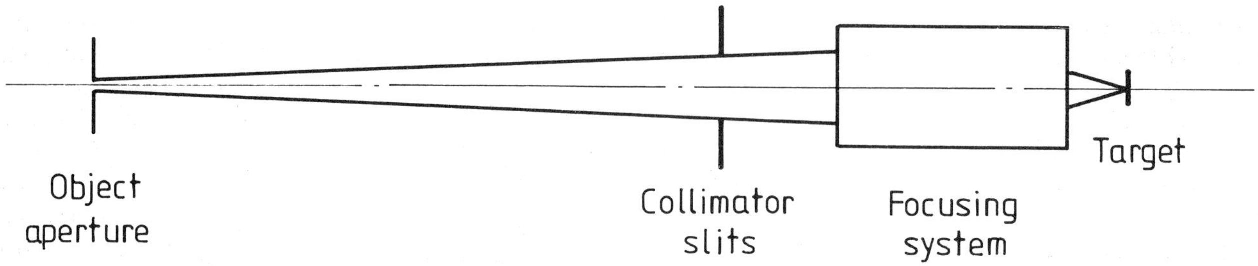

Figure 6.1 Schematic diagram of a microprobe system.

Table 6.1 The geometry and relevant beam-optical parameters of the Heidelberg doublet [NO77], the Oxford coupled triplet [WA82] and the Harwell Russian Quadruplet [CO72].

	Heidelberg Doublet	Oxford Triplet	Harwell Quadruplet
Beam energy	2 MeV	4 MeV	3 MeV
Quadrupole length (geometric)	40 mm	200 mm	180.5 mm
Quadrupole length (effective)	42.7 mm	214 mm	201.6 mm
Gap between adjacent quadrupoles	40 mm	46 mm	45 mm
Pole gap	5 mm	30 mm	38.4 mm
Object distance	1.84 m	5.786 m	3.5 m
Image distance	100 mm	160 mm	210 mm
Assumed fractional beam-momentum spread (δ)	0.05%	0.025%	0.05%
1st quadrupole excitation (k)	14.84 m^{-1}	5.987 m^{-1}	3.120 m^{-1}
2nd quadrupole excitation (k)	−18.99 m^{-1}	−5.987 m^{-1}	−4.801 m^{-1}
3rd quadrupole excitation (k)	—	6.532 m^{-1}	4.801 m^{-1}
4th quadrupole excitation(k)	—	—	−3.120 m^{-1}
Vertical demagnification (y)	−26.4	−15	−5.3
Horizontal demagnification (x)	4.6	68	−5.3
Major chromatic aberration coefficients			
$<x\|\theta\delta>$	36.5	−256	63.7
$<x\|\phi\delta>$	39.6	783	103.7
Spherical aberration coefficients			
$<x\|\theta^3>$	−3.1	87.9	−1.1
$<y\|\phi^3>$	−1.1	−438	−4.6
$<x\|\theta\phi^2>$	−1.6	111	−5.8
$<y\|\phi\theta^2>$	−7.9	−428	−5.6

All geometries have been extracted from the relevant publications. All beam-optical parameters have been calculated using OXRAY.

figurations. The geometries and relevant beam-optical parameters of each system are tabulated in table 6.1.

Since the effects of the parasitic aberrations on the spot size can be dependent on the input beam divergence angles, nominal values for the input divergence angles have been chosen where relevant. It must be stressed, therefore, that this chapter is not intended to be a comparison between the three chosen systems, but rather an attempt to compare the effect of different parasitic aberrations for each system. A quadrupole alignment technique is described in which the effects of horizontal, vertical, tilt and rotational misalignments within the focusing system are minimised.

The chapter concludes with a brief description of the ways in which a focused beam may be scanned over the sample, with the emphasis on the beam-scanning system used at Oxford.

6.1 Fringing fields

Whether or not the shape of the longitudinal field profile plays an important part in probe formation is an important question when considering the design and construction of individual elements within the focusing system. In general, the fringing field of a quadrupole lens introduces both an axial component of the magnetic field and higher-order field harmonics into the quadrupole end-fields over a region approximately equal to the radius of the pole gap (see §2.2). One would expect the deviation from the pure quadrupole field to be greatest for thin lenses, i.e. lenses with a relatively high bore to length ratio. The effect of the shape of the longitudinal field profile on the attainable spot size is difficult to categorise, since any effect will be dependent on the quadrupole configuration used in the probe formation and the geometry of the elements within this configuration.

A two-part investigation has been carried out in an attempt to estimate:

(1) The effect of the form of the fringing field on the spot size for the Oxford coupled triplet configuration, with the effective lengths of the lenses in each case remaining constant.

(2) The effect on the spot size of varying the quadrupole bore to length ratio for the Oxford coupled triplet configuration i.e. varying the fringe-field region to quadrupole effective length ratio, with the lens excitations adjusted for each case to maintain the first-order focusing properties.

(1) OXRAY has been used to investigate the effect of using several fringing-field profiles in the coupled triplet configuration of the Oxford microprobe. The fringe-field profiles used in the calculations, together with the respective fringe-field coefficients as defined in §2.2 are shown in figure 6.2. The effective length for each selected profile is constant. The primary effect of introducing fringe fields into a focusing element is to produce an astigmatic image caused by weak focusing due to the z-field, which fortunately can be remedied by adjusting the excitation of the quadrupole lenses within the focusing system. In a practical approach to investigating the effect of the shape of the fringing fields on the spot size, it is sensible to observe the change in the third-order spherical aberration coefficients (and the introduction of any higher-order coefficients) with each selected profile after first adjust-

ing the quadrupole pole-tip fields so as to maintain a stigmatic image at a fixed image distance.

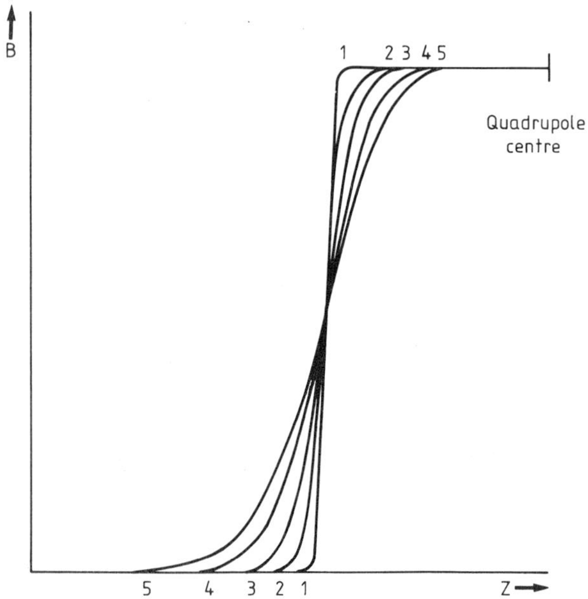

Figure 6.2 Fringe-field profiles of the quadrupole element used in the third-order beam-optics calculations. Profile 1: $c_1 = 20$, profile 2: $c_1 = 3.5$, profile 3: $c_1 = 2.0$, profile 4: $c_1 = 1.2$, profile 5: $c_1 = 0.9$; $c_0 = -6.0$, $c_2 = c_3 = c_4 = 0$ in all cases. Fringe-field coefficients are defined in §2.2. Profile 2 represents the field profile of the Oxford quadrupole.

A comparison of the results of calculations of the third-order properties for the selected profiles is shown in table 6.2. Several assumptions have been made in deriving these results. The fringing fields have been calculated using the expressions in chapter 2 and are hence derived from a pure quadrupole field with no parasitic harmonic components included. It is also assumed that there is no magnetic coupling between adjacent lenses, although any fringe field overlap is added algebraically.

Table 6.2 Tabulation of spherical aberration coefficients for the Oxford coupled triplet with various fringe-field profiles (see figure 6.2).

Field Profile	$\langle x\|\theta\phi^2\rangle$	$\langle x\|\theta^3\rangle$	$\langle y\|\phi^3\rangle$	$\langle y\|\theta^2\phi\rangle$
1	106	84	−421	−511
2	105	84	−447	−495
3	109	88	−462	−500
4	114	95	−488	−514
5	118	100	−514	−546

Profile 2 represents the field profile of the Oxford quadrupole. The field profiles refer to all quadrupole elements. All units are in mrad or μm.

From table 6.2 it is clear that, for the configuration chosen, the effect of the nature of the fringing field on the third-order spherical aberration coefficients appears to be small. No other significant aberration coefficients were introduced.
(2) OXRAY has also been used to investigate the effects of using various longitudinal field profiles in the coupled triplet configuration of the Oxford microprobe. The longitudinal field profiles used in

Table 6.3 Tabulation of spherical aberration coefficients for the Oxford coupled triplet with several longitudinal field profiles (see figure 6.3).

Field Profile	L_E/d	$\langle x\|\theta\phi^2\rangle$	$\langle x\|\theta^3\rangle$	$\langle y\|\phi^3\rangle$	$\langle y\|\theta^2\phi\rangle$
A	7.13	105	84	−447	−495
B	5.33	80	83	−350	−357
C	4.00	69	89	−333	−320
D	3.33	67	100	−336	−304
E	2.33	66	125	−388	−309

L_E is the effective length of the quadrupole lens and d is the pole gap. Profile A represents the field profile of the Oxford quadrupole. The field profiles refer to all quadrupole elements. All units in mrad or μm.

the calculations are shown in figure 6.3. These profiles have realistic fringing fields ($c_0 = -6$, $c_1 = 3.5$, $c_2 = c_3 = c_4 = 0$) and different effective lengths. In order that a reasonable comparison may be made for the selected profiles within each configuration, the pole-tip fields of each quadrupole element have been adjusted so that a stigmatic image is formed at a constant distance from the effective field boundary of the last quadrupole element.

Table 6.3 shows the results of calculations of the third-order properties for the selected elemental longitudinal profiles. Results in this table indicate that as the quadrupole elements tend towards

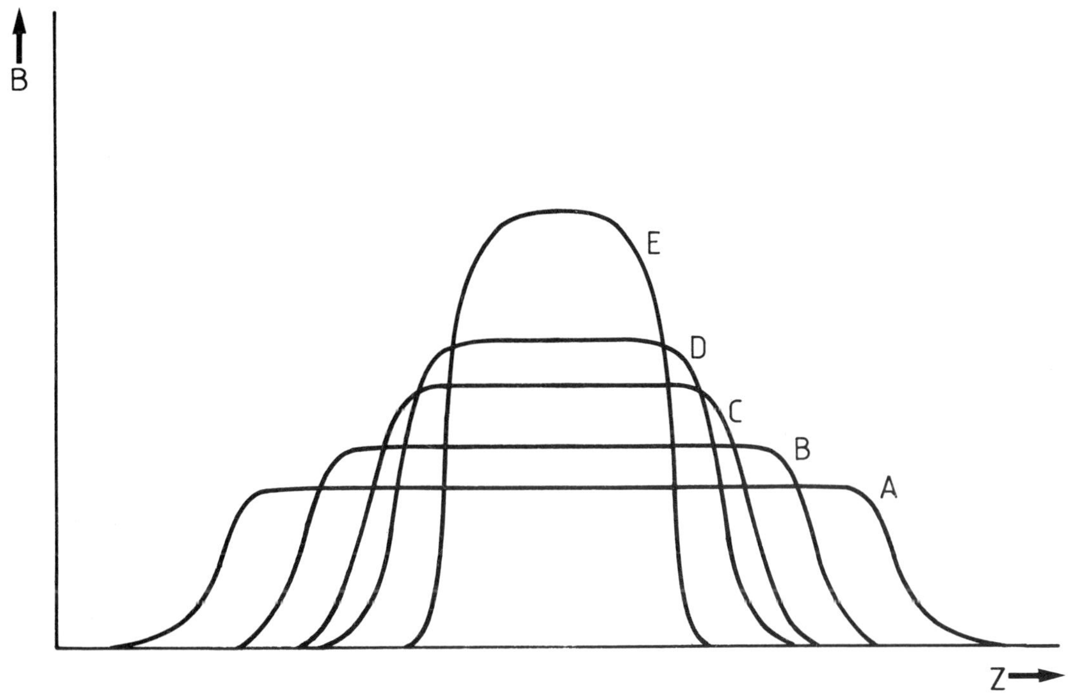

Figure 6.3 Longitudinal field profiles of the quadrupole element used in the third-order beam-optics calculations.
Profile A: effective length/bore diameter ratio = 7.13
Profile B: effective length/bore diameter ratio = 5.33
Profile C: effective length/bore diameter ratio = 4.00
Profile D: effective length/bore diameter ratio = 3.33
Profile E: effective length/bore diameter ratio = 2.33
Profile A represents the field profile of the Oxford quadrupole. Fringe field coefficients (see §2.2) are $c_0 = -6.0$, $c_1 = 3.5$, $c_2 = c_3 = c_4 = 0$, for all profiles.

shorter lengths, i.e. as the ratio of the fringe-field region to the main quadrupole region becomes larger, then a slight change in the third-order aberration coefficients becomes apparent. No other significant aberration coefficients were introduced for the configurations investigated.

In conclusion, calculations indicate that, for the Oxford coupled triplet system, the magnitudes of the third-order spherical aberration coefficients remain relatively insensitive to the effective length to bore radius ratio and fringe-field characteristics of the quadrupole element used in the focusing system.

6.2 Quadrupole translational misalignment

The translational misalignment of a quadrupole lens or lenses in a focusing system introduces parasitic aberrations which degrade the spot size. The primary effect of introducing a translational misalignment is the appearance of a first-order parasitic aberration of the form $<x|U_n>$ and $<y|V_n>$, where U_n and V_n are the x- and y-translations of quadrupole n from the beam axis. These particular first-order aberrations displace the image by an amount proportional to the displacement of a lens element from the beam axis.

OXRAY has been used to investigate the effects of such lens misalignments in three systems, the Oxford coupled triplet, the Harwell Russian quadruplet and the Heidelberg doublet. Table 6.4 shows the calculated first-order translational aberration coefficients for these three systems. As an example, a y-displacement of the second lens in the Harwell Russian quadruplet system of 10 μm causes a displacement of the image in the image plane of 43 μm from the beam axis. The first-order translational aberration coefficients do not cause a degradation of the image profile since they are independent of the object coordinates. There are, however, numerous higher-order aberrations introduced by translational misalignment. These aberration coefficients are of the form $<x|\theta^j\phi^k U^l V^m>$ and do degrade the spot size. Because of the complexity of presenting these effects in tabular form, the results are presented as scatter plots.

Table 6.4 First-order translational aberration coefficients for the Oxford coupled triplet, the Harwell Russian quadruplet and the Heidelberg doublet.

	Oxford Triplet	Harwell Quadruplet	Heidelberg Doublet
$<x\|U_1>$	−0.48	1.25	3.29
$<x\|U_2>$	−0.42	−2.28	−2.07
$<x\|U_3>$	1.90	2.87	
$<x\|U_4>$		−0.64	
$<y\|V_1>$	−3.74	−1.42	−0.65
$<y\|V_2>$	8.07	4.30	1.69
$<y\|V_3>$	−3.27	−2.27	
$<y\|V_4>$		−0.58	

U_n and V_n are the x- and y-translations of quadrupole n.

Figures 6.4 (a)–(e) show scatter plots of the calculated beam spot profile for the Heidelberg doublet for first-element x-displacements of 0, 100, 200, 500, 1000 μm respectively. The x and y object divergences are a nominal 1 mrad and the quadrupole pole-tip fields have been adjusted in each case to reduce any induced astigmatism to zero.

Figures 6.5 (a)–(e) show scatter plots of the calculated beam-spot profile for the Oxford coupled triplet for first-element x-displacements of 0, 100, 200, 500 and 1000 μm respectively. The x and y object divergences are a nominal 0.3 mrad and the quadrupole pole-tip fields have been adjusted in each case to reduce any induced astigmatism to zero.

Figures 6.6 (a)–(e) show scatter plots of the calculated beam-spot profile for the Harwell Russian quadruplet for first-element x-displacements of 0, 100, 200, 500 and 1000 μm respectively. The x and y object divergences are a nominal 1 mrad and the quadrupole pole-tip fields have been adjusted in each case to reduce any induced astigmatism to zero.

The scatter plots for each system have been calculated for zero object dimensions. The series of plots therefore depict the change in the image profile in relation to the intrinsic spherical aberration of each system.

As can be seen, especially in the case of the Oxford coupled triplet, the first lens element need only be displaced a small amount from the beam axis for higher-order parasitic aberrations to become comparable with the spherical aberration. Further calculations involving the displacement of other lens elements indicate that some elements are more sensitive than others concerning degradation of the beam-spot profile. For these systems at least, it seems that the elements must be translationally aligned to within 100 μm so that beam-broadening due to quadrupole misalignment is not a major effect. Luckily, this is normally within the tolerances with which these systems can be aligned (see §6.7).

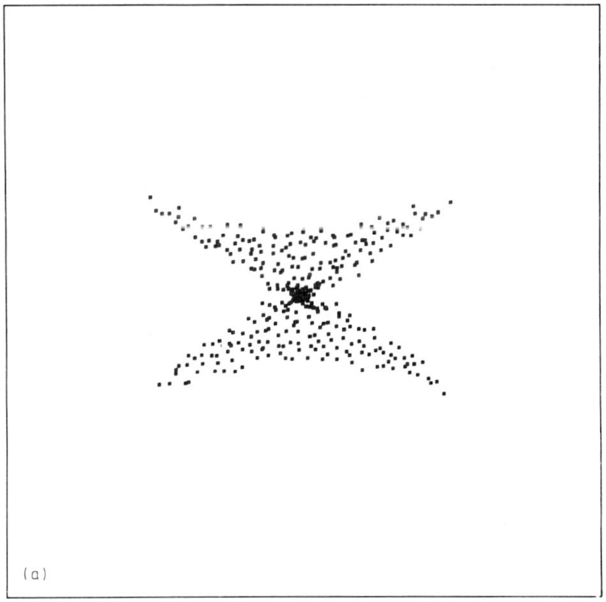

Figure 6.4 Scatter plots of the calculated beam-spot profiles for the Heidelberg doublet. Plots (a)–(e) result from first element horizontal displacements of 0, 100, 200, 500 and 1000 μm respectively. The input divergences are a nominal 1 mrad and the quadrupole pole-tip fields have been slightly adjusted in each case to reduce any induced astigmatism to zero. The plots each represent 4 μm × 4 μm in the image plane.

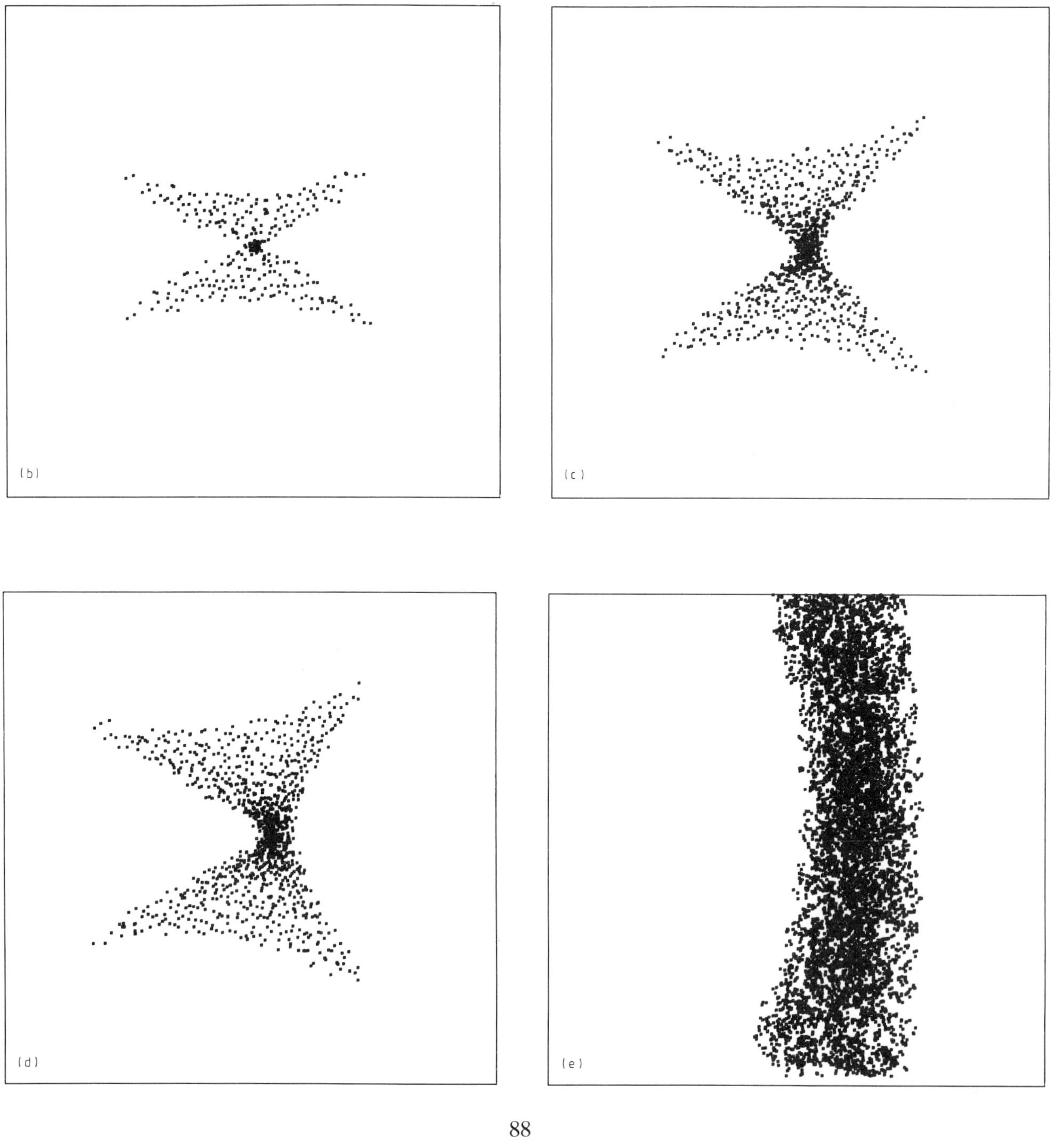
(b)
(c)
(d)
(e)

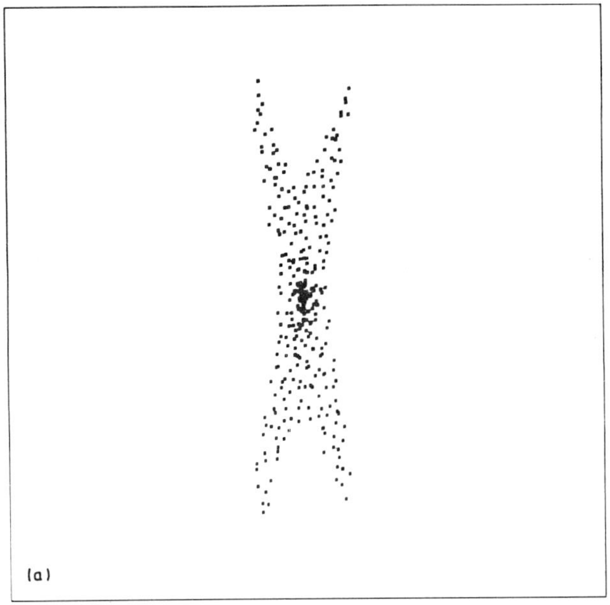

Figure 6.5 Scatter plots of the calculated beam-spot profiles for the Oxford coupled triplet. Plots (a)–(e) result from first element horizontal displacements of 0, 100, 200, 500 and 1000 μm respectively. The input divergences are a nominal 0.3 mrad and the quadrupole pole-tip fields have been slightly adjusted in each case to reduce any induced astigmatism to zero. The plots each represent 10 μm × 10 μm in the image plane.

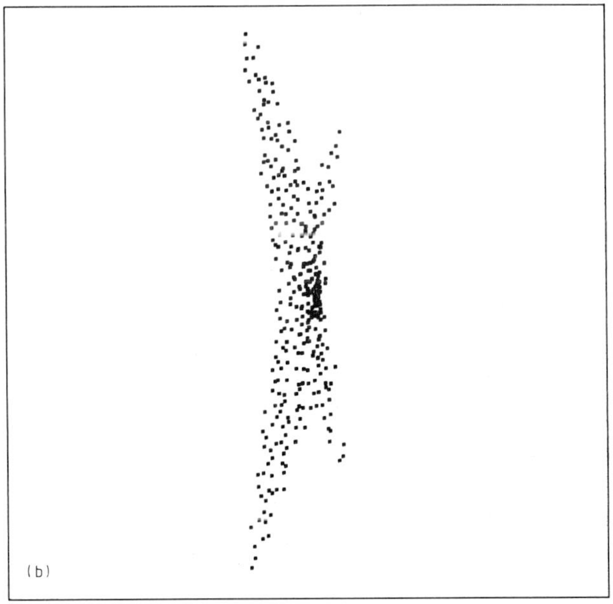

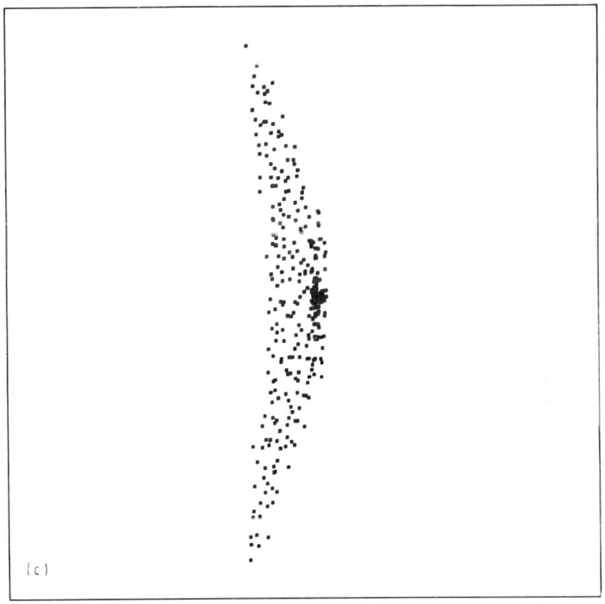

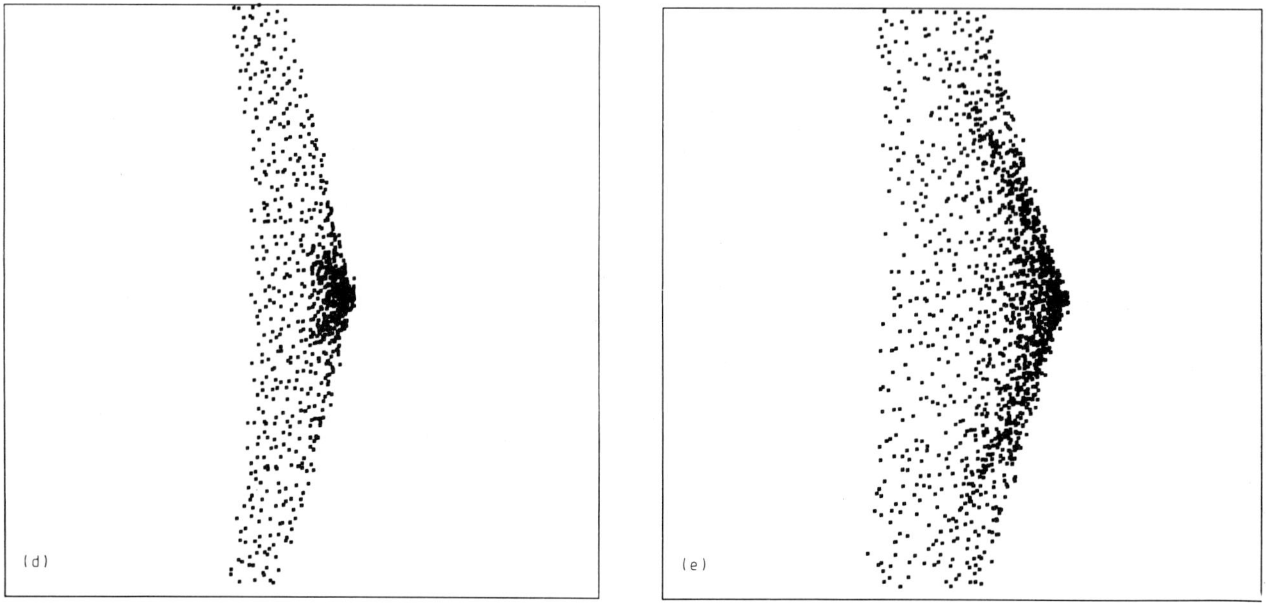

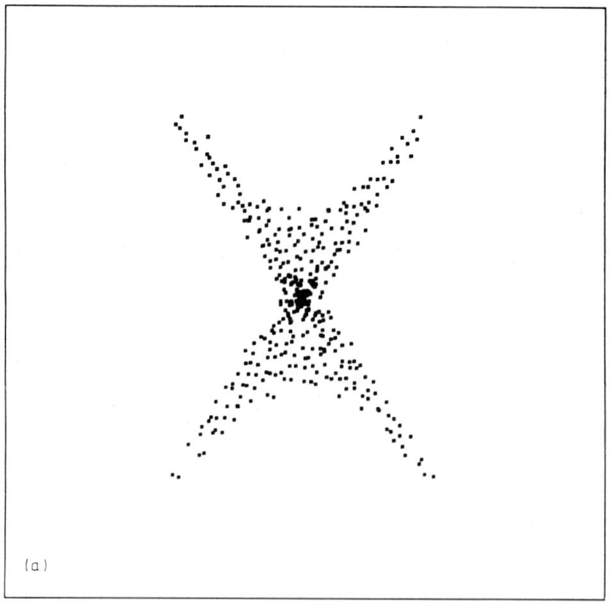

Figure 6.6 Scatter plots of the calculated beam spot profiles for the Harwell Russian quadruplet. Plots (a)–(e) result from first element horizontal displacements of 0, 100, 200, 500 and 1000 μm respectively. The input divergences are a nominal 1 mrad and the quadrupole pole-tip fields have been slightly adjusted in each case to reduce any induced astigmatism to zero. The plots each represent 4 μm × 4 μm in the image plane.

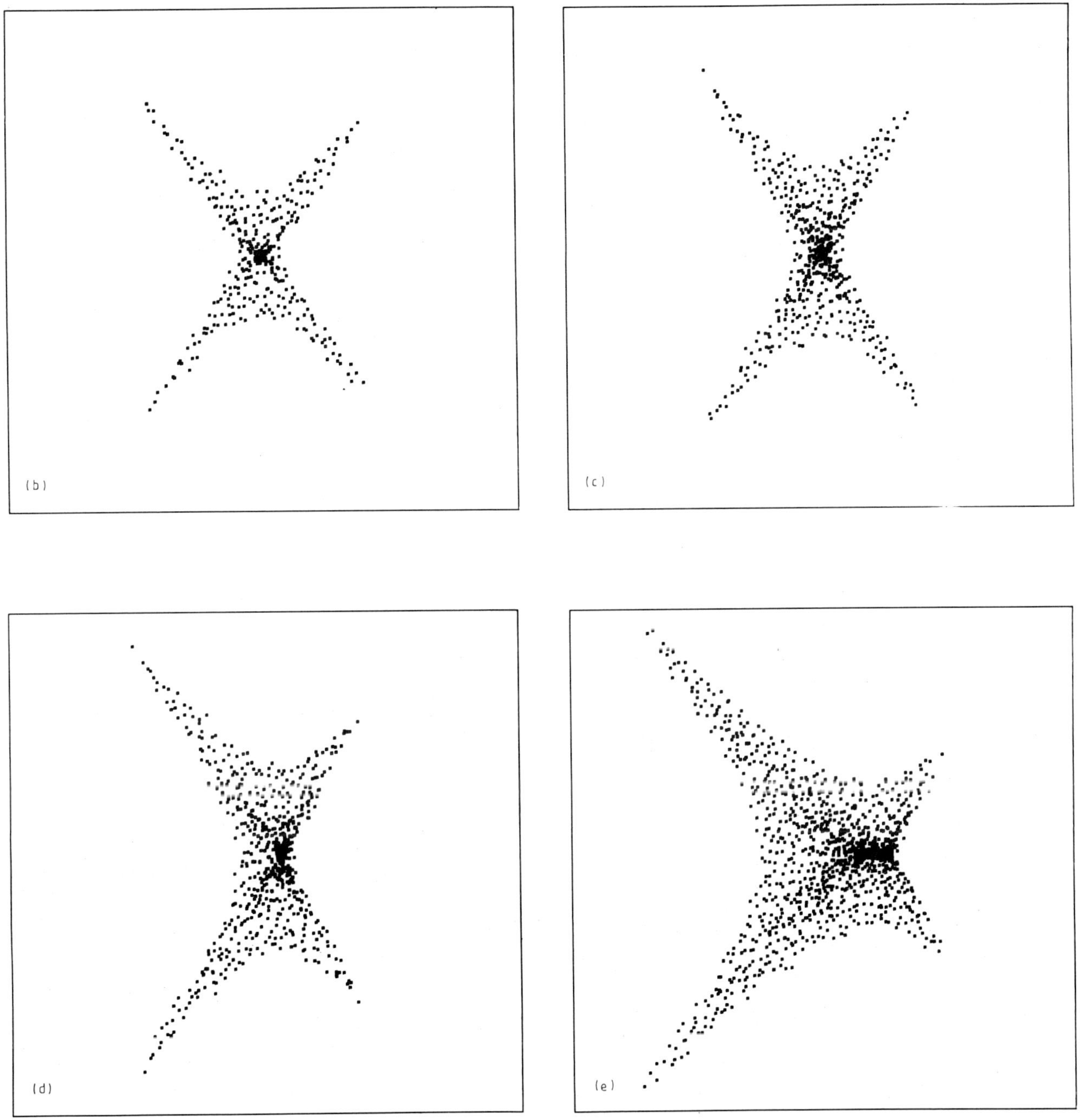
(b)
(c)
(d)
(e)

6.3 Quadrupole tilt misalignment

In the previous section, the effect on the spot profile of displacing the quadrupole lens from the beam axis was considered. Tilting a lens within a focusing system appears to have a similar, although less pronounced, effect. The major effect of introducing a tilt in one of the focusing elements is to introduce a first-order parasitic aberration of the form $<x|\alpha_n>$ or $<y|\beta_n>$ where α_n, β_n are the tilts in the x- and y-planes measured in mrad. Table 6.5 shows the calculated first-order tilt aberration coefficients for three chosen systems: the Oxford coupled triplet, the Harwell Russian quadruplet and the Heidelberg doublet. As an example, a 1 mrad tilt in the element of the Heidelberg doublet in the x-plane causes a displacement of the image in the image plane of 4.65 μm.

Tilting a lens element within a focusing system also introduces numerous higher-order aberrations of the form $<x|\theta^j\phi^k\alpha^l\beta^m>$, which degrade the focused-beam spot size. As an indication of the relative insensitivity of the effect of tilt misalignment on the spot size, scatter plots of calculated beam-spot profiles for the three chosen systems are shown in Figures 6.7–9.

Figures 6.7 (b)–(e) show the effect of tilting each element 20 mrad in the x- and y-planes for the Heidelberg doublet. The x and y object divergences are a nominal 1 mrad and the quadrupole pole-tip fields have been adjusted to minimise any induced astigmatism.

Figures 6.8 (b)–(g) show the effect of tilting each element 20 mrad in the x- and y-planes for the

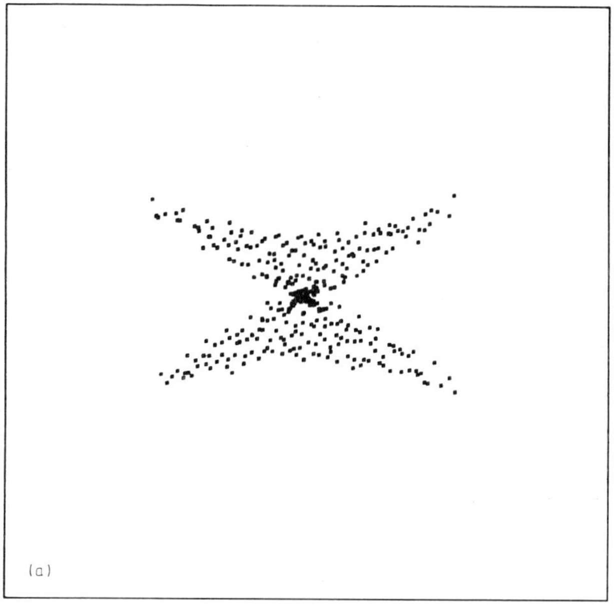

Figure 6.7 Scatter plots of the calculated beam-spot profiles for the Heidelberg doublet. Plot (a) depicts the profile of the aligned system. Plots (b) and (c) result from each element in turn being tilted 20 mrad in the x-plane. Plots (d) and (e) result from each element in turn being tilted 20 mrad in the y-plane. The input divergences are a nominal 1 mrad and the quadrupole pole-tip fields have been adjusted to minimise any induced astigmatism. The plots each represent an area of 4 μm × 4 μm in the image plane.

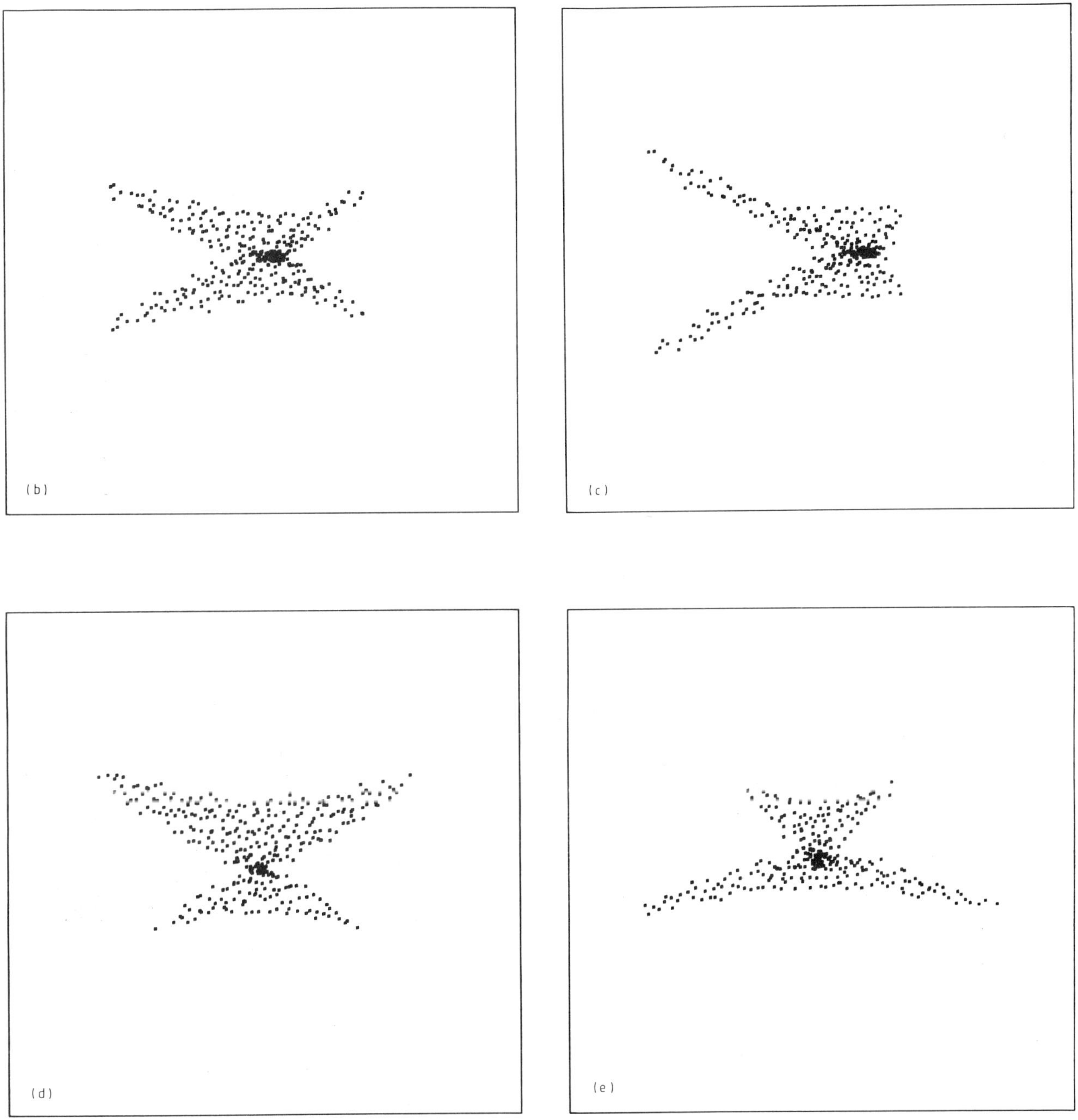
(b)
(c)
(d)
(e)

Oxford coupled triplet. The x- and y-divergences are a nominal 0.3 mrad and the lens pole-tip fields have been adjusted to minimise any induced astigmatism.

Figures 6.9 (b)–(i) show the effect of tilting each element 20 mrad in the x- and y-planes for the Harwell Russian quadruplet. The x- and y-divergences are a nominal 1 mrad and the pole-tip fields have been adjusted to minimise any induced astigmatism.

As in the previous section, these scatter plots have been calculated for zero object dimensions and therefore show the effect of each lens tilt in relation to the system spherical aberration. Scatter plots of these systems perfectly aligned, figures 6.7 (a), 6.8 (a) and 6.9 (a), are shown to aid the comparison.

As can be seen from these plots, a 20 mrad tilt in any lens element for each system does little more than 'rearrange' the halo caused by spherical aberration. To put these results into perspective, the 20 cm long quadrupole lenses forming the Oxford coupled triplet require a displacement of ±2 mm at the quadrupole lens ends to cause such an angular tilt. This is well within the tolerances with which this system can be aligned.

Table 6.5 First-order tilt aberration coefficients for the Oxford coupled triplet, the Harwell Russian quadruplet and the Heidelberg doublet.

	Oxford Triplet	Harwell Quadruplet	Heidelberg Doublet
$\langle x\|\alpha_1\rangle$	8.18	−3.07	−2.21
$\langle x\|\alpha_2\rangle$	−17.7	−1.18	4.65
$\langle x\|\alpha_3\rangle$	−2.29	−3.74	
$\langle x\|\alpha_4\rangle$		8.59	
$\langle y\|\beta_1\rangle$	−50.2	−5.8	−0.52
$\langle y\|\beta_2\rangle$	−8.4	−9.48	−0.39
$\langle y\|\beta_3\rangle$	80.4	22.7	
$\langle y\|\beta_4\rangle$		−5.0	

α_n and β_n are the x-plane and y-plane angular tilts of quadrupole n. Units are in μm and mrad.

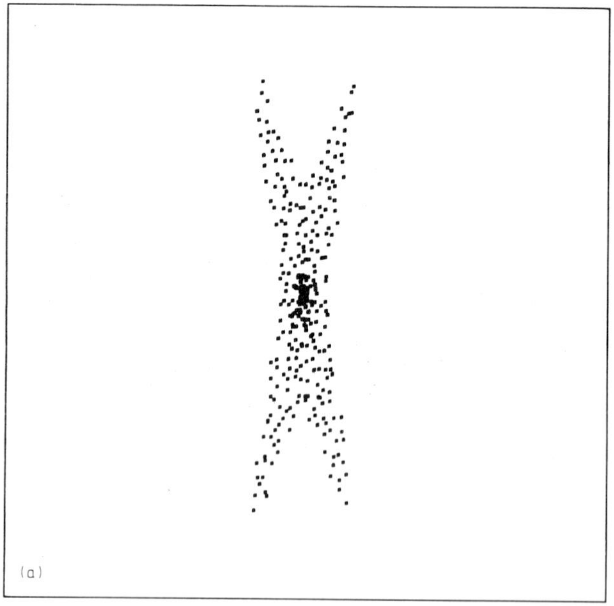

Figure 6.8 Scatter plots of the calculated beam-spot profiles for the Oxford triplet. Plot (a) depicts the profile of the aligned system. Plots (b)–(d) result from each element in turn being tilted 20 mrad in the x-plane. Plots (e)–(g) result from each element in turn being tilted 20 mrad in the y-plane. The input divergences are a nominal 0.3 mrad and the quadrupole pole-tip fields have been adjusted to minimise any induced astigmatism. The plots each represent an area of 8 μm × 8μm in the image plane.

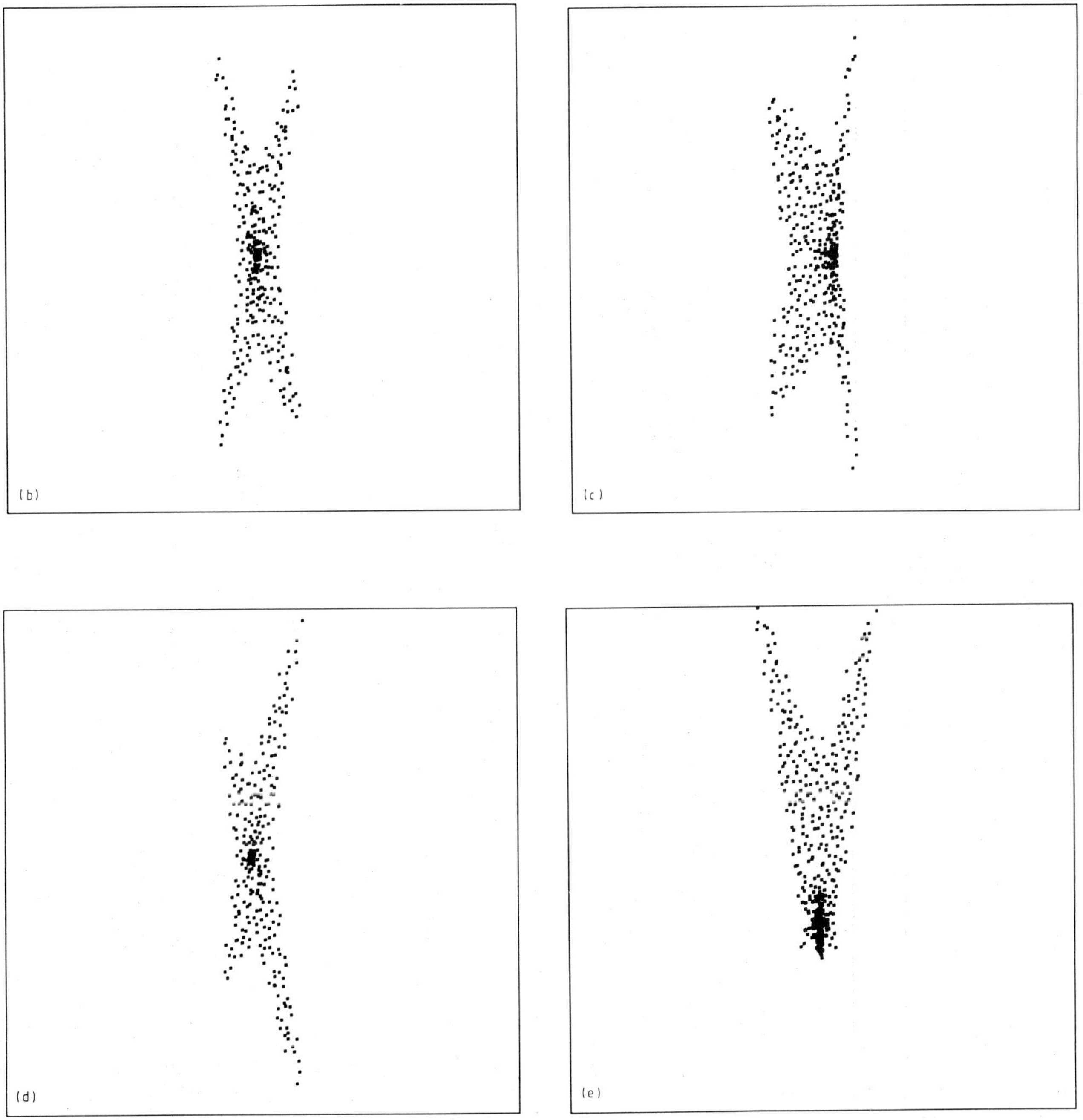
(b)
(c)
(d)
(e)

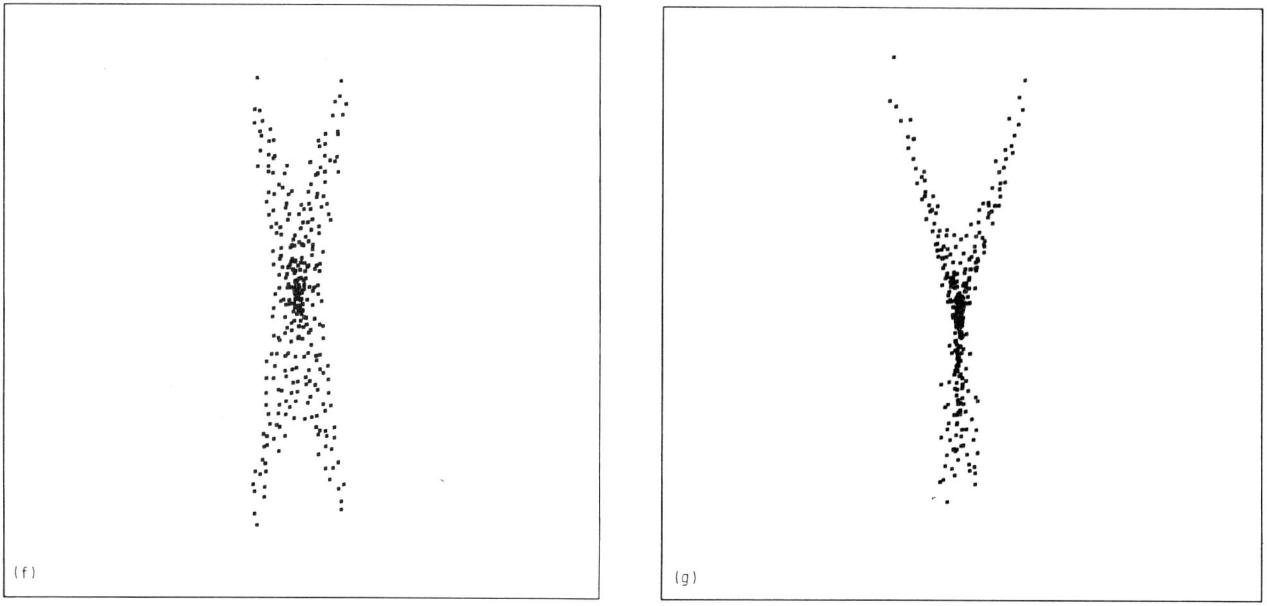

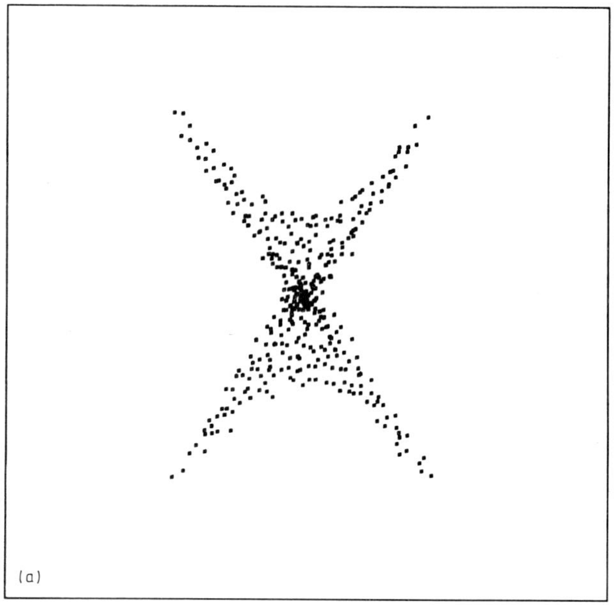

Figure 6.9 Scatter plots of the calculated beam-spot profiles for the Harwell Russian quadruplet. Plot (a) depicts the profile of the aligned system. Plots (b)–(e) result from each element in turn being tilted 20 mrad in the x-plane. Plots (f)–(i) result from each element in turn being tilted 20 mrad in the y-plane. The input divergences are a nominal 1 mrad and the quadrupole pole-tip fields have been adjusted in each case to minimise any induced astigmatism. The plots each represent an area of 8 μm × 8μm in the image plane.

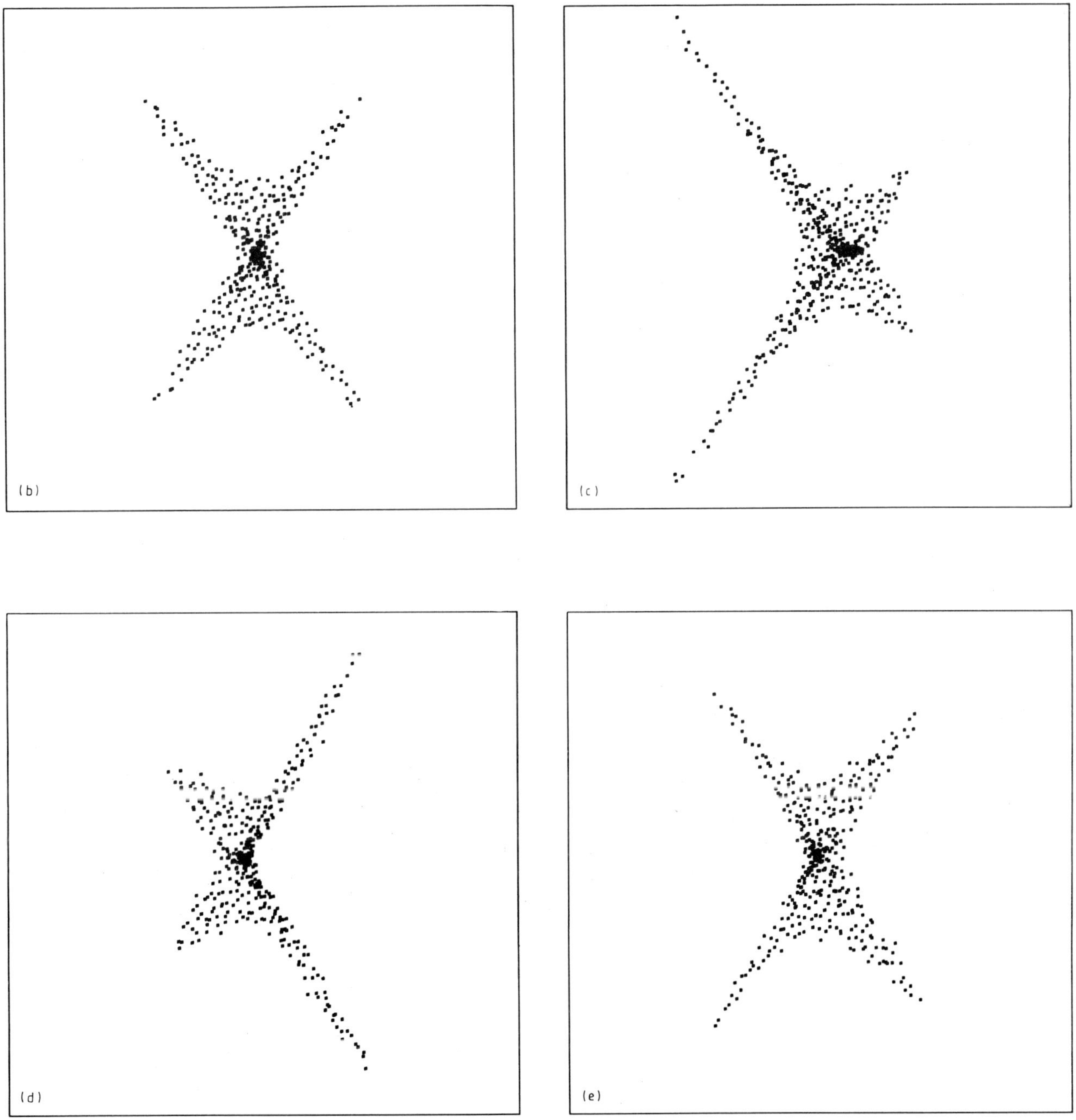
(b)
(c)
(d)
(e)

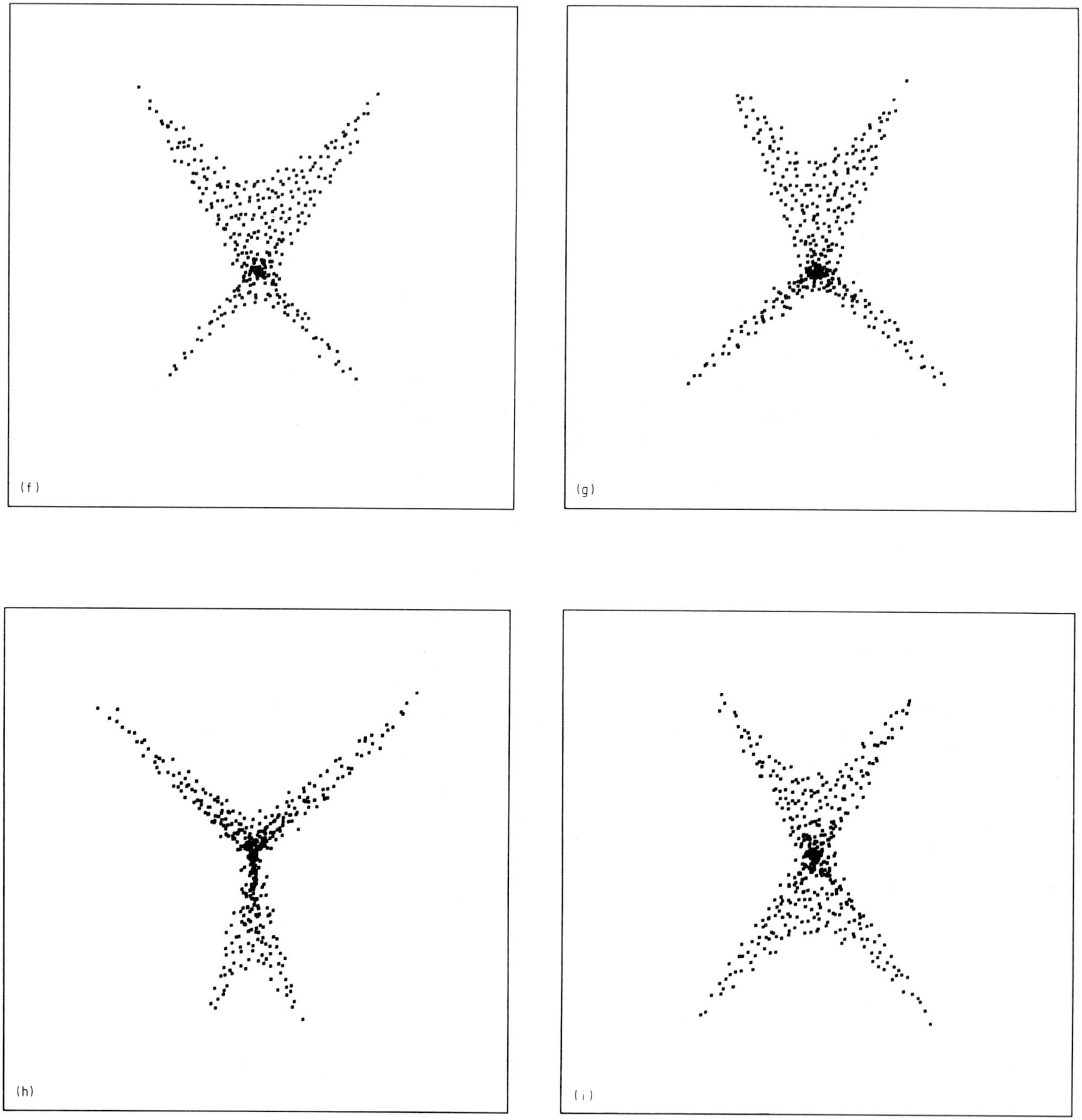
(f)
(g)
(h)
(i)

6.4 Quadrupole rotation misalignment

The rotational misalignment of a quadrupole lens in a focusing system can affect the beam-spot size in a dramatic way [GR82] and [CO72]. The primary effect of introducing rotational misalignment into a focusing system is the introduction of second-order parasitic aberration coefficients of the form $<x|\phi\rho_1>$, $<x|\phi\rho_2>$, ..., $<x|\phi\rho_n>$ and $<y|\theta\rho_1>$, $<y|\theta\rho_2>, \ldots, <y|\theta\rho_n>$ where $\rho_1, \rho_2, \ldots, \rho_n$ are the rotational misalignments of quadrupole lenses 1,2, ..., n. Table 6.6 shows the calculated rotational second-order coefficients for the Oxford coupled triplet, the Harwell Russian quadruplet and the Heidelberg doublet. As an example, a rotational misalignment of 1 mrad in the second lens of the Heidelberg doublet, for a nominal input beam divergence of 1 mrad in each plane, causes a point image to be increased to a rectangle of dimensions 12.02 μm in the x-direction and 2.09 μm in the y-direction. An interesting point to note is the coupling between planes introduced by the rotational misalignment. For example, the coefficient $<x|\phi\rho_1>$ shows that the increase in the image size in the x-direction is proportional to the entrance angle in the y-plane for a fixed quadrupole-lens rotation.

Table 6.6 Rotational aberration coefficients of the Oxford coupled triplet, the Harwell quadruplet and the Heidelberg doublet.

	Oxford Triplet	Harwell Quadruplet	Heidelberg Doublet
$<x\|\phi\rho_1>$	6.55	−9.57	−12.02
$<x\|\phi\rho_2>$	13.54	22.35	12.02
$<x\|\phi\rho_3>$	−20.12	−14.87	
$<x\|\phi\rho_4>$		2.09	
$<y\|\theta\rho_1>$	−29.3	−9.60	−2.10
$<y\|\theta\rho_2>$	−60.67	22.35	2.09
$<y\|\theta\rho_3>$	89.60	−14.87	
$<y\|\theta\rho_4>$		2.09	

ρ_n is the rotation in mrad of element n.
All units in mrad or μm.

From the results in table 6.6 it is apparent that for the attainment of small spot sizes, a high degree of rotational alignment between lenses is required. For example, in order to attain a spot size of 1 μm for a nominal input divergence of 0.3 mrad using the Oxford coupled triplet, the third element needs to be rotationally aligned to within 30 μrad. For a cylindrical quadrupole of diameter 30 cm, this alignment accuracy represents a positional accuracy at the circumference of 5 μm!

There are, however, several properties of the rotational aberration coefficients which reduce the practical problems associated with the need for such precise rotational alignment. Blower [BL83] has shown that:

(1) The algebraic sum of the second-order parasitic aberration coefficients corresponding to each plane is zero, i.e. $\Sigma<x|\phi\rho_n> = 0$ and $\Sigma<y|\theta\rho_n> = 0$ simultaneously.
(2) There is a fixed constant of proportionality between coefficients relating to the same quadrupole element, and this constant of proportionality is equal to the ratio of the magnification in each plane, i.e.

$$\frac{<x|\phi\rho_1>}{<y|\theta\rho_1>} = \frac{<x|\phi\rho_2>}{<y|\theta\rho_2>} = \ldots$$

$$= \frac{<x|\phi\rho_n>}{<y|\theta\rho_n>} = \frac{M_x}{M_y} = \frac{D_y}{D_x} .$$

Note that in the case of the Russian quadruplet, which has orthomorphic properties, i.e. equal demagnification in each plane, this constant is unity.

The consequences of these relationships are that any increase in spot size due to the rotational misalignment of quadrupole lens elements in a multi-elemental system can be eliminated by the rotational adjustment of one lens only, i.e. the second-order parasitic rotational aberration coefficients can be made to cancel in both planes simultaneously. An obvious practical advantage, therefore, is that only one quadrupole lens within a focusing system is required to have a high degree of rotational alignment capability. A second advantage is that the quadrupole lens chosen for the second-order rotational aberration correction can be that element which is least sensitive to the misalignment, thereby reducing the degree of rotational accuracy required in that element.

The above conclusions only hold providing the second-order rotational aberrations are dominant and any higher-order aberrations introduced are negligible. Calculations using OXRAY have shown that higher-order coefficients of the form $<x|\phi^3\rho_n>$, $<x|\theta^2\phi\rho_n>$, $<y|\theta^3\rho_n>$, $<y|\theta^2\phi\rho_n>$, ... can be significant at high ρ_n. A relevant question is then: 'what degree of misalignment can be tolerated within a system of quadrupole elements in order that the rotation of one lens element can reduce the resultant increase in spot size to negligible proportions?'

Once again we can only consider specific systems. Consider the Oxford coupled triplet with the first element rotated +10 mrad and the second lens rotated −10 mrad from perfect alignment. Calculations show that a rotation of the third element by −3.43 mrad cancels the effects of the second-order parasitic aberrations and that any residual increase in the spot size due to the effects of the higher-order coefficients is small compared with the intrinsic spherical aberration of this system.

Consider the Harwell Russian quadruplet with the first element rotated 10 mrad, the second element left aligned and the third element rotated −10 mrad. Calculations show that the net effect of rotating the fourth quadrupole lens by −25.2 mrad reduces the resultant increase in spot size to values small compared with the intrinsic spherical aberration of this system.

These results indicate, at least for the two systems considered, that the degree of rotational alignment necessary within a system, such that the rotational misalignment effects can be eliminated using one lens is well within the tolerances of even the most crudely aligned systems.

An alternative approach to the physical rotation of a single quadrupole lens in the correction of system rotational misalignment is the introduction of a secondary perturbing quadrupole field, either magnetic or electrostatic, which may be used to compensate for rotational misalignment in a mechanically rigid system. For further details, the reader should consult [CR67], [KO77] and [MA81b].

Because quadrupole lens rotational misalignment affects the spot size in such a dramatic way, and yet can be corrected relatively easily, it becomes useful when optimising a lens system if it can be easily determined whether or not the spot size is limited by rotational misalignments.

Optimising a lens system to produce small spot sizes involves the adjustment of the lens excitations to minimise astigmatism. A common way of observing the beam-spot profile during the optimisation is to allow the beam to impinge on a glass or quartz slide situated in the image plane and to view the resultant fluorescent pattern using a microscope. Any beam-spot broadening caused by rotational misalignment can be identified through a characteristic set of beam-spot profiles as the lens excitations are adjusted to minimise the spot size. Figures 6.10 (a)–(e) show scatter plots of the calculated beam-spot profiles for the Harwell Russian quadruplet. The second element has been rotated 0.1 mrad and the entrance x- and y-divergences are a nominal 1 mrad respectively. Each plot represents the effects due to the introduced rotational misalignment mixed with varying degrees of astigmatism. As can be seen, if minimising the beam-spot size by adjusting the lens excitations results in a 'rotation' of the beam profile, then these calculations indicate that rotational misalignment is a major contributory factor in limiting the optimisation of the system. This identifying characteristic results from the partial cancellation of the aberrations introduced by rotational misalignment by astigmatism. This effect has certainly been observed experimentally at Oxford during the optimisation of the Oxford coupled triplet.

Throughout the above discussion we have ignored the possibilities of magnetic field rotational misalignment occurring within the lens element, caused, for example, by poor construction or a twisting of the pole pieces, due to stressed iron. This effect is not investigated here although, since the spatial resolution of a focusing system seems to be so sensitive to rotational misalignments, it is possible that this effect may be important at sub-micron spot sizes.

Figure 6.10 Scatter plots of the calculated beam-spot profiles for the Harwell Russian quadruplet showing a characteristic 'rotation' of the spot profile due to mixing of rotational misalignment and astigmatism during optimisation. The second element of the quadruplet has been rotated 0.1 mrad and the input ray divergences are a nominal 1 mrad.

The aberration coefficients determining each profile are as follows:

Profile	Rotational aberration coefficients		Astigmatism coefficients	
	$\langle x\|\phi\rangle$	$\langle y\|\theta\rangle$	$\langle x\|\theta\rangle$	$\langle y\|\phi\rangle$
a	2.23	2.23	2.23	2.23
b	2.23	2.23	1.12	1.12
c	2.23	2.23	0	0
d	2.23	2.23	−1.12	−1.12
e	2.23	2.23	−2.23	−2.23

The plots each represent an area 4 μm × 4 μm in the image plane.

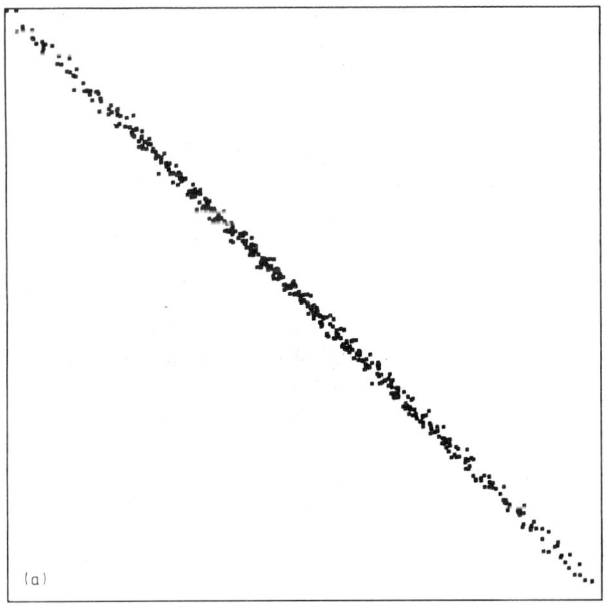

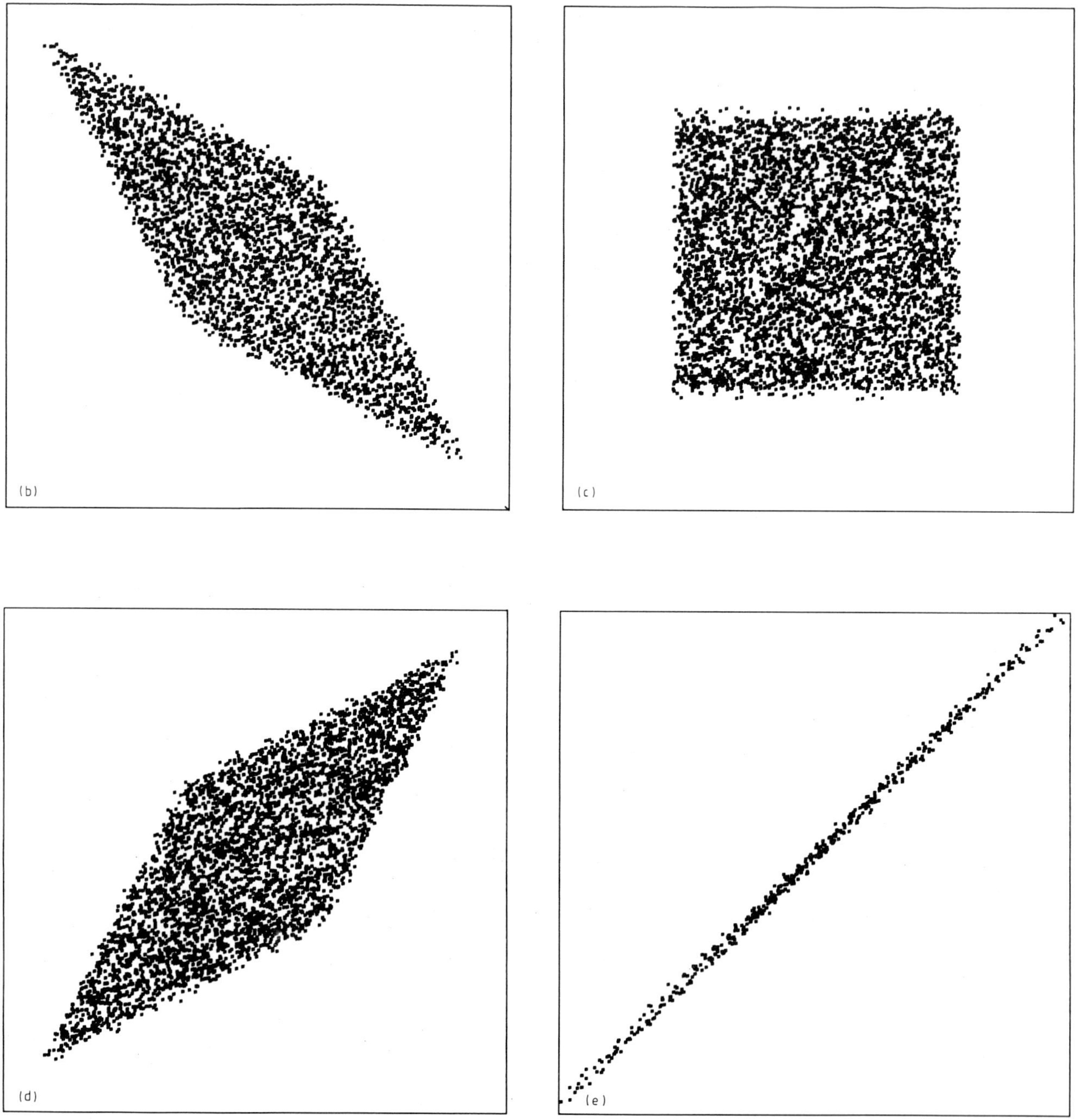
(b)
(c)
(d)
(e)

6.5 Harmonic contamination

As noted in §2.2, higher multipoles in a quadrupole lens can arise from the physical construction of the magnet. Both the pole-tip contour and mechanical asymmetries of the pole pieces introduce various multipole fields. The most important property of these contaminant fields is that a harmonic field of order $2N$ (where $N=2$ for a quadrupole, $N=3$ for a sextupole, etc) varies with the aperture radius as r^{N-1}, and hence the ratio of this field to the basic quadrupole field varies with radius as r^{N-2}. Thus, the further away the beam particle is from the quadrupole axis, the greater the influence of the contaminant field. As a result, it would be expected that any increase in the spot size would be caused by the extreme rays through the focusing system. It would also be expected that the effects of higher-order harmonic contamination in focusing systems could be reduced appreciably by reducing the entrance divergence of the particle beam.

For a particular probe-forming system, the entrance divergences are usually chosen so that the effects of intrinsic (and parasitic) aberrations are reduced to acceptable levels. The design and construction of the quadrupole lens should therefore be such that for these input divergences, the effects of harmonic contamination on the spot size are small compared with the intrinsic aberrations.

As an attempt to estimate the relative importance of harmonic contamination, we again can only consider specific systems. The effects on the spot size of a 1% harmonic constituent in the last element of the Heidelberg doublet, the Oxford triplet and the Harwell Russian quadruplet have been investigated. The harmonics considered in the calculations are those which are known to be predominant in quadrupole design, i.e. sextupole, octupole, decapole, dodecapole and 20-pole. The results are shown as scatter plots in figures 6.11–13. A harmonic contamination of 1% has been chosen, since this value represents a realistic upper limit, at

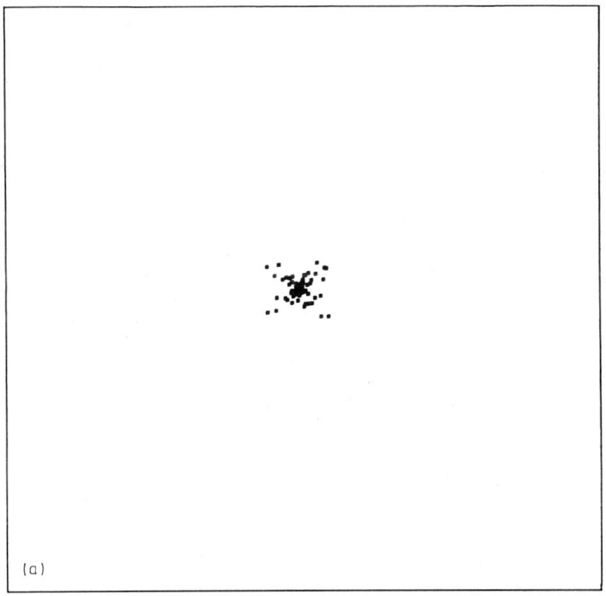

Figure 6.11 Scatter plots of the calculated beam-spot profiles showing the effect of harmonic contamination in the Heidelberg doublet. Plot (a) depicts the spot profile of the aligned system. Plots (b)–(f) result from the addition into the last element quadrupole field of a 1% contaminant level of sextupole, octupole, decapole, dodecapole and 20-pole harmonics respectively. The beam input divergences are a nominal 1 mrad. The plots each represent an area in the image plane of 20 μm × 20 μm.

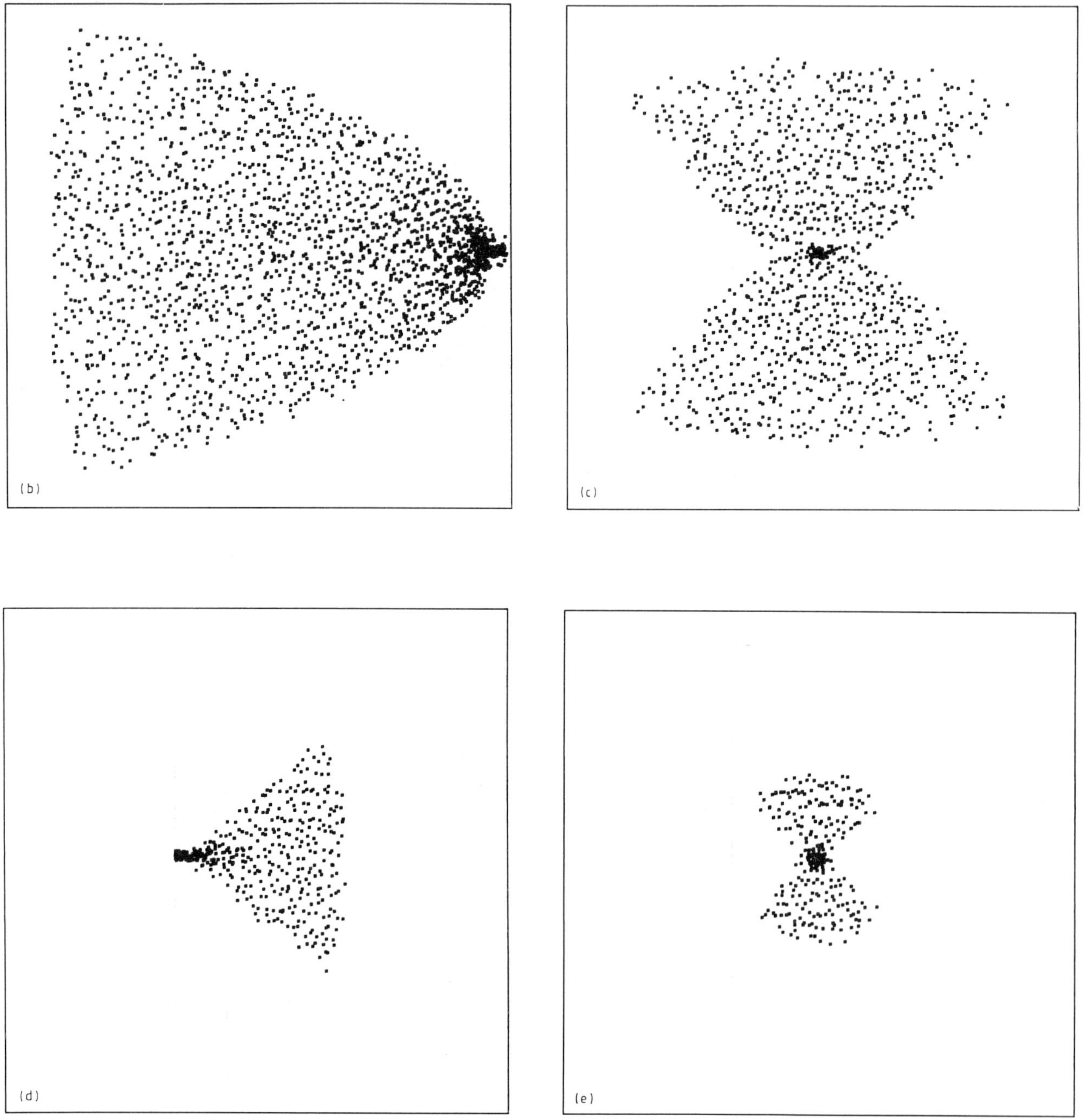
(b)
(c)
(d)
(e)

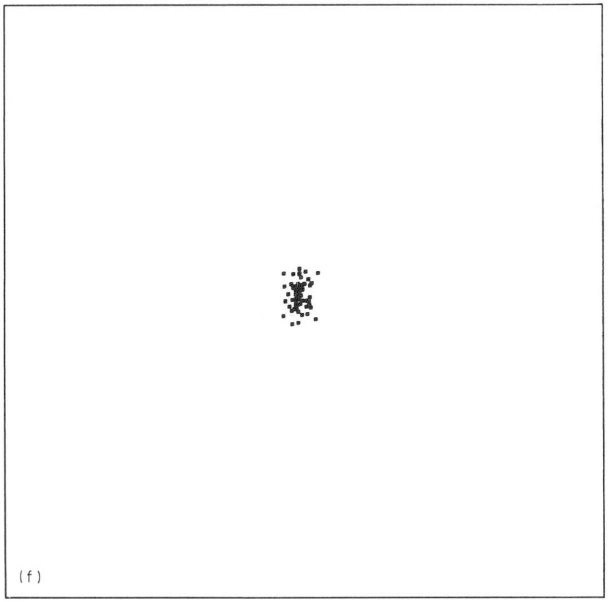

least for quadrupoles designed for beam transport of nuclear particles [LO72]. It must be remembered, however, that for quadrupoles of smaller bore radius than those used in nuclear particle-beam transport, construction and machining needs to be more precise, in order that this upper limit is realistic.

Some general comments can be formed from the results depicted in figures 6.11–13.

(1) As expected, the sextupole component produces the largest effect on the spot size. For each of the three systems considered, with the input conditions as shown, the effect of 1% sextupole harmonic contamination in the last element is large compared with the spherical aberration. Since a sextupole component is generated by a mechanical asymmetry of the poles [CO65], great care must be taken to construct the quadrupole lens with precise fourfold symmetry. The calcu-

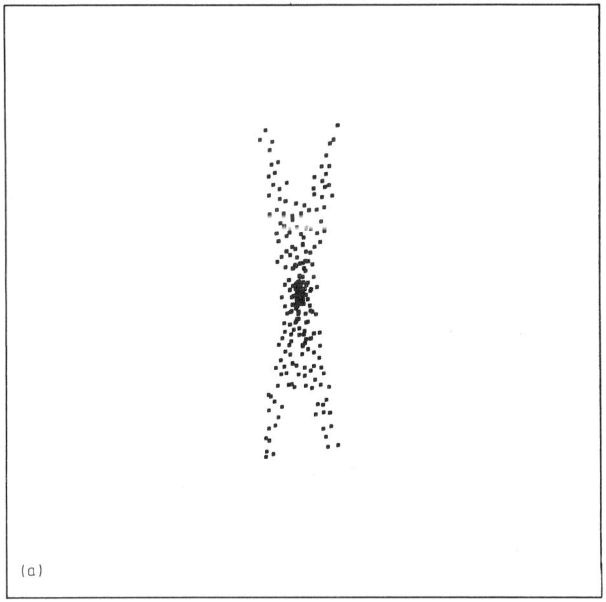

Figure 6.12 Scatter plots of the calculated beam-spot profiles showing the effect of harmonic contamination in the Oxford triplet. Plot (a) depicts the spot profile of the aligned system. Plots (b)–(f) result from the addition into the last element quadrupole field of a 1% contaminant level of sextupole, octupole, decapole, dodecapole and 20-pole harmonics respectively. The beam input divergences are a nominal 0.3 mrad. The plots each represent an area in the image plane of $10\,\mu m \times 10\,\mu m$.

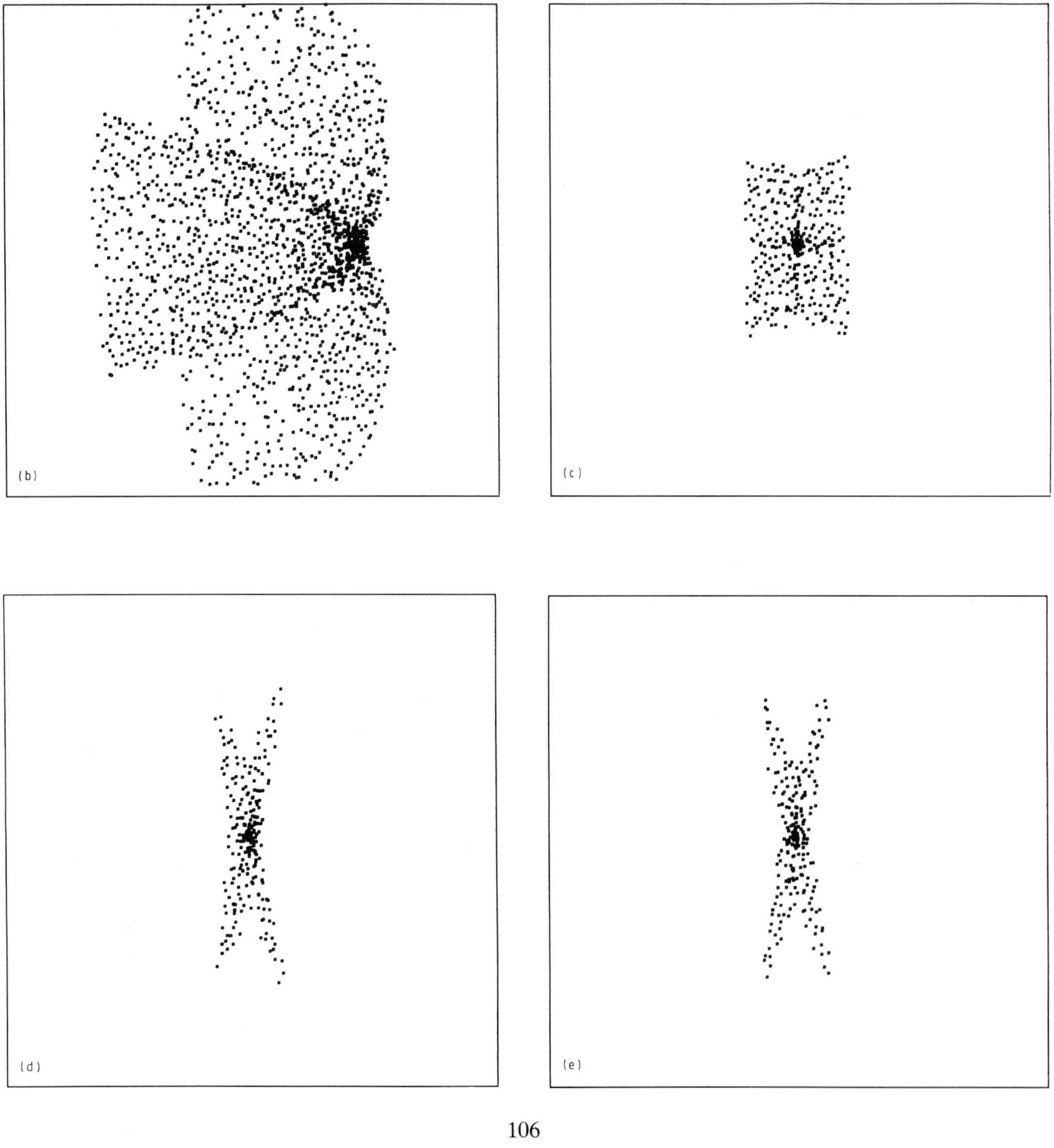
(b)
(c)
(d)
(e)

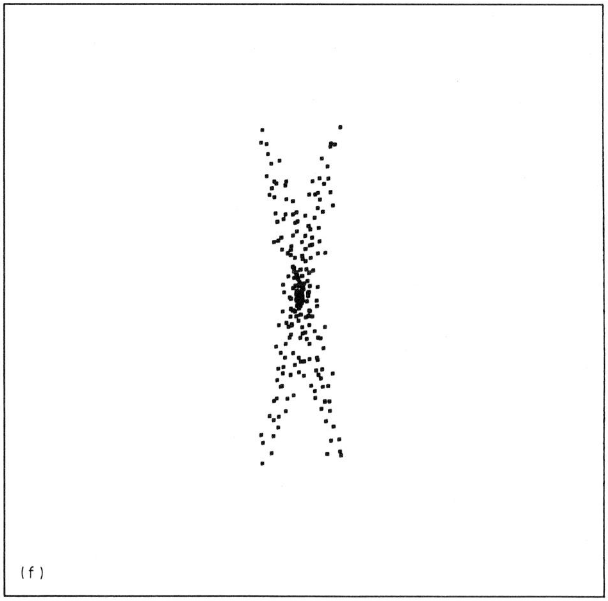

lations also show that the sextupole contamination introduces second-order aberration coefficients of the form $<x|\phi^2>$, $<y|\phi^2>$, $<x|\theta\phi>$, $<y|\theta^2>$ and $<y|\phi^2>$.

(2) The primary effect of octupole contamination is to cause a change in the values of the third-order spherical aberration coefficients. In the case of the Heidelberg doublet, the third-order spherical aberration coefficients have increased drastically, for the Harwell quadruplet a minimal effect is noticed, and for the Oxford triplet the net effect is to reduce the spot size. This is in accordance with a proposal by Scherzer [SC47], that specific values of an octupole field can eliminate spherical aberrations in quadrupole probe-forming systems. Since an octupole component can be generated by either a mechanical asymmetry of the poles or a radial movement of two opposite poles

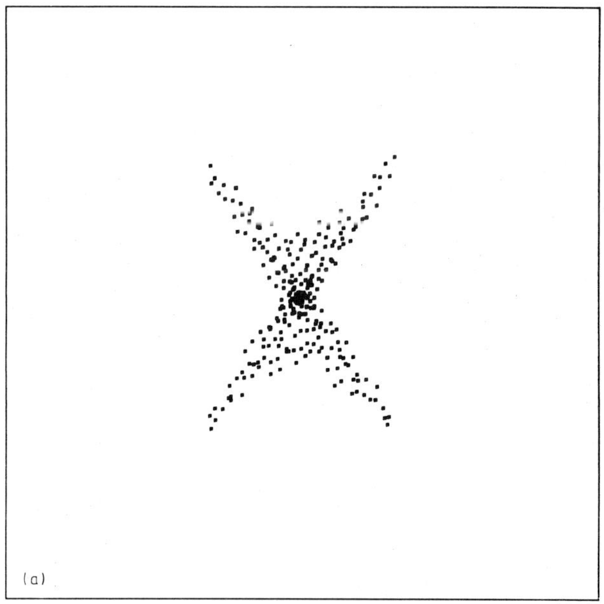

Figure 6.13 Scatter plots of the calculated beam-spot profiles showing the effect of harmonic contamination in the Harwell Russian quadruplet. Plot (a) depicts the spot profile of the aligned system. Plots (b)–(f) result from the addition into the last element quadrupole field of a 1% contaminant level of sextupole, octupole, decapole, dodecapole and 20-pole harmonics respectively. The beam input divergences are a nominal 1.0 mrad. The plots each represent an area in the image plane of 5μm × 5μm.

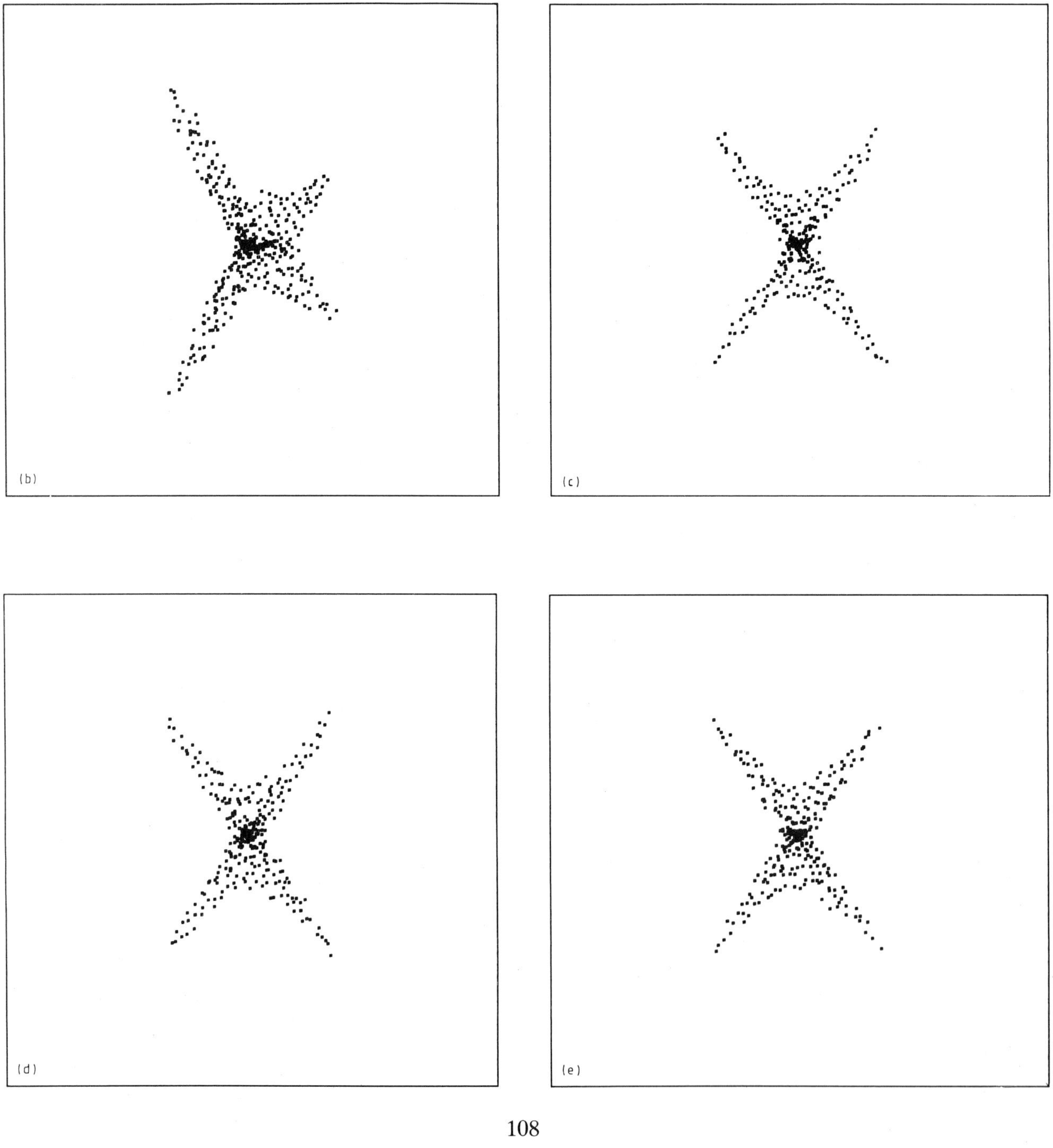
(b)
(c)
(d)
(e)

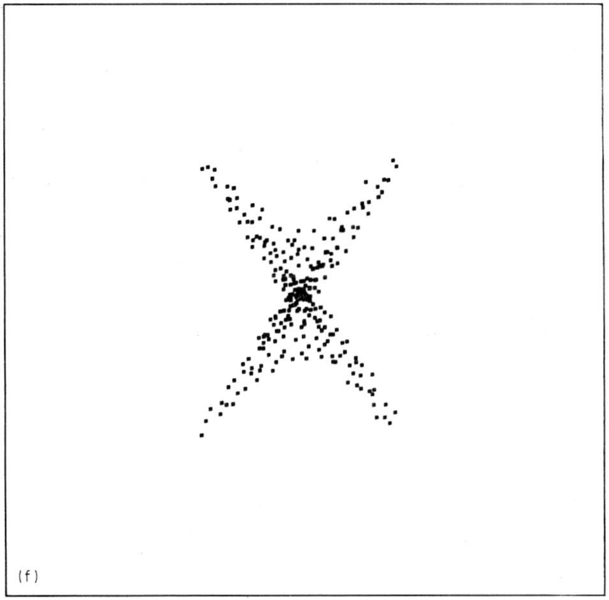

(f)

[CO65], then again the need for precise fourfold symmetry in indicated.

(3) The decapole component, again produced by pole asymmetry, has very little effect on the Harwell quadruplet or the Oxford triplet probe-forming properties, whereas for the Heidelberg doublet the effect is pronounced. This is probably a consequence of the fact that the Heidelberg quadrupole lens has a very small bore diameter of 5 mm compared with the Harwell lens (d = 38.4 mm) and the Oxford lens(d = 30 mm). The path of the beam in the Heidelberg doublet configuration for an input divergence of 1 mrad passes much closer to the pole tip so that the effects of the harmonic contamination are much increased.

(4) The major source of dodecapole and 20-pole contamination comes from the mechanical approximation of the pole shape to that of a perfect hyperbola. The Heidelberg doublet configuration is the only system in which a pronounced effect is noticed, again probably as a consequence of the small lens design and large input divergence.

Although the tests carried out above were limited to specific harmonic phase angles ($\gamma_n = 0$, see equation (2.17)), and only the last element in a few selected systems was considered, a generalised conclusion can be indicated. It seems likely that, in a system where the beam passes close to the pole tip of a lens, the design and construction of the lens is crucial to the performance of the probe. By reducing the entrance divergence of the beam, which reduces transmitted beam intensity, the effects of the harmonic contamination on the spot size can be reduced appreciably. Sextupole contamination, produced by poor lens construction, seems to be the major effect, rather than contamination caused by the pole shape. The effects of poor magnet iron or field saturation are difficult to quantify or calculate. If any such effects could be envisaged as causing a local harmonic fluctuation within the lens, then the above general comment could still apply.

6.6 Power supply stability

It is important in any probe-forming system for the quadrupole lens to have sufficient stability in excitation so as not to affect the minimum attainable spot size. The effect of changes in excitation can be expressed in terms of second-order parasitic aberrations of the form $\langle x|\theta\varepsilon_1\rangle$, $\langle y|\phi\varepsilon_1\rangle$,

$\langle x|\theta\varepsilon_2\rangle$, $\langle y|\phi\varepsilon_2\rangle$, ..., $\langle y|\phi\varepsilon_n\rangle$, where θ and ϕ are the input divergences and ε_n is the percentage change in excitation of quadrupole n.

Calculations of these aberration coefficients for six chosen microprobe systems have been published [GR82] and are reproduced here. The calculated aberration coefficients are listed in table 6.7.

As an example, consider an excitation instability of 0.01% in the second lens of the Melbourne quadruplet. From table 6.7 it is readily seen that for an input divergence angle of 0.3 mrad the spot size is increased by (228.4 × 0.3 × 0.01) μm = 0.7 μm in the x-direction and (443.1 × 0.3 × 0.01) μm = 1.3 μm in the y-direction. The results listed in table 6.7 indicate, at least for these six chosen microprobe systems, that current stabilities of better than 1 part in 10^4 may be required.

Table 6.7 Quadrupole excitation aberration coefficients for six existing microprobe systems.

Coefficient	Harwell Triplet [CO81]	Harwell Quadruplet [CO72]	Heidelberg Doublet [NO77]	Karslruhe Doublet [HE79]	Melbourne Quadruplet [LE79]	Oxford Triplet [WA82]
$\langle x\|\theta\varepsilon_1\rangle$	13.5	−42.2	12.0	−79.3	−123.8	21.7
$\langle x\|\theta\varepsilon_2\rangle$	−20.4	59.6	−49.0	9.1	228.4	−23.4
$\langle x\|\theta\varepsilon_3\rangle$	155.4	−94.0			−335.8	261.8
$\langle x\|\theta\varepsilon_4\rangle$		11.9			77.0	
$\langle y\|\phi\varepsilon_1\rangle$	76.7	55.0	−52.8	7.7	135.7	268.5
$\langle y\|\phi\varepsilon_2\rangle$	−657.8	−211.2	13.2	−69.1	−443.1	−1281.3
$\langle y\|\phi\varepsilon_3\rangle$	135.1	60.3			204.6	200.4
$\langle y\|\phi\varepsilon_4\rangle$		−9.3			−70.1	

ε_n is the percentage change in excitation of quadrupole n.
All other units are in μm or mrad.

6.7 Quadrupole system alignment procedures

It has already been indicated in §§6.4, 6.2, 6.3 that rotational, translational and to a lesser extent tilt misalignment between quadrupoles within a probe-forming system can lead to degradation of the minimum attainable spot size. It is therefore important that these misalignments are minimised. The quadrupole lens alignment procedure described below is the one used at Oxford [WA82], which is a variation of the technique first used at Harwell [CO70].

The magnetic elements can be aligned optically to an accuracy of about 150 μm by means of a telescope viewing an optical target which can be accurately located both at the front and the rear of each quadrupole lens. This procedure is taken as a first step and serves to reduce tilt misalignment to acceptable tolerances (see §6.3).

Since the beam axis does not necessarily coincide with the optical axis (due to the Earth's magnetic field or other stray fields) then a second procedure must be adopted to reduce any translational misalignment to tolerable proportions. With the quadrupoles dormant, the particle beam is allowed to drift through the object aperture and impinge on a thin quartz or glass slide situated at the target position. By using a microscope to observe the fluorescent pattern produced by the beam as it strikes the quartz target, the beam axis can be defined by adjusting the collimator slits until the beam is centred at the crosswires of the microscope. By energising the first quadrupole, a vertical line focus can be produced at the target position, and with a horizontal adjustment of the quadrupole lens the image of the line focus can be moved so as to be observed to coincide with the centre of the crosswires. In a similar way, reversal of the current through the quadrupole produces a horizontal line focus and by adjusting the lens in the vertical direction the image of the line focus can be moved to coincide with the crosswire centre. With the quadrupole in this position, no steering of the beam takes place, and provided the magnetic axis of the lens is parallel with the defined beam axis then these axes are now coincident. Using this technique of 'dipole minimisation' the magnetic axis of each quadrupole can be aligned to the beam axis to an accuracy of less than 20 μm, well within the tolerances required for translational alignment (see §6.2). It should be remembered, however, that the beam axis, if affected by stray fields, may be energy-dependent, so the above procedure is valid only for particles of the same rigidity. If several beams of different energies are used, then either Helmholtz coils can be used to eliminate stray fields [BO78a] or the quadrupoles can be realigned at each energy. At Oxford, since the total alignment procedure takes less than an hour, coils were not used.

The final alignment, and by far the most important in the attainment of micron spot sizes, is the removal of rotational misalignment between lenses. The problem and identification of the effects of rotational misalignment has been dealt with in §6.4. This section indicated that the effects due to rotational misalignment can be effectively eliminated by the rotation of one quadrupole lens or by the introduction of a secondary perturbing quadrupole field, either magnetic or electrostatic, to compensate for rotational misalignment in a rigidly fixed system.

6.8 Beam Scanning

In chapter 1 it was shown that focusing a high-energy particle beam to small dimensions not only improves the minimum detectable concentration using techniques such as PIXE, RBS and PRA, but also allows two- or even three-dimensional elemental maps of the sample to be constructed by scanning the sample and observing the reaction products (see for example [WA82]). Although the scanning is sometimes carried out by mechanically moving the sample relative to a fixed beam, it is usually more convenient to deflect the beam over a stationary sample using electrostatic or magnetic fields.

Very little work has been carried out on the effects on the final spot size of deflecting the beam away from the axis of a lens system, but since many groups have successfully incorporated beam scanning into their focusing systems it can be assumed that increases in beam-spot size due to deflection do not present major problems. It is beyond the scope of this book to consider in detail the various systems reported and the reader is directed towards a recent summary by Legge [LE82b] or the original papers of the various groups as listed in table 1.3.

It is convenient from a beam optics point of view to divide the microprobe system into three regions where electrostatic or magnetic deflecting fields may be inserted: these are the object space (pre-lens deflection), the lens system itself, and the image space (post-lens deflection). Of these, post-lens deflection is the most desirable, since the path of the beam through the lens is unchanged by deflection and so the focusing properties are not affected. The disadvantage of this system, however, is that the space between the lens and the target is usually restricted and so large deflections are difficult to achieve experimentally. The amplitude of the deflection may be increased by moving the deflecting field back into the last lens (for example, Cookson has used electrostatic deflector plates within the bore of the final lens [CO72]). Since deflection in the final lens has a minimal effect on the overall trajectory, the aberrations introduced are small, but again there may be practical problems with generating the fields within the lens, especially with lenses of small bore radius. The third alternative, pre-lens deflection, is experimentally very convenient since the object distance in a microprobe system is usually large, but because the deflected beam enters the lens system off axis, the resulting aberrations may be very large. The scanning system used in the Oxford microprobe has been investigated experimentally and theoretically and is described briefly below.

Initially, the Oxford microprobe utilised post-lens deflection [WA82] using an electrostatic deflecting field between the final lens and the target. At the maximum available field of 1kV cm^{-1}, the largest area which may be scanned with 4 MeV protons is 60 μm × 60 μm and, as would be expected with post-lens deflection, there is no noticeable increase in spot size over this area. Although this system has proved useful in small-sample scanning (e.g. biological cells) it was felt that the ability to scan over much larger areas would be beneficial not only for studying large samples but also for the identification and positioning of small samples. A second scanning system was also developed which has some novel and interesting features.

The beam spot is rastered over the target using a fast vertical (y-plane) scan and a slow horizontal (x-plane) scan. The fast y-deflection is produced by two air-cored coils situated before the lens system and arranged so that the beam always crosses the axis at the entrance principal plane (see figure 6.14). The effect of this is to ensure that the beam leaving the system always appears to diverge from the axis at the exit principal plane. Since in the Oxford system the principal planes in the y-direction coincide with the extremities of the lenses (as shown in the diagram), this arrangement minimises the deviation from the axis of the deflected beam within the lens and so reduces the aberrations due to deflection [HE78b]. In the x-plane, however, the entrance and exit principal planes occur in the centre of the lens system and behind the image respectively. As a result, using this technique will still cause large deviations within the lens, with consequent large aberrations at the extremities of the scan. In order to avoid this, therefore, the horizontal deflection takes place within the final lens, but rather than place a separate coil within the bore of the lens, the dipole field is created by unbalancing the currents fed to the coils of the quadrupole lens. This system minimises the aberration due to deflection, and areas ranging from 20 μm × 20 μm to 4 mm × 4 mm may be scanned, although

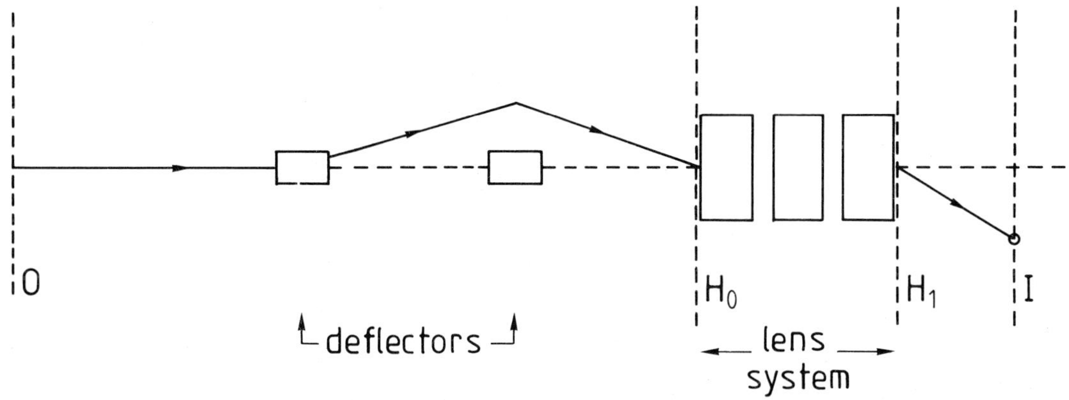

Figure 6.14 Schematic diagram of the path of the central ray in the Oxford coupled triplet using pre-lens deflection with two coils. O and I are the object and image planes; H_0 and H_1 are the entrance and exit principal planes. The two deflectors are arranged so that the central ray always crosses the axis at H_0, in which case the deflected beam will always diverge from the axis at H_1 [HE78b].

due to the large inductance of the coils in the final lens, the maximum undistorted scan rate is limited to one complete scan in twenty seconds. The effects of hysteresis in the iron of the lens are avoided by ensuring that data is only collected when the beam is scanning in one direction. Calculations, which have been supported by experimental evidence, have indicated that for an undeflected spot of 1 μm diameter, the broadening of the beam at the extremities of a 4 mm × 4 mm scan is less than 10 μm.

In the light of the foregoing discussion it is worth pointing out that, for large scan sizes, the resolution of a microprobe elemental map may be limited by the resolution of the display. For example, in the case of the Oxford microprobe, the elemental maps are digitally reconstructed and displayed as an array with a maximum size of 512 × 512. Thus for a 4 mm × 4 mm scan, each element in the picture corresponds to an area of 8 μm × 8 μm and so it is only at the extremities of the scan that the increase in spot size due to off-axis deflection will be noticeable.

7

Alternative approaches to the production of ion microbeams

This chapter is primarily concerned with the description of several new techniques for producing small probe diameters from MeV particle beams using focusing elements other than magnetic quadrupoles. At the time of writing, most of these devices are at an early stage of development and therefore have yet to realise their full potential. Nevertheless, those described show novel and interesting features and are well worth investigating. Also in this chapter is a description of possible methods of further reducing the spot sizes achieved by magnetic quadrupole focusing, using techniques which reduce or eliminate the major aberrations limiting the spatial resolution.

As most of these techniques will only be briefly and simply summarised, the reader is asked to follow references to the work described for further information and to realise that much of the work will have progressed since the time of writing.

7.1 The plasma lens

The focusing action of a plasma lens is derived from the field produced by a uniform, stable, axially-symmetric negatively-charged plasma. Following Mobley [MO79], if the plasma is confined by a cylindrical electrode of radius R, coaxial with the beam, held at potential V in a solenoidal magnetic field, such that the plasma density is uniform and that zero potential is produced on axis, then the radial field $E(r)$ is given by

$$E(r) = 2Vr/R^2.$$

For particles of energy E_0, the focal length of the lens in the thin lens approximation can be estimated to be

$$f = E_0R^2/VL,$$

where L is the length of the cylindrical electrode. The thick lens focal length is calculated to be

$$f = \frac{L}{2} + R\left(\frac{E_0}{V}\right)^{1/2} \cot\left[\frac{L}{R}\left(\frac{V}{E_0}\right)^{1/2}\right].$$

Note that, in the above expressions, the focal length is independent of particle mass or charge, offering an immediate advantage of this type of lens over magnetic quadrupoles. Also, because of its cylindrical symmetry, the plasma lens is inherently stigmatic, unlike the quadrupole lens with its focusing and defocusing action in the x- and y-planes.

Stable operation of plasma lenses, first discussed by Borries and Ruska in 1932 [BO32] and Gabor in 1947 [GA47], has been reported by at least four groups [ZH69, MO79, MA80a, LE79b].

In 1969, Zhukov and co-workers demonstrated the stable operation of the plasma lens for low-energy (keV) ions [ZH69].

At Brookhaven, successful operation of five different geometric configurations has been described [MO79]. The Brookhaven group found that there were no beam-current dependent effects for primary proton currents of up to 200 mA at proton energies of 40 keV.

The University of California at San Diego group [MA80a and references therein] have investigated the plasma lens both theoretically and experimentally. Long electron-trapping times of up to twenty minutes have been observed in ultra-high vacuum conditions. Theoretical work on electron diffusion and mobility within the electron trap indicates that the electrons behave collectively like a plasma.

The experimental group at Livermore [LE79b] has been investigating plasma lenses for some years, and at the time of writing has achieved focal spot sizes of 30 μm for 3 MeV ions [LE82], with a trapped-electron density within the lens approaching 10^{10} cm^{-3}.

Successful development of the focusing properties of the plasma lens to attain spot sizes competing with magnetic quadrupole lens systems (1 μm) will require a continuing research and development programme. Investigations carried out into plasma stabilisation, plasma end-effects and alignment tolerances will go some way towards realising the potential of the plasma lens. In particular, if the shaping of the plasma end-profiles to reduce or eliminate aberrations [LE75] can be achieved experimentally, then the plasma lens could well lead to exciting developments in the field of proton and heavy-ion microprobes.

7.2 The electrostatic coaxial lens and associated developments

This section is mainly concerned with the work carried out by a research group in Sydney on novel types of electrostatic focusing utilising (a) the electrostatic field present between two coaxial cylinders and (b) the electrostatic field present between an array of transverse charged rods.

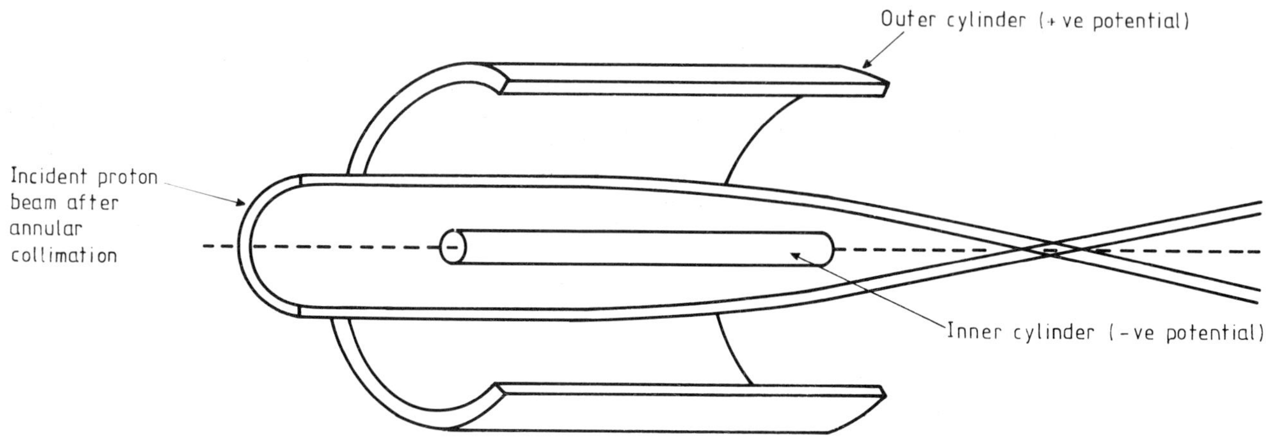

Figure 7.1 Schematic representation of the focusing action of a single convergent coaxial lens element.

The focusing action of the coaxial lens is shown in figure 7.1. The idea of using an axial lens element was first proposed by Gabor in 1946 [GA46] who suggested that such a lens may be useful in correcting spherical aberration in conventional lenses.

The basic coaxial lens element developed by the Sydney group consists of a stainless steel tube of 8mm inside radius with an axial wire of 100μm radius [KR79]. With such a lens, a collimated beam of MeV protons can be brought to a focus in 100 mm with a lens voltage of the order of several kV. The advantages of such a lens system are its simplicity of design and construction, its cylindrical symmetry and, being electrostatic, the mass-independence of its focusing action.

The field between the inner coaxial wire and the outer cylinder, however, has a $1/r$ dependence and therefore particles closer to the axis experience a stronger focusing action, resulting in very high spherical aberration. It is necessary, therefore, to collimate the incoming particle beam by an annular entrance aperture, thereby ensuring that the particles enter the lens at the same radial distance. This automatically reduces the amount of beam available for focusing. In spite of annular collimation, however, the minimum spot size achievable is still dominated by spherical aberration.

Further development of the coaxial lens [KR80] has involved using two coaxial lens elements in tandem in a diverging/converging sequence. This has the effect of reducing the spherical aberration. With this doublet, beam-spot sizes of 40μm have been achieved for a 425 keV proton beam.

Multi-elemental systems may offer an even greater reduction in aberrations although difficulties associated with lens alignment necessarily become more crucial. Dalglish [DA81] states that a system with five tandem elements is able to match or exceed the performance of conventional quadrupole multiplets, although no details are given in his paper.

Dalglish also describes in this publication a different approach to the problem of focusing charged-particle beams. He proposes that the use of a lens system containing a more suitable basic element than the coaxial lens should produce a good result when developed in a similar way to the

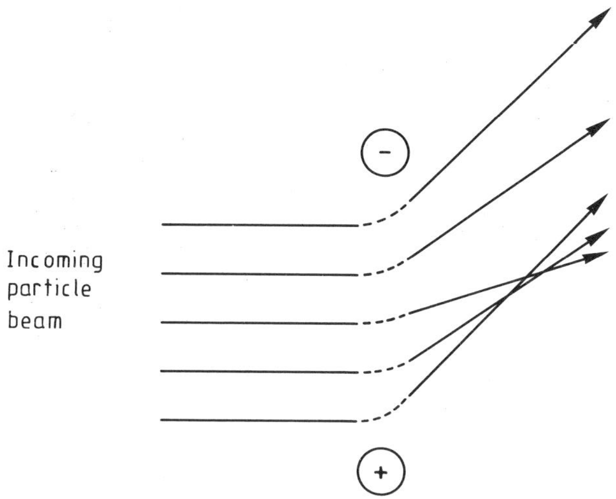

Figure 7.2 Diagrammatic representation of the focusing action of the twin transverse rod configuration. The focusing action is convergent on one side of the symmetry plane and divergent on the other.

multi-elemental coaxial lens system. In a preliminary theoretical investigation, the element chosen takes the form of a twin transverse rod configuration. Figure 7.2 shows a diagrammatic representation of the trajectories of a particle beam traversing the region between the rods and a resulting focusing action in one plane. The addition of a second offset independent twin rod element, to form a configuration which he terms the Caledonian quadruplet, has theoretical properties which predict that an input particle beam of 1mm size can be focused to a line of 1 μm width. It is further stated that since the coupling action of the Caledonian quadruplet planar focusing lens to a plane at right angles to the lens is small, then the addition of a second orthogonal lens should enable a spot size of 1 μm diameter to be formed from an input particle beam of 1mm dimensions.

If the theoretical predictions for this system can be experimentally verified, then such a simple system will provide an exciting new direction in the field of ion microprobe development. At the time of writing no experimental results regarding minimum achievable spot sizes were available.

7.3 Superconducting solenoids

The properties of a simple solenoid used for focusing charged particles are well known and this type of lens has been used extensively in electron microscopy [HA77] and β-ray spectroscopy [SI65]. Typical focal lengths for focusing MeV electrons are in the region of 50 cm with field strengths at the centre of the solenoid of about 0.5T.

To focus MeV ions using solenoids requires an increase in the field strength proportional to the ratio of the ion to electron rigidities (or momenta) and such high fields (of the order of 10T) can only be achieved with a superconducting solenoid. The advantages of a working superconducting solenoid lens in the development of scanning proton microprobes are numerous:

(1) Because of its close analogy with conventional solenoids, the beam-optical properties are very well understood.

(2) Due to the cylindrical symmetry of such a device, the focusing properties are inherently stigmatic.

(3) The spherical aberrations (and to a lesser extent the chromatic aberrations) asso-

ciated with solenoidal focusing are, in general, smaller than those for quadrupole multiplets of the same first-order focusing properties [MO70, MO71, KO80]. This indicates that beam densities for a given spot size can be correspondingly larger, due to the increased phase space acceptance of the solenoid system.

(4) The solenoid is a single element and, unlike quadrupole multiplets, does not suffer from inter-element misalignments, although the minimum spot size achievable is very sensitive to lens tilt and displacement from the beam axis [KO80]. Additionally, the focusing of a particle beam requires the adjustment of only one parameter, the solenoid current.

Two superconducting high-energy microprobe systems have been constructed at the time of writing: at the Massachusetts Institute of Technology [GR78b] and the Los Alamos Scientific Laboratory [MA80b, MA81c]. Both systems were designed to achieve spot sizes of 1 μm for MeV ion beams and are described briefly here.

The MIT solenoid is 10 cm long and has a bore diameter of 1.6 cm. The object and image distances, measured from the centre of the lens, are 4.05 m and 160 mm respectively giving a system demagnification of approximately 25. A maximum field of 6.5T on the axis of the solenoid is attainable for a coil current of 23A. The focusing parameters of this system have been calculated using the Enge and Kowalski raytrace program [EN70].

The Los Alamos solenoid is 5 cm long and has a maximum field strength at the centre of the coil of 8T for a current of 69.56A. Focusing parameters have been calculated using the bell-shaped model of the form

$$B = B_0 \left[1 + (z/a)^2\right]^{-1},$$

where a=6.5 cm, and is a good approximation to the experimentally measured field [MA81c]. The solenoid is situated 1.14 m from the object slits and the beam is brought to a focus 110 mm from the lens giving a demagnification of approximately 10.

A comparison of the aberration coefficients for these two systems with those of quadrupole systems is not straightforward because of differing formalisms.

Koyama-Ito and Grodzins of MIT have used conventional definitions for the chromatic and spherical aberration coefficients as applied to cylindrical lenses [KO80]. The radius of a focused spot due to a particular aberration only is given by

$$R_{CH} = C_{CH} M \theta \Delta E/E$$
$$R_S = C_S M \theta^3$$

where C_{CH} and C_S are the chromatic and spherical aberration coefficients, M is the system magnification, θ is the half-angle input divergence in radians and $\Delta E/E$ is the fractional energy spread. Values of $C_S = 3.08 \times 10^7$ cm and $C_{CH} = 9.41 \times 10^3$ cm are calculated by Koyama-Ito and Grodzins for the MIT superconducting solenoid system.

Maggiore of Los Alamos has used a different formalism in his calculations [MA81c]:

$$R_{CH} = C_{CH}\, \alpha\, \Delta E/E$$
$$R_S = \tfrac{1}{4} C_S\, \alpha^3$$

where C_{CH} and C_S are the chromatic and spherical aberration coefficients, α is the half-angle of convergence in the focused spot in radians and $\Delta E/E$ is the fractional energy spread. Values of C_{CH}=9.3 cm and C_S=14 cm are calculated by Maggiore for the Los Alamos superconducting system.

In order to attempt a comparison between the aberration coefficients for the superconducting solenoid systems, these coefficients are converted to be compatible with the notation used throughout the book using the equations

$$R_{CH} = C_{CH}\, \theta\, \delta \qquad (7.1)$$
$$R_S = C_S\, \theta^3 \qquad (7.2)$$

where θ is the half-angle input divergence in mrad and δ is the fractional momentum spread of the beam expressed as a percentage.

Table 7.1 gives the spherical aberration coefficients for the Los Alamos solenoid and the MIT solenoid expressed in the above notation, together with the aberration coefficients of the Heidelberg quadrupole doublet, the Oxford triplet and the Harwell quadruplet. It must be remembered however that any comparisons must take into account the cross terms present in the quadrupole multiplet aberration coefficients.

aberration figures. Calculated values of the phase space acceptance of solenoid systems [KO80, MA81c] show that at least twice the beam density should be obtainable within a 1 μm focused spot than can be presently achieved with quadrupole multiplet systems.

Initial experimental results for superconducting solenoid systems have been disappointing. At the time of writing, the MIT solenoid has not achieved a spot size below 10 μm [GR81b] and the Los Alamos system has failed to reach spot sizes below 5 μm [MA81c]. It is thought that the relatively large energy spreads of the beams delivered by the MIT and Los Alamos accelerators are primarily responsible for the poor performances of the solenoid systems and it seems likely therefore that the full potential of the solenoids has yet to be realised.

Table 7.1 Aberration coefficients for the Los Alamos solenoid, the MIT solenoid, the Heidelberg quadrupole doublet, the Oxford triplet and the Harwell quadruplet.

	MIT Solenoid	LASL Solenoid		Heidelberg Doublet	Oxford Triplet	Harwell Quadruplet
Demagnification D	25	10	D_x	−4.6	68	−5.3
			D_y	−26.4	−15	−5.3
Spherical aberration C_S	12	0.035	$\langle x\|\theta^3\rangle$	−3.1	87.9	−1.1
			$\langle y\|\phi^3\rangle$	−1.1	438	−4.6
			$\langle x\|\theta\phi^2\rangle$	−1.6	111	−5.8
			$\langle y\|\phi\theta^2\rangle$	−7.9	−428	−5.6
Chromatic aberration C_{CH}	76	18	$\langle x\|\theta\delta\rangle$	36.5	−256	63.7
			$\langle y\|\phi\delta\rangle$	39.6	783	103.7

All units are in μm and mrad. C_S and C_{CH} are defined by equations (7.1) and (7.2). $\langle x|\theta^3\rangle$, $\langle y|\phi^3\rangle$, . . . are defined in § 1.10.

It can be seen from table 7.1 that a general comparison of the spherical and chromatic aberration coefficients for the solenoid systems with those of the quadrupole multiplets is somewhat inconclusive. However, the table does indicate that for similar demagnifications the solenoids show smaller

7.4 The electrostatic quadrupole triplet

Augustyniak and co-workers at the Bell Laboratories in New Jersey have developed a high-energy ion microprobe operating at MeV energies based on a symmetric CDC electrostatic triplet configuration [AU78]. It is included in this chapter because of its convenient and simple design, enabling the complete focusing system to be built inside a standard four-inch tee-section commonly available in nuclear accelerator installations, and also because of the mass-independence of its focusing action, allowing the formation of heavy-ion probes.

The lens elements are made up from four titanium-plated quartz rods accurately positioned with fourfold symmetry. The lengths of the electrodes are 30 mm, 44 mm, and 30 mm respectively with a bore diameter of 2 mm. These parameters indicate the degree of miniaturisation with which this system has been constructed.

For an excitation of approximately 15 kV on the electrodes, 1.5 MeV singly-charged particles can be focused to a spot within 10 cm of the end of the lens system. Particle beams of about 5nA have been focused to spot sizes of 15 μm in diameter, adequate for many microprobe applications. The New Jersey group has estimated that this figure of 15 μm represents a realistic minimum achievable spot size, since the focused spot at this dimension is heavily dominated by aberrations, thought to be caused by imperfections in the ends of the quartz rod electrodes. In this particular system, however, several types of aberration may prevail.

Chromatic aberration, perhaps the most dominant of the intrinsic aberrations affecting the spot size in present magnetic quadrupole systems, is greater by a factor of two in electrostatic focusing systems (see §3.5). Aberrations due to rotational misalignment, which if uncorrected represent the largest source of parasitic aberration in quadrupole systems, will no doubt be present to some extent in the electrostatic system, because of its mechanically rigid design. Lastly, because of the very small dimensions of the system, effects due to harmonic contamination of the quadrupole field may greatly influence the spot size.

That a 15 μm spot size has been achieved with the miniature triplet is perhaps indicative of the precision to which this system has been constructed.

7.5 Achromatic quadrupole lenses

Since the real and parasitic aberrations of a quadrupole multiplet focusing system are in general dependent on the beam input angles, the effects of aberrations on the focused spot size may be minimised by reducing the beam input divergence, although at the expense of the transmitted beam current. A limit to the improvement in beam-spot size that may be carried out in this manner is set by the lower-order aberrations, since the effects of these aberrations remain relatively large at small input angles. In present magnetic probe-forming systems the effect of chromatic aberration, which is proportional to the input divergence of the beam and the beam energy spread, is a major limitation in the production of usable micron and sub-micron beams [GR82].

The effects of chromatic aberration can also be reduced by minimising the energy spread in the incident beam, usually by improving the energy stabilisation of the accelerator. Typical energy spreads in beams of MeV ions are in the region of one part in 10^3 although one part in 10^4 can be achieved using sophisticated energy stabilisation techniques. The 1 μm spot size achieved with the Oxford coupled triplet for 4 MeV protons [WA82] required an energy stability in the beam of five parts in 10^4.

The construction of an achromatic focusing system is therefore important for the attainment of sub-micron spot sizes. It is interesting to note that although it is not possible to construct achromatic lenses which have cylindrical symmetry [SC36], quadrupole probe-forming systems can be constructed which have achromatic properties.

There are two ways of constructing an achromatic system of quadrupoles, both of which make use of the different chromatic aberration properties of magnetic and electrostatic lenses (see §3.5). The first method is to combine a series of spatially-separated electrostatic and magnetic lenses into a probe-forming system. In this type of configuration, however, the condition for achromatic behaviour can be satisfied in one plane only [KE62]. Preliminary calculations at Oxford using OXRAY have shown that a specific geometry of triplet (CDC) with quadrupole elements mixed (EME : electrostatic/magnetic/electrostatic) has zero chromatic aberration in one plane.

A second method, first discussed by Kel'man and Yavor in 1962 [KE62], makes use of quadrupole elements which have superimposed electrostatic and magnetic fields. If the excitations of these quadrupole lenses are adjusted such that the magnetic force on a charged particle is everywhere opposite to the electrostatic force and has twice its magnitude, then to first order the entire trajectory of a charged particle can be achromatic. A system of combined quadrupole lenses can therefore be constructed which has a focal length in both planes independent of ion energy, to first order.

A schematic diagram of this type of quadrupole lens is shown in figure 7.3. To satisfy the condition for achromatic behaviour, the electrostatic and magnetic poles are spaced at 45° intervals.

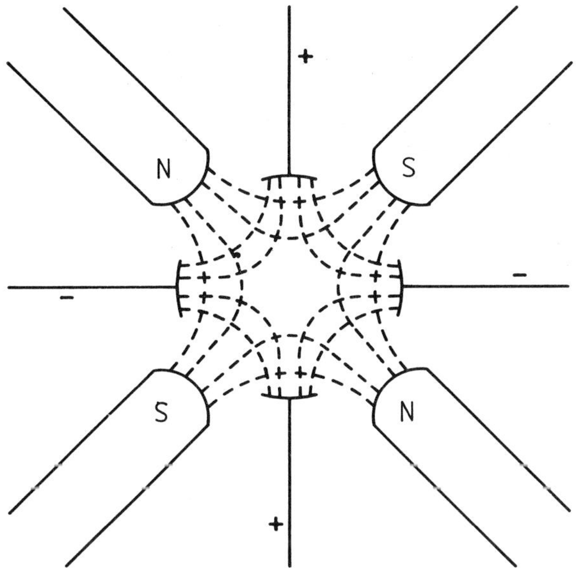

Figure 7.3 Schematic diagram of a combined magnetic/electrostatic lens.

A doublet of achromatic quadrupole lenses has been constructed by F W Martin of Microscope Associates Inc and R Goloskie of the Worcester Polytechnic Institute, Massachusetts [MA82]. Each lens element is of length 6 cm and diameter 14 cm, and contains a magnetic quadrupole of maximum pole-tip field 1.0T and bore 3.4 mm, and an electrostatic quadrupole of maximum pole potential 3 kV and bore 1.05 mm. Calculations have predicted the maximum achromatic operating energy for protons to be 29 MeV.

Achromatic operation of the doublet has been tested using protons of energy ranging from 400 to 800 keV [MA82]. For a focused beam-spot size of 15 μm at 600 keV, tests have shown that over the full energy range an increase in spot size of only 10 μm was observed. This can be compared to an increase of 85 μm for an energy change of 150 keV when the same system is operated in the magnetic mode only. The chromatic aberration coefficient in one plane was determined by Martin and Goloskie to be 20 cm for the magnetic mode, reducing to less than 1.2 cm for achromatic operation. The formalism used in the aberration coefficient determination was

$$R_{CH} = C_x \, \Delta E/E \; 2\alpha$$

where R_{CH} is the distance off-axis of a ray in the focal plane, C_x is a chromatic aberration coefficient in one plane, $\Delta E/E$ is the fractional beam energy spread and α is the half-angle of convergence at the focus.

A spot size of 4 μm in one direction has been measured using the achromatic doublet. The transmitted beam current at this resolution is not given, however. The limitation in resolution is thought to be due to parasitic aberrations resulting from harmonic contamination of the quadrupole field introduced by mechanical imperfections.

The small size of the system and the complexity of the magnetic and electrostatic pole arrangements will make the reduction of parasitic aberrations more difficult than in conventional pure magnetic quadrupole systems. Nevertheless, the system has achieved an order of magnitude reduction in chromatic aberration, which is at least one step in the right direction toward sub-micron high-energy particle beams.

7.6 Spherical aberration correction using octupoles

In the previous section, correction of chromatic aberration using combined electrostatic and magnetic quadrupoles was discussed. The other major intrinsic aberration of quadrupole probe-forming systems is the third-order aberration or spherical aberration. An introduction to the elimination of image aberrations using multipole elements is given by Wollnik [WO72]. In this paper he indicates how the introduction of an octupole lens (shown schematically in figure 7.4) into a system of quadrupoles influences the third- and higher-order aberrations without affecting the first-order focusing properties. As early as 1947, Scherzer [SC47] suggested that the third-order spherical aberration of axially symmetric lenses may be corrected using octupole elements, and subsequently Archard [AR60] outlined a proposal for the construction of a multi-elemental system containing quadrupole and octupole elements which would be free of spherical aberration. Most of the work carried out in the field of spherical aberration correction has been directed towards achieving higher resolving powers in electron microscopes [SE66].

In particular, a series of investigations by a Russian group on the correction of spherical aberration using octupoles has outlined analytical methods of calculating octupole positions and strengths [FI68b, YA72, LY74 and references therein]. More recently, Jamieson [JA79a, JA82] has theoretically investigated spherical aberration correction in the Melbourne Russian quadruplet configuration [LE78] using a computer program PRAM based on the analytical work of the Russians.

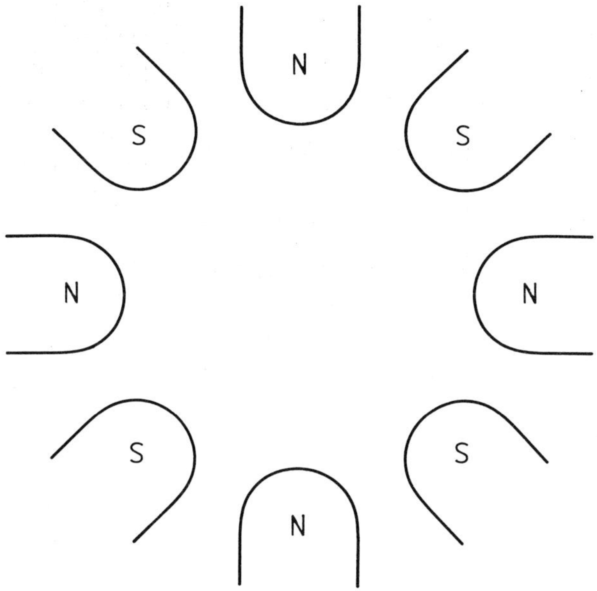

Figure 7.4 Schematic representation of an octupole lens.

These calculations have indicated that by using three independently excited octupoles, positioned QOQOQOQ (where Q represents a quadrupole and O an octupole), third-order spherical aberration can be eliminated, although fifth-order effects are increased.

A relevant question when discussing the practical application of octupole correctors is whether or not the degree of parasitic aberrations introduced by the octupole elements will negate the improvement in spatial resolution caused by the reduction in spherical aberration. As has been indicated in chapter 6, parasitic aberrations caused by element misalignments and deviation from hyperbolic pole-tip shape can be a major concern. The introduction of a series of octupole lenses into a quadrupole multiplet system will make the minimisation of such parasitic aberrations even more complex.

Additionally, the practical difficulties of optimising three independently excited octupoles in an attempt to minimise spherical aberration will undoubtedly introduce added operational complexities.

7.7 Ion source development

As discussed in §7.5, the effect on the spot size of both intrinsic and parasitic aberrations can be minimised by reducing the entrance divergence of the particle beam at the expense of the transmitted beam intensity. An obvious limit to the minimum spot size achieved using this method is set by the amount of useful beam that can be focused down. Experience at Oxford with the coupled triplet probe has indicated that a particle beam as low as 100 pA is acceptable for trace element studies at the parts per million level in thin-section biological samples using the PIXE method. Although this figure is obviously very approximate, since it is dependent on detector area, detector to target distance, sensitivity requirements, etc, it nevertheless represents a reasonable indication of the smallest amount of transmitted beam which allows useful trace element mapping to be carried out. Since the average maximum transmitted beam density for most quadrupole probe-forming systems in operation at present is of the order of 100 pA μm^{-2} or less, then this implies that sub-micron spot sizes will not be reached by reducing the beam entrance divergences while still retaining adequate beam transmissions. An obvious approach to achieving sub-micron spot sizes therefore is in the field of ion-source development.

Most of the ion sources used in the production of ions prior to acceleration have been developed for nuclear structure research, especially the negative ion sources for tandem Van de Graaff accelerators. These sources, by nature of their development, are not optimised for maximum brightness over the much-reduced angular acceptance required by quadrupole multiplet systems. As discussed in §1.7, the brightness of conventional duoplasmatron sources and RF sources is in the region of 10^9 A m^{-2} sr^{-1}. Experience with the Oxford microprobe has shown that a conventional direct-extract off-axis duoplasmatron source [LA65] has sufficient brightness such that 80 pA of 4 MeV protons can be focused to a 1 μm spot [WA82]. It is likely that the development of existing sources to maximise brightness over angular ranges compatible with the acceptances of microprobe systems will lead to increased transmitted beam and consequently smaller spot sizes.

An alternative approach to the problem is to make use of recent advances in new ion-source technology. Two new types of ion source have been developed, the field ionisation source and the liquid metal source, both of which have higher brightness characteristics compared with existing RF and duoplasmatron sources.

The phenomenon of field ionisation has been known for some years and involves the ionisation of atoms and molecules by tunnelling of the electrons in very high electric fields. The high fields are achieved by applying a potential in the region of 10 keV between a sharp iridium or tungsten tip of radius typically 50 nm and a nearby cathode aperture. For a given potential difference, fields of tens of volts per nanometre can occur over small regions of the tip surface and, depending on the tip temperature and the gas pressure around the tip, positive ions such as hydrogen, oxygen, neon, etc can be produced with a source brightness of approximately 10^{11} A m^{-2} sr^{-1}, two orders of magnitude greater than conventional sources. It is beyond the scope of this book to describe the detailed work carried out on the development of field-ionisation sources. The reader is asked to consult articles by Gomer [GO61] for a discussion on field ionisation, Hanson and Siegel [HA79] and Orloff and Swanson [OR78] and references therein. The field-ionisation source has also been used in ion microscopes (see Levi-Setti and Fox [LE80a] and references therein). A proposed design by Cho *et al* [CH78] for an ion microprobe using a field-ionisation source also makes interesting reading.

There are practical disadvantages associated with the field-ionisation source when proposed as an injector of ions for a nuclear accelerator. The high brightness characteristics and stable operation of the source occur when the tip is cooled, preferably to liquid helium temperatures. Incorporating liquid helium cooling into a terminal of a nuclear accelerator such as a single-ended Van de Graaff machine will undoubtedly create enormous practical difficulties. A second point to note is that the relatively small total beam current available from the present generation of field-ionisation sources (in the picoamp to nanoamp range) indicates the need for efficient transmission of the beam through the accelerator and effective stabilisation of the beam using small ion currents.

The second type of source, the liquid metal source, operates on the principle of electric field ionisation of a molten metal film on a wetted needle of micron dimensions. The application of kV potential differences between the needle and a cathode aperture creates an intense electric field sufficient to produce metallic positive ion emission by field ionisation. Brightness characteristics of better than 10^{10} A m^{-2} sr^{-1} have been achieved with this type of source over a wide range of metal ions. The advantages of the liquid metal ion source over the gas field-ionisation source are that no tip cooling is required, no gas loading of the source occurs and primary currents in the range 10–100 μA can be achieved. For further information the reader is asked to consult relevant recent articles [CL78, RI78, SW79] and references therein.

An obvious disadvantage of the liquid metal source is that, in general, metallic ions are not suitable for the analytical techniques described in chapter 1. Development of a lithium source based on this principle, however, may prove to be useful. In order to obtain high cross-sections for x-ray production, at least 12 MeV lithium ions would be needed and this would require a relatively large tandem accelerator. Also, because the liquid metal source delivers positive ions, charge exchange to produce negative ions must take place prior to acceleration. How much the higher angular density of the beam would be affected by this need for charge exchange and stripping will require further investigation.

Although the two sources described here have serious operational disadvantages in the production of particle beams suitable for the analytical techniques described in chapter 1, one important aspect of these sources warrants further discussion. A feature common to both sources is that the effective size is small, i.e. the region from which the ions are emitted is, in general, much less than 1 μm across. An important consequence of this characteristic is that it is possible to use a direct, demagnified image of the source as the ion probe. These sources therefore have important applications in the fields of ion microscopy and micro ion implantation [LE80c, WI80]. It is possible then, for these applications, that low demagnification (and therefore low aberration) quadrupole systems may become useful in sub-micron high-energy ion-probe formation.

Appendix 1

First-, second- and third-order imaging properties of quadrupole probe-forming systems

For each type of lens, i.e. doublet, triplet or quadruplet, there are different lens configurations, and for each configuration there are four possible spacings: 0, 1, 2 or 5 quadrupole lengths. For each spacing of each configuration there are two plots and one table. The two plots appear on facing pages and the relevant table appears either immediately before or immediately after the plots. The configurations are arranged in the following order:

Doublet	CD
Triplet	CDC
	CDC
	CDC
Quadruplet	CDDC
	CDCD
	CDDC
	CDCD
	CDDC
	CDCD

The spacings of each configuration are arranged in ascending order. An asterisk in a table of spherical aberration coefficients indicates that at least one of the four coefficients has a value greater than 10^{10}.

Lens Configuration 20C0D

KL (QUAD 1)	KL (QUAD 2)	Normalised image distance	Normalised spherical aberration coefficients			
			x/θ^3QL	$x/\theta\phi^2$QL	y/ϕ^3QL	$y/\theta^2\phi$QL
1.014	1.445	0.5	−9.2 (3)	−1.10 (5)	−5.41 (4)	−1.14 (4)
0.950	1.220	1.0	−8.1 (3)	−8.72 (4)	−4.07 (4)	−1.26 (4)
0.905	1.095	1.5	−7.8 (3)	−7.18 (4)	−3.17 (4)	−1.42 (4)
0.871	1.014	2.0	−8.5 (3)	−6.71 (4)	−2.70 (4)	−1.68 (4)
0.844	0.957	2.5	−7.0 (3)	−6.50 (4)	−2.89 (4)	−1.68 (4)
0.823	0.907	3.0	−7.0 (3)	−6.30 (4)	−2.61 (4)	−1.71 (4)
0.803	0.879	3.5	−7.0 (3)	−6.24 (4)	−2.60 (4)	−1.68 (4)
0.786	0.851	4.0	−7.0 (3)	−6.32 (4)	−2.37 (4)	−1.87 (4)

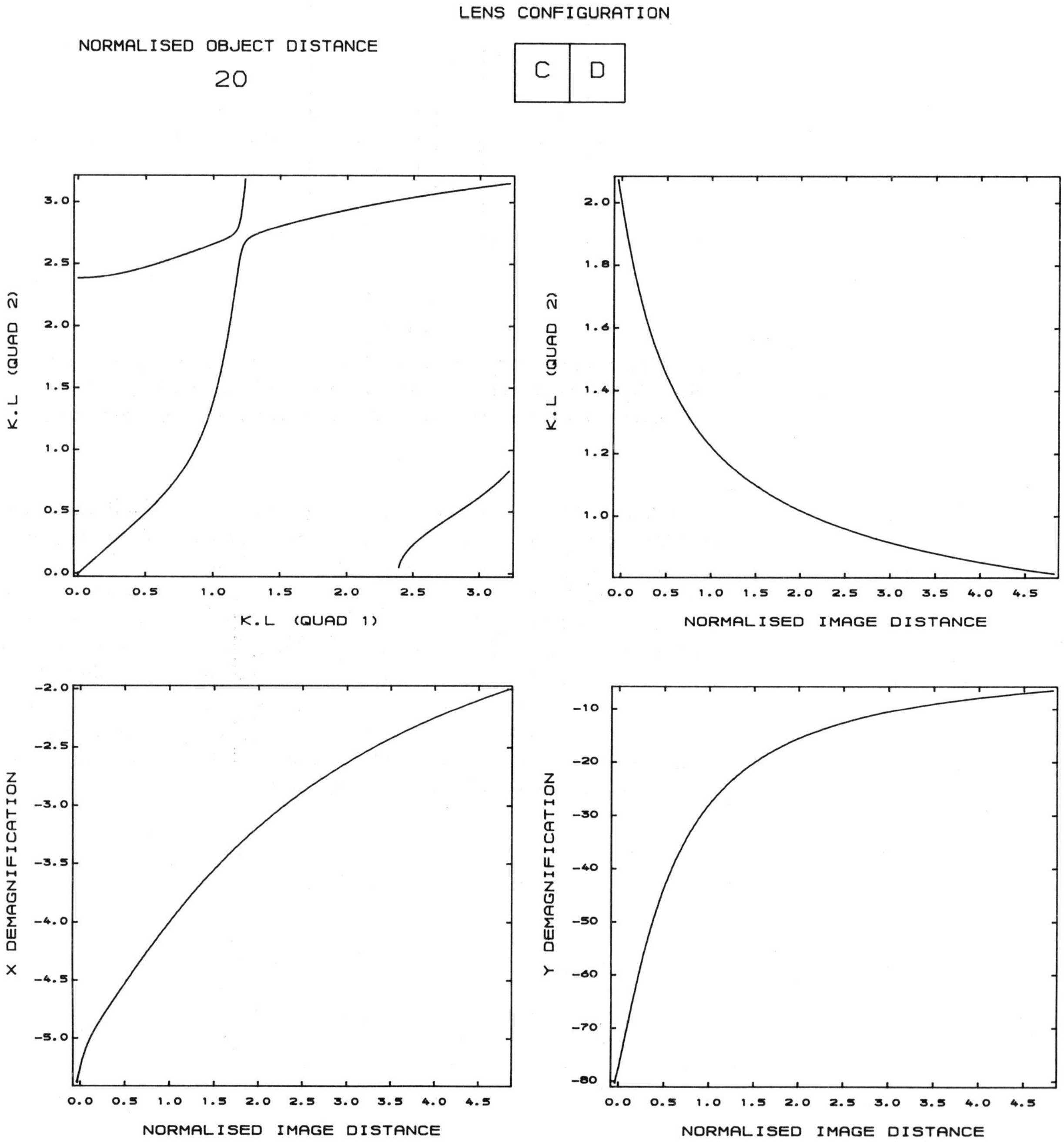
LENS CONFIGURATION
NORMALISED OBJECT DISTANCE
20
C
D
K.L (QUAD 2)
K.L (QUAD 1)
NORMALISED IMAGE DISTANCE
X DEMAGNIFICATION
Y DEMAGNIFICATION

LENS CONFIGURATION

NORMALISED OBJECT DISTANCE

20

C	D

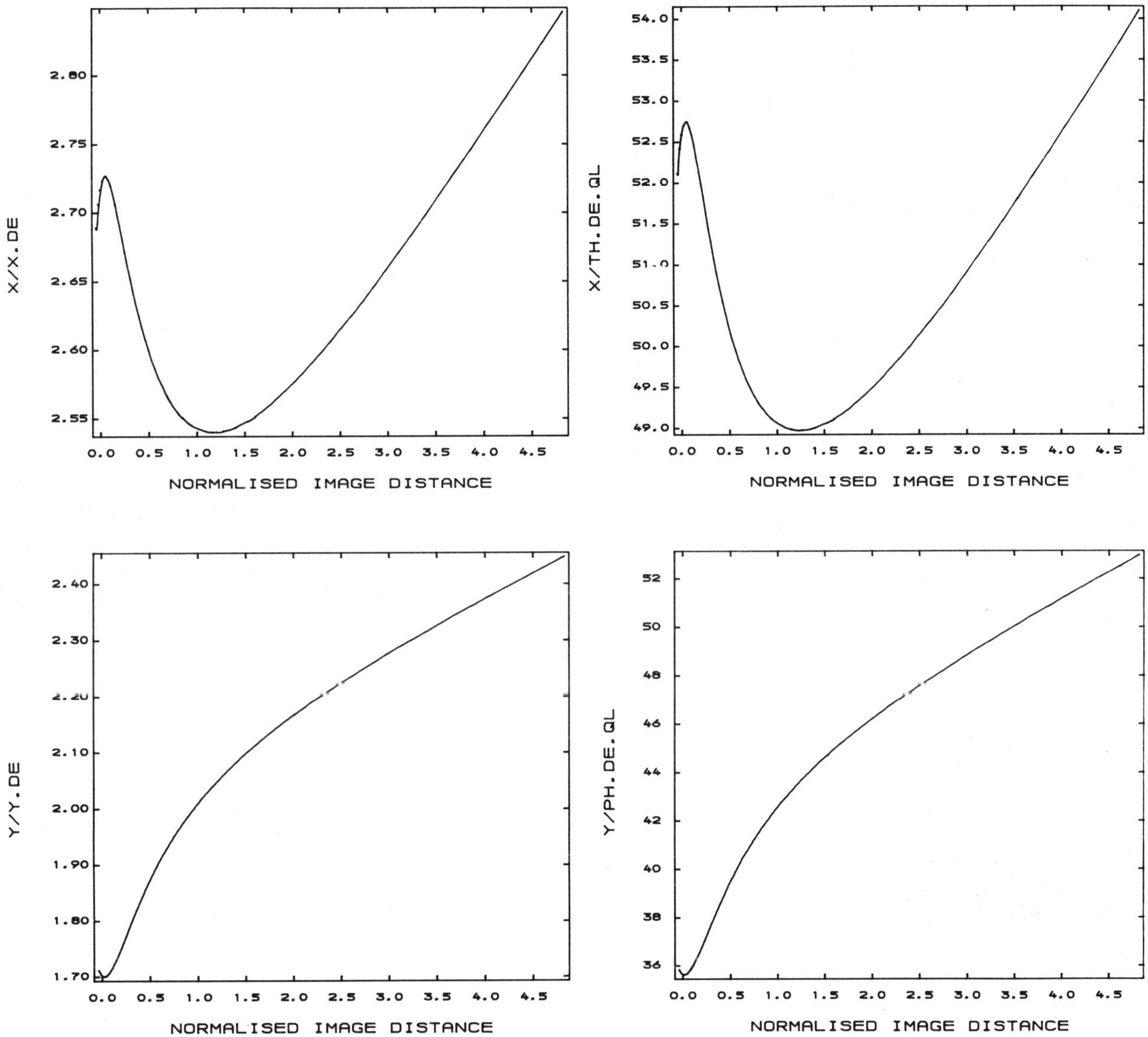

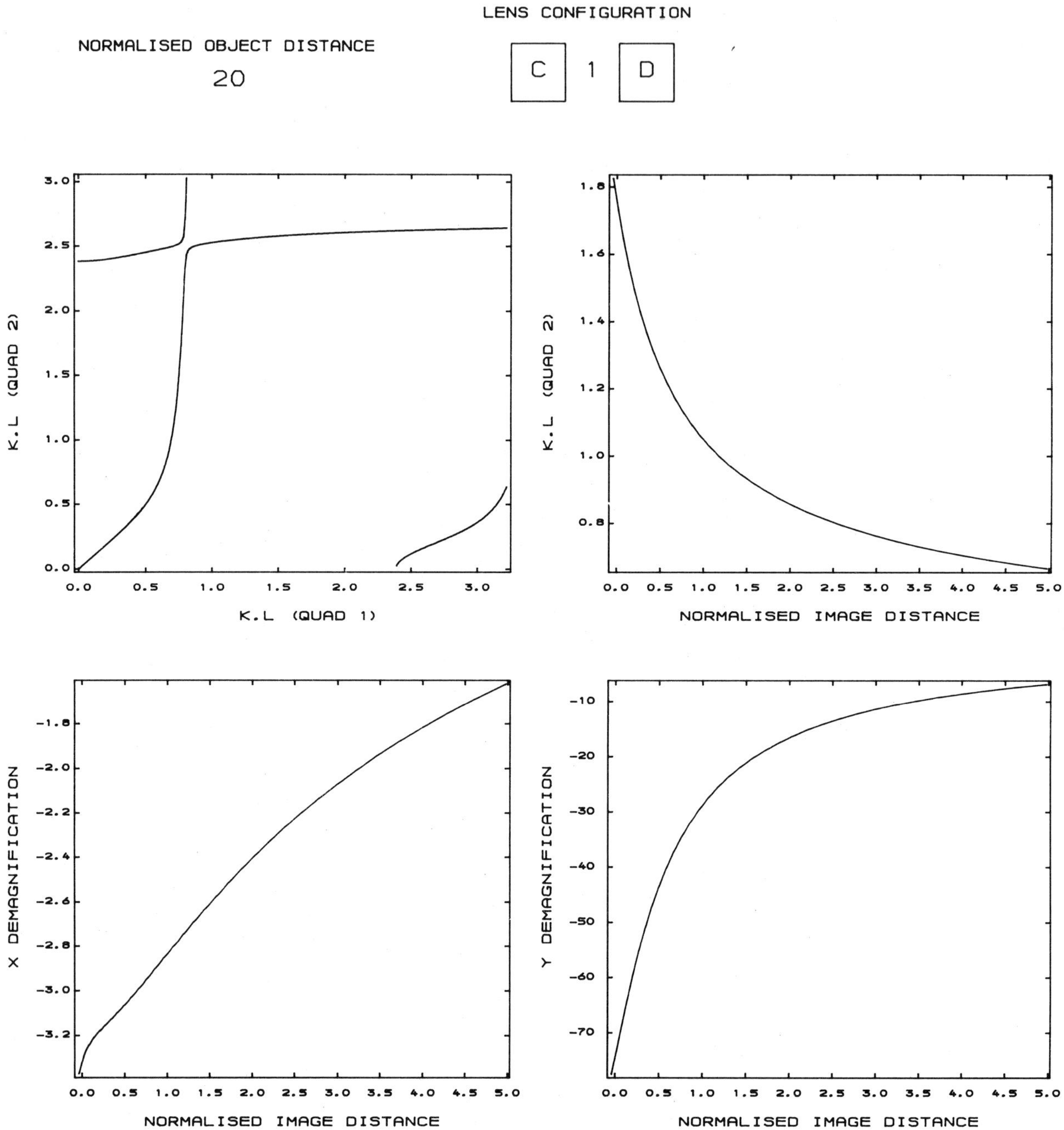
LENS CONFIGURATION
NORMALISED OBJECT DISTANCE
20
C
1
D
K.L (QUAD 2)
K.L (QUAD 1)
NORMALISED IMAGE DISTANCE
X DEMAGNIFICATION
Y DEMAGNIFICATION
0.0
0.5
1.0
1.5
2.0
2.5
3.0
0.8
1.2
1.4
1.6
1.8
3.5
4.0
4.5
5.0
-1.8
-2.0
-2.2
-2.4
-2.6
-2.8
-3.0
-3.2
-10
-20
-30
-40
-50
-60
-70

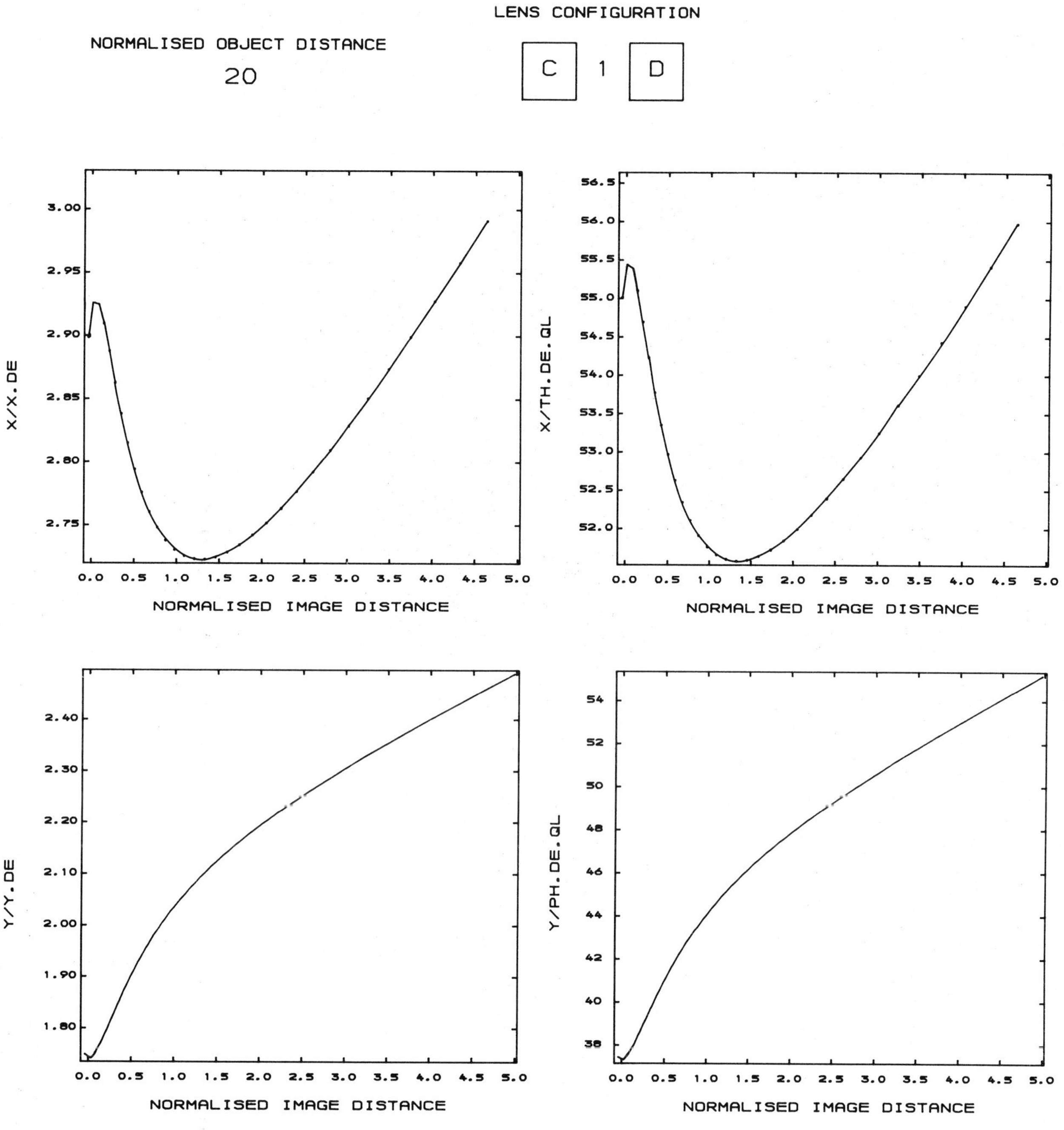
LENS CONFIGURATION
NORMALISED OBJECT DISTANCE
20
C
1
D
X/X.DE
X/TH.DE.QL
Y/Y.DE
Y/PH.DE.QL
NORMALISED IMAGE DISTANCE
NORMALISED IMAGE DISTANCE
NORMALISED IMAGE DISTANCE
NORMALISED IMAGE DISTANCE

Lens Configuration 20C1D

KL (QUAD 1)	KL (QUAD 2)	Normalised image distance	Normalised spherical aberration coefficients			
			x/θ^3QL	$x/\theta\phi^2QL$	y/ϕ^3QL	$y/\theta^2\phi QL$
0.726	1.256	0.5	−4.7 (3)	−5.00 (4)	−4.28 (4)	−3.5 (3)
0.698	1.046	1.0	−2.7 (3)	−4.09 (4)	−2.95 (4)	−4.0 (3)
0.676	0.930	1.5	−1.4 (3)	−3.71 (4)	−2.30 (4)	−5.2 (3)
0.658	0.854	2.0	−2.6 (3)	−3.41 (4)	−2.13 (4)	−5.3 (3)
0.643	0.800	2.5	−3.6 (3)	−3.30 (4)	−1.86 (4)	−4.5 (3)
0.630	0.760	3.0	−2.5 (3)	−3.17 (4)	−1.58 (4)	−5.5 (3)
0.619	0.728	3.5	−0.6 (3)	−3.15 (4)	−1.45 (4)	−6.3 (3)
0.609	0.702	4.0	−3.0 (3)	−2.94 (4)	−1.38 (4)	−6.6 (3)

Lens Configuration 20C2D

KL (QUAD 1)	KL (QUAD 2)	Normalised image distance	Normalised spherical aberration coefficients			
			x/θ^3QL	$x/\theta\phi^2$QL	y/ϕ^3QL	$y/\theta^2\phi$QL
0.609	1.198	0.5	−0.3 (3)	−3.46 (4)	−4.33 (4)	−1.9 (3)
0.592	0.990	1.0	−1.9 (3)	−2.82 (4)	−2.92 (4)	−2.6 (3)
0.578	0.873	1.5	−3.2 (3)	−2.35 (4)	−2.28 (4)	−2.5 (3)
0.566	0.797	2.0	−1.4 (3)	−2.30 (4)	−1.90 (4)	−2.4 (3)
0.556	0.744	2.5	−2.6 (3)	−2.22 (4)	−1.71 (4)	−3.1 (3)
0.547	0.703	3.0	−2.8 (3)	−2.14 (4)	−1.39 (4)	−3.1 (3)
0.539	0.671	3.5	−1.8 (3)	−1.95 (4)	−1.30 (4)	−3.6 (3)
0.532	0.645	4.0	−1.8 (3)	−2.19 (4)	−1.21 (4)	−2.9 (3)

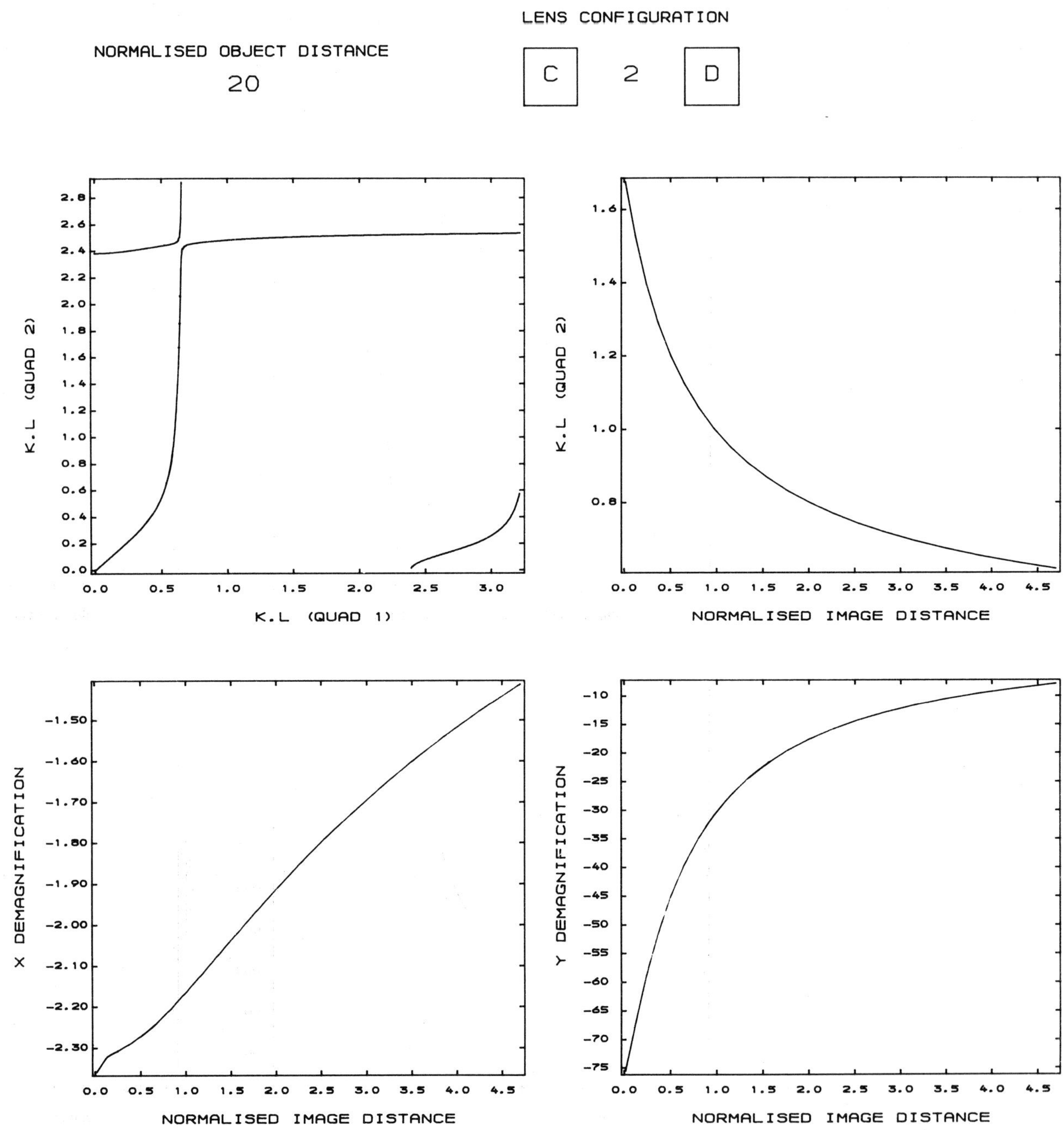

LENS CONFIGURATION
NORMALISED OBJECT DISTANCE
20
C
2
D
K.L (QUAD 2)
K.L (QUAD 1)
NORMALISED IMAGE DISTANCE
X DEMAGNIFICATION
Y DEMAGNIFICATION

LENS CONFIGURATION

NORMALISED OBJECT DISTANCE

20

C 2 D

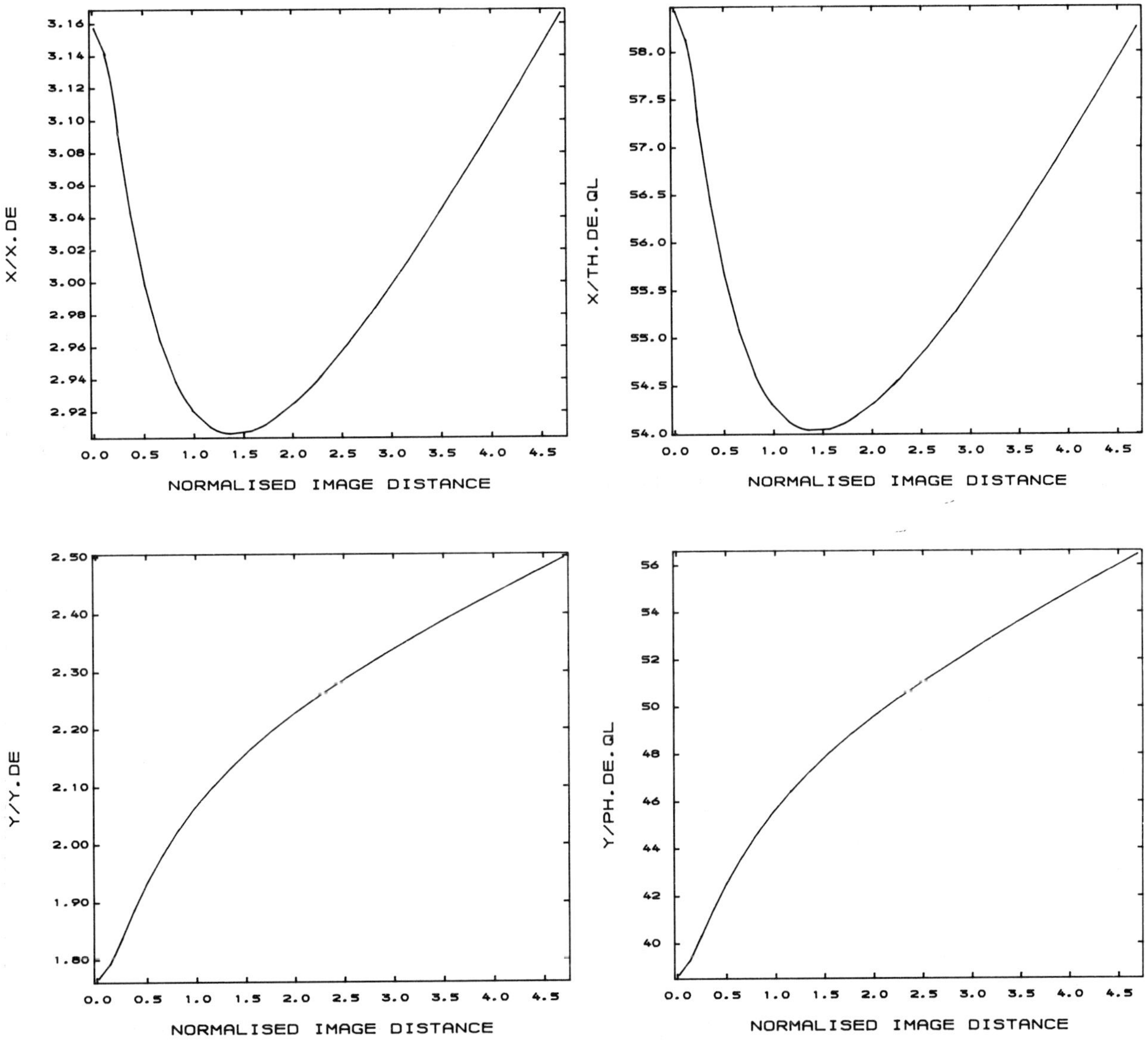

LENS CONFIGURATION

NORMALISED OBJECT DISTANCE 20

C 5 D

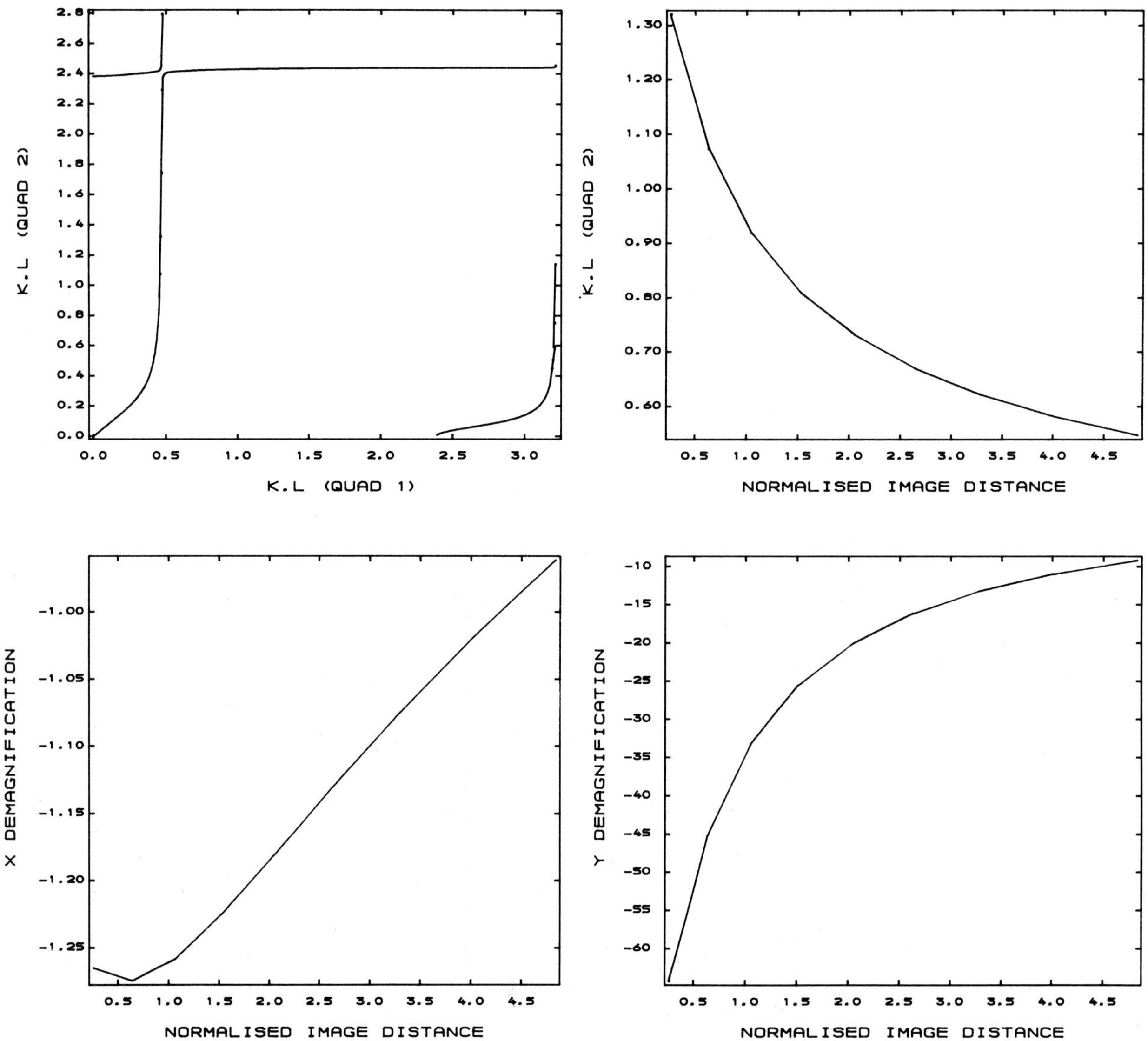

LENS CONFIGURATION

NORMALISED OBJECT DISTANCE

20

C 5 D

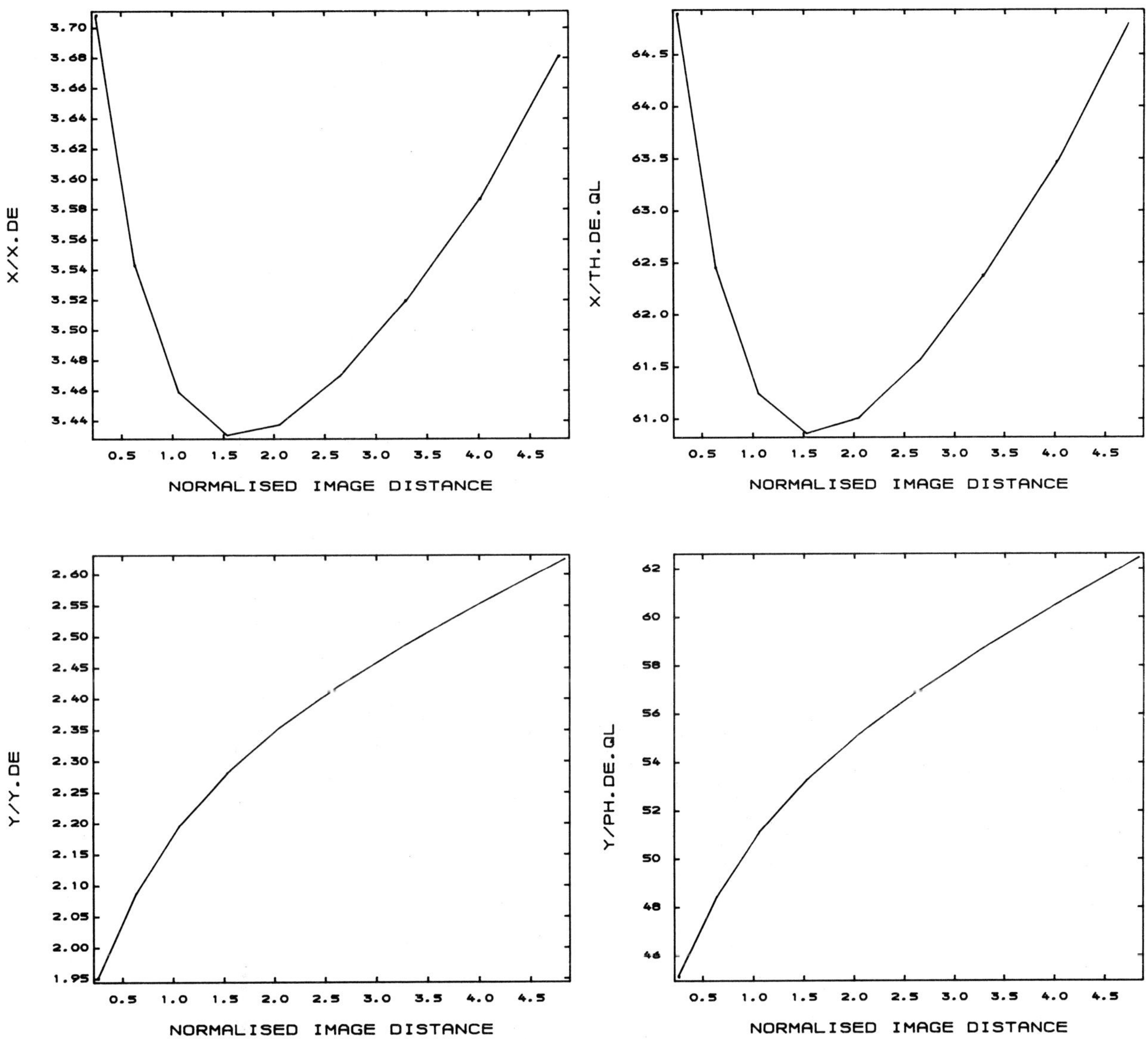

Lens Configuration 20C5D

KL (QUAD 1)	KL (QUAD 2)	Normalised image distance	Normalised spherical aberration coefficients			
			x/θ^3QL	$x/\theta\phi^2QL$	y/ϕ^3QL	$y/\theta^2\phi QL$
0.461	1.140	0.5	−0.4 (3)	−1.92 (4)	−5.68 (4)	−0.5 (3)
0.454	0.931	1.0	−1.5 (3)	−1.50 (4)	−3.72 (4)	−1.2 (3)
0.449	0.812	1.5	−0.5 (3)	−1.49 (4)	−2.80 (4)	−0.3 (3)
0.443	0.735	2.0	−0.7 (3)	−1.32 (4)	−2.18 (4)	−1.0 (3)
0.439	0.680	2.5	−2.9 (3)	−1.22 (4)	−1.78 (4)	−1.0 (3)
0.435	0.638	3.0	1.1 (3)	−1.28 (4)	−1.56 (4)	−1.1 (3)
0.431	0.606	3.5	−0.6 (3)	−1.17 (4)	−1.45 (4)	−0.6 (3)
0.427	0.579	4.0	−1.4 (3)	−1.23 (4)	−1.26 (4)	−1.0 (3)

Lens Configuration 20C0D0C

KL (QUAD 1)	KL (QUAD 3)	Normalised image distance	Normalised spherical aberration coefficients			
			x/θ^3QL	$x/\theta\phi^2QL$	y/ϕ^3QL	$y/\theta^2\phi QL$
1.565	1.670	0.5	4.26 (5)	5.50 (5)	−2.45 (6)	−2.75 (6)
1.499	1.457	1.0	2.92 (5)	5.27 (5)	−2.03 (6)	−1.95 (6)
1.458	1.349	1.5	2.43 (5)	5.36 (5)	−1.92 (6)	−1.66 (6)
1.430	1.285	2.0	2.16 (5)	5.63 (5)	−1.89 (6)	−1.51 (6)
1.409	1.242	2.5	2.06 (5)	6.92 (5)	−1.90 (6)	−1.45 (6)
1.393	1.211	3.0	1.97 (5)	6.26 (5)	−1.97 (6)	−1.44 (6)
1.381	1.188	3.5	1.97 (5)	6.61 (5)	−2.04 (6)	−1.41 (6)
1.371	1.170	4.0	2.00 (5)	6.95 (5)	−2.08 (6)	−1.47 (6)

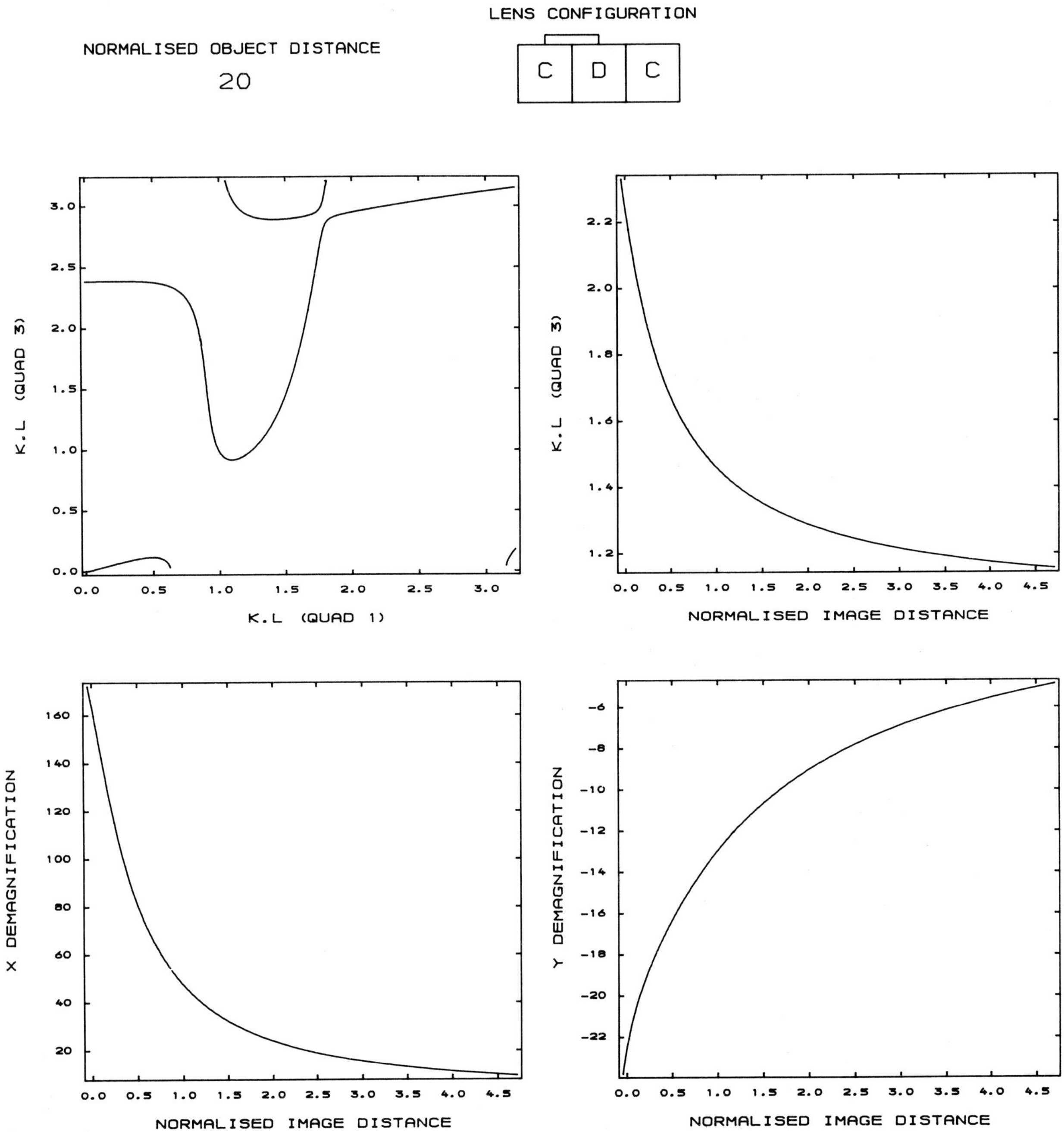
LENS CONFIGURATION
NORMALISED OBJECT DISTANCE
20
C
D
C
K.L (QUAD 3)
K.L (QUAD 1)
NORMALISED IMAGE DISTANCE
X DEMAGNIFICATION
Y DEMAGNIFICATION

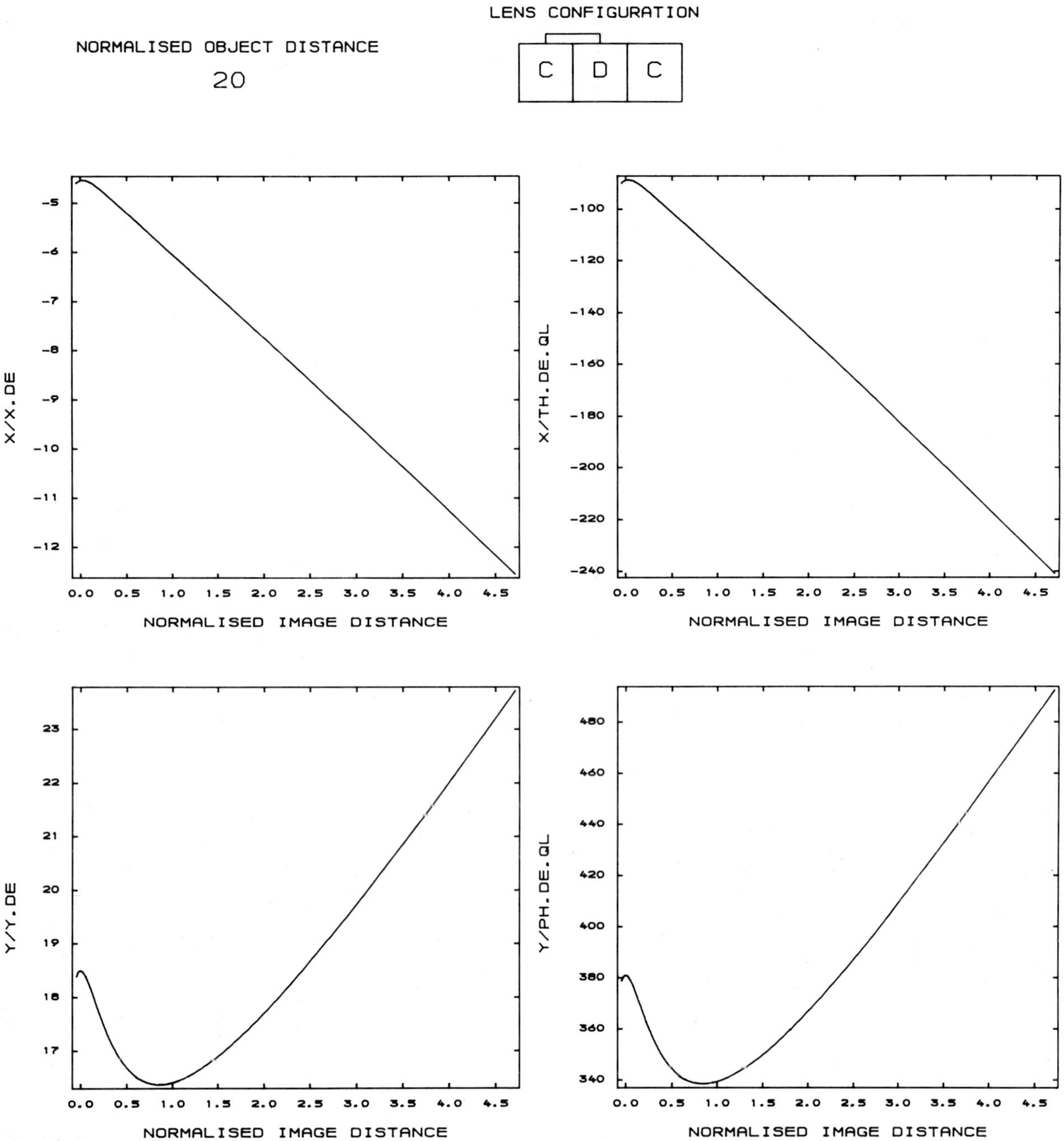
LENS CONFIGURATION
NORMALISED OBJECT DISTANCE
20
C
D
C
X/X.DE
X/TH.DE.QL
Y/Y.DE
Y/PH.DE.QL
-5
-6
-7
-8
-9
-10
-11
-12
-100
-120
-140
-160
-180
-200
-220
-240
23
22
21
20
19
18
17
480
460
440
420
400
380
360
340
0.0
0.5
1.0
1.5
2.0
2.5
3.0
3.5
4.0
4.5
NORMALISED IMAGE DISTANCE

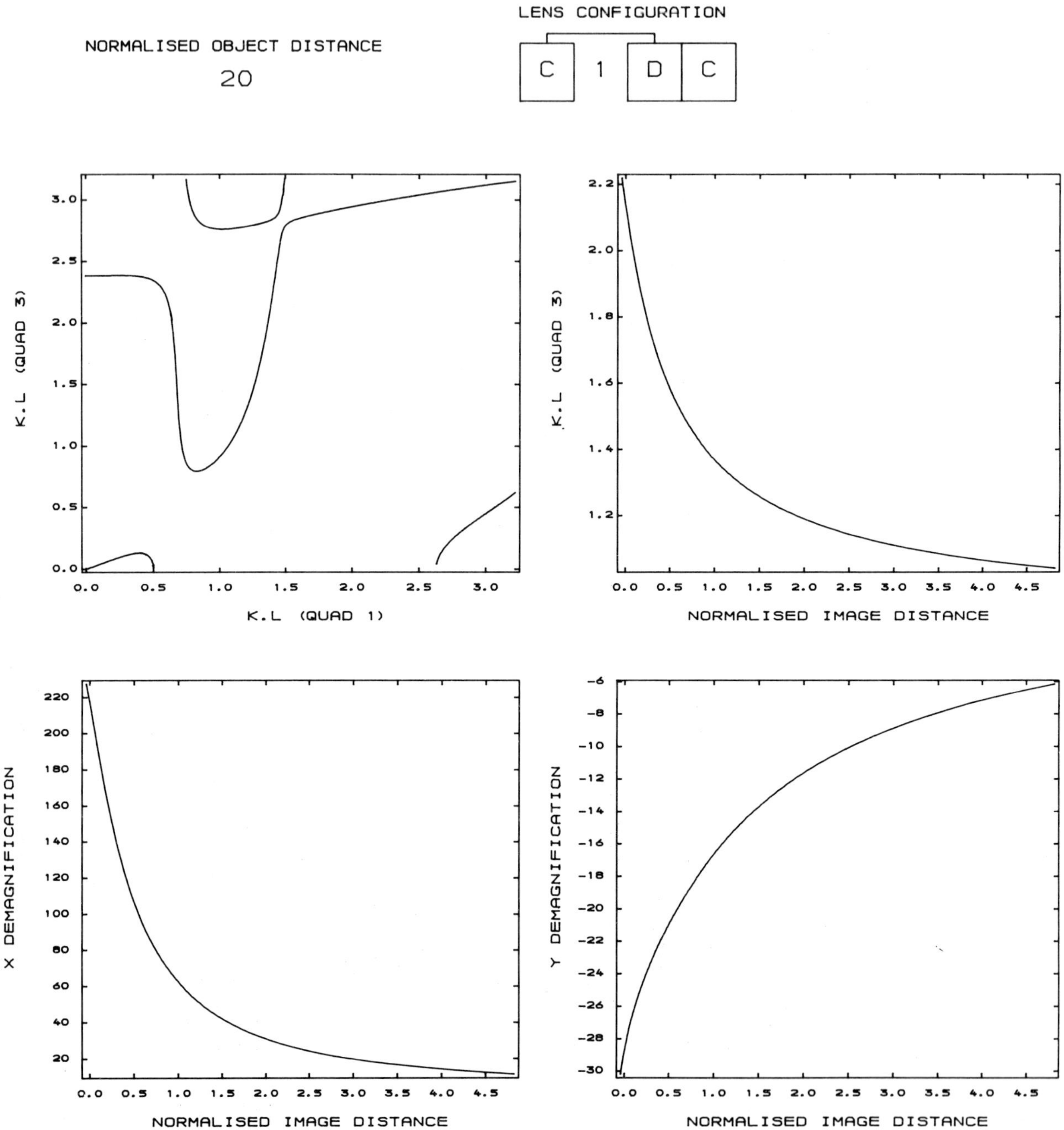

LENS CONFIGURATION
NORMALISED OBJECT DISTANCE
20
C
1
D
C
K.L (QUAD 3)
K.L (QUAD 1)
NORMALISED IMAGE DISTANCE
X DEMAGNIFICATION
Y DEMAGNIFICATION

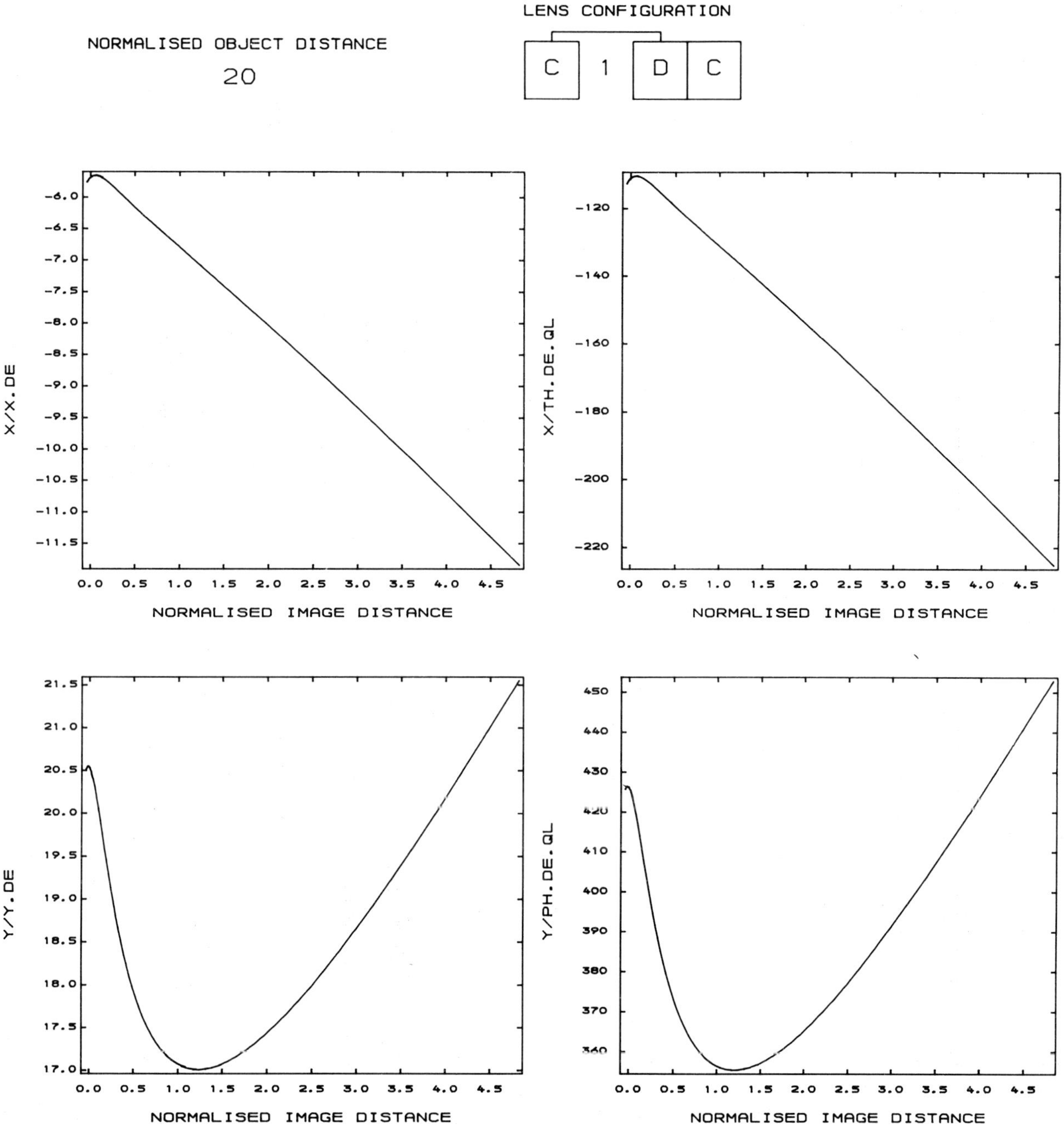
LENS CONFIGURATION
NORMALISED OBJECT DISTANCE
20
C
1
D
C
X/X.DE
X/TH.DE.QL
Y/Y.DE
Y/PH.DE.QL
NORMALISED IMAGE DISTANCE
NORMALISED IMAGE DISTANCE
NORMALISED IMAGE DISTANCE
NORMALISED IMAGE DISTANCE

Lens Configuration 20C1D0C

KL (QUAD 1)	KL (QUAD 3)	Normalised image distance	Normalised spherical aberration coefficients			
			x/θ^3QL	$x/\theta\phi^2QL$	y/ϕ^3QL	$y/\theta^2\phi QL$
1.281	1.581	0.5	8.90 (5)	8.90 (5)	−2.48 (6)	−4.56 (6)
1.223	1.366	1.0	5.65 (5)	7.97 (5)	−1.94 (6)	−3.01 (6)
1.185	1.254	1.5	4.29 (5)	7.71 (5)	−1.71 (6)	−2.39 (6)
1.158	1.186	2.0	3.58 (5)	7.68 (5)	−1.61 (6)	−2.07 (6)
1.138	1.140	2.5	3.21 (5)	7.75 (5)	−1.57 (6)	−1.89 (6)
1.122	1.107	3.0	3.00 (5)	7.88 (5)	−1.56 (6)	−1.79 (6)
1.109	1.082	3.5	2.81 (5)	8.10 (5)	−1.57 (6)	−1.72 (6)
1.098	1.062	4.0	2.67 (5)	8.29 (5)	−1.59 (6)	−1.68 (6)

Lens Configuration 20C2D0C

KL (QUAD 1)	KL (QUAD 3)	Normalised image distance	Normalised spherical aberration coefficients			
			x/θ^3QL	$x/\theta\phi^2$QL	y/ϕ^3QL	$y/\theta^2\phi$QL
1.189	1.540	0.5	1.85 (6)	1.64 (6)	−3.64 (6)	−8.60 (6)
1.131	1.323	1.0	1.11 (6)	1.40 (6)	−2.66 (6)	−5.40 (6)
1.092	1.209	1.5	8.06 (5)	1.30 (6)	−2.26 (6)	−4.10 (6)
1.064	1.137	2.0	6.52 (5)	1.25 (6)	−2.04 (6)	−3.43 (6)
1.043	1.089	2.5	5.60 (5)	1.23 (6)	−1.93 (6)	−3.03 (6)
1.026	1.053	3.0	4.99 (5)	1.22 (6)	−1.88 (6)	−2.77 (6)
1.012	1.026	3.5	4.53 (5)	1.23 (6)	−1.84 (6)	−2.60 (6)
1.001	1.005	4.0	4.29 (5)	1.23 (6)	−1.82 (6)	−2.50 (6)

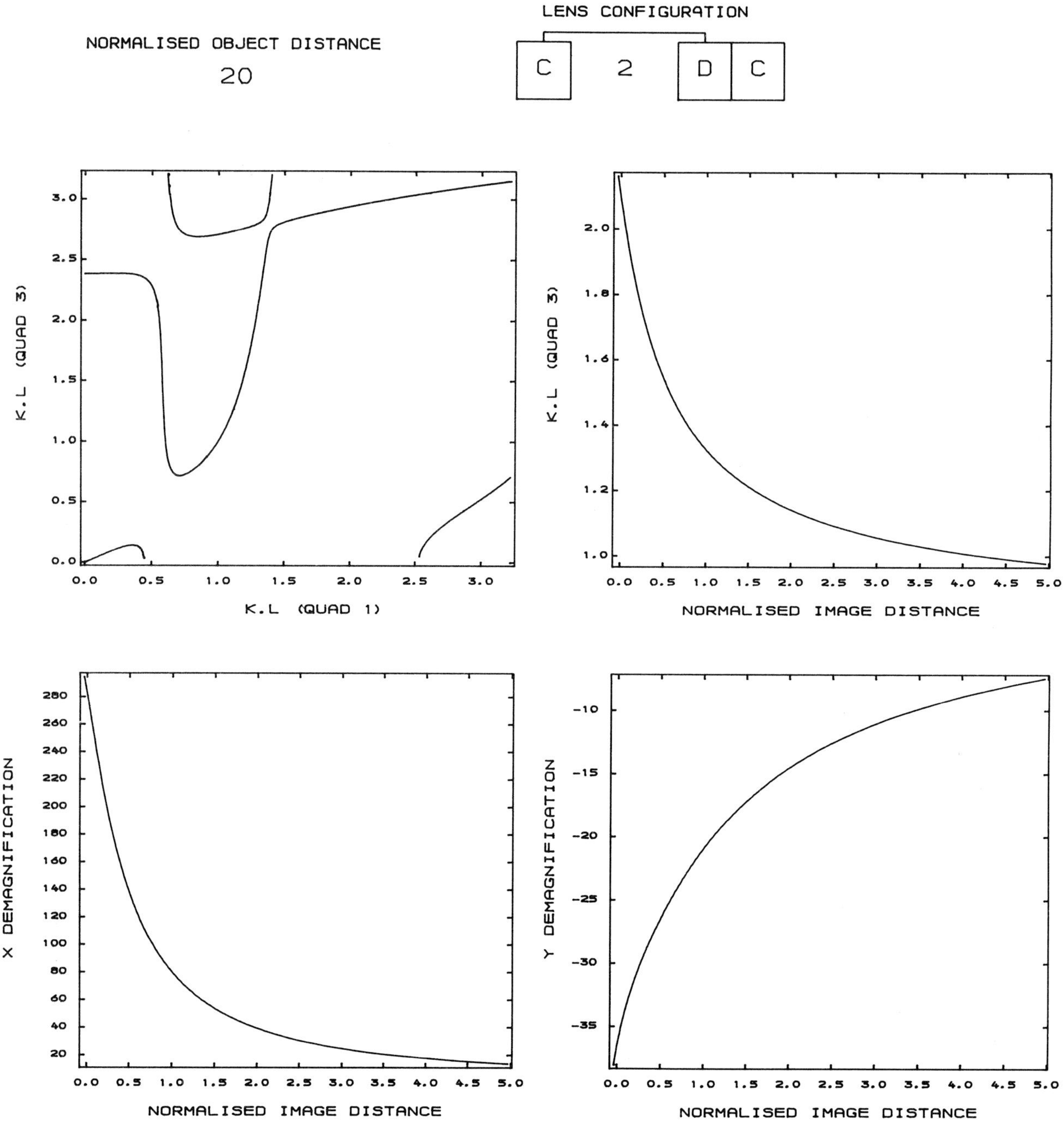
LENS CONFIGURATION
NORMALISED OBJECT DISTANCE
20
C
2
D
C
K.L (QUAD 3)
K.L (QUAD 1)
K.L (QUAD 3)
NORMALISED IMAGE DISTANCE
X DEMAGNIFICATION
NORMALISED IMAGE DISTANCE
Y DEMAGNIFICATION
NORMALISED IMAGE DISTANCE

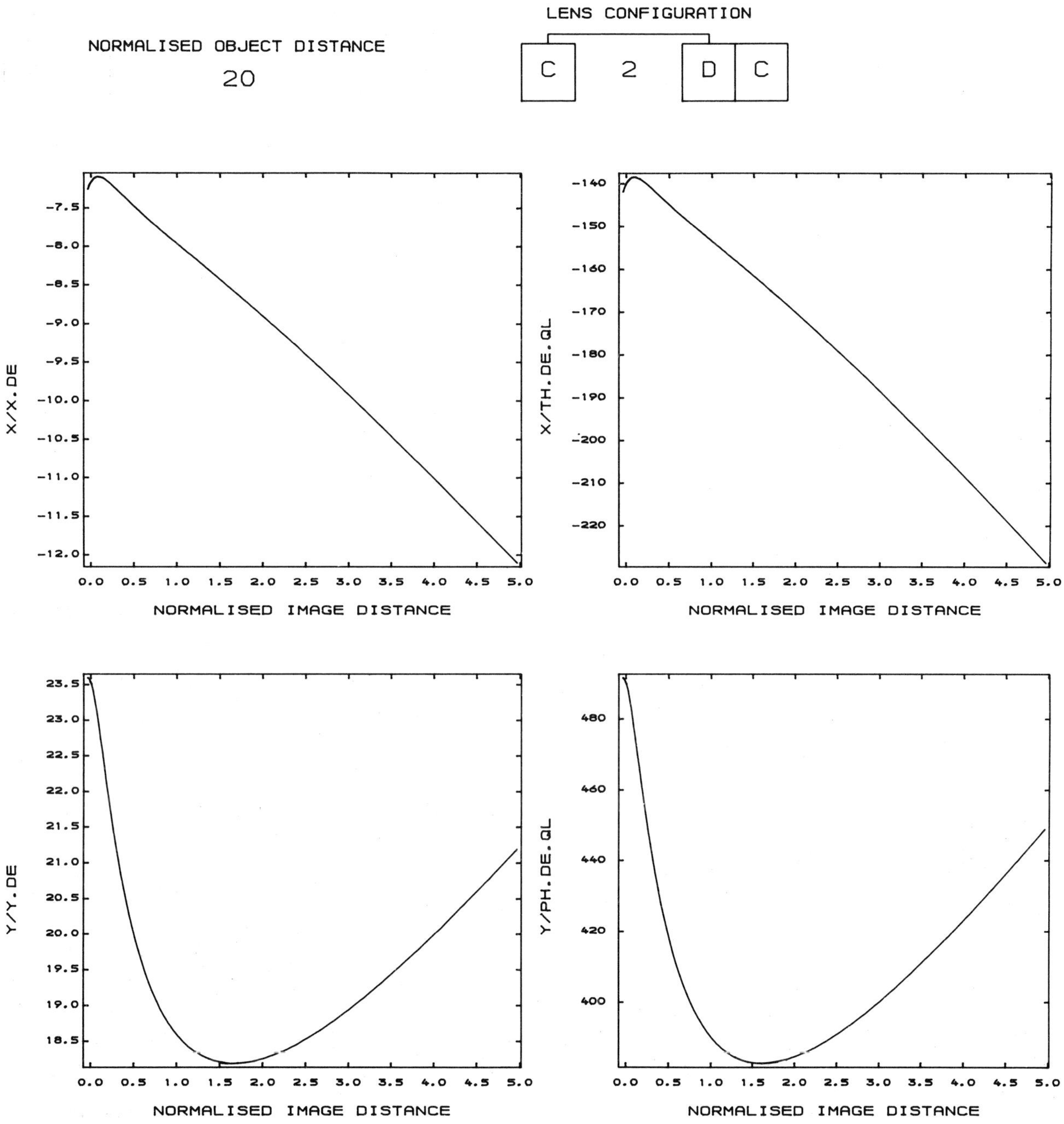
NORMALISED OBJECT DISTANCE
20
LENS CONFIGURATION
C
2
D
C
X/X.DE
X/TH.DE.QL
Y/Y.DE
Y/PH.DE.QL
NORMALISED IMAGE DISTANCE

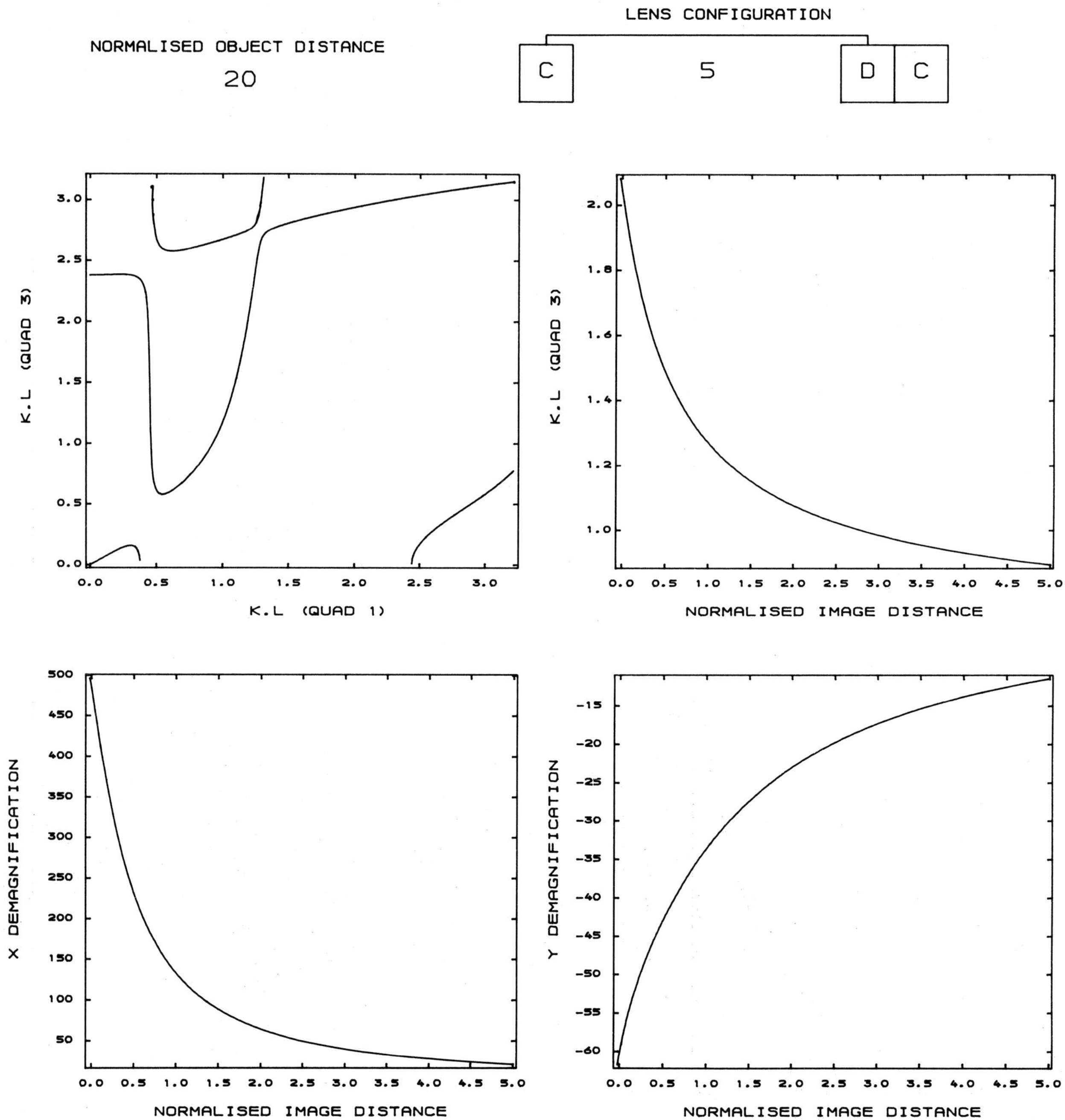
NORMALISED OBJECT DISTANCE
20
LENS CONFIGURATION
C
5
D
C
K.L (QUAD 3)
K.L (QUAD 1)
NORMALISED IMAGE DISTANCE
X DEMAGNIFICATION
Y DEMAGNIFICATION

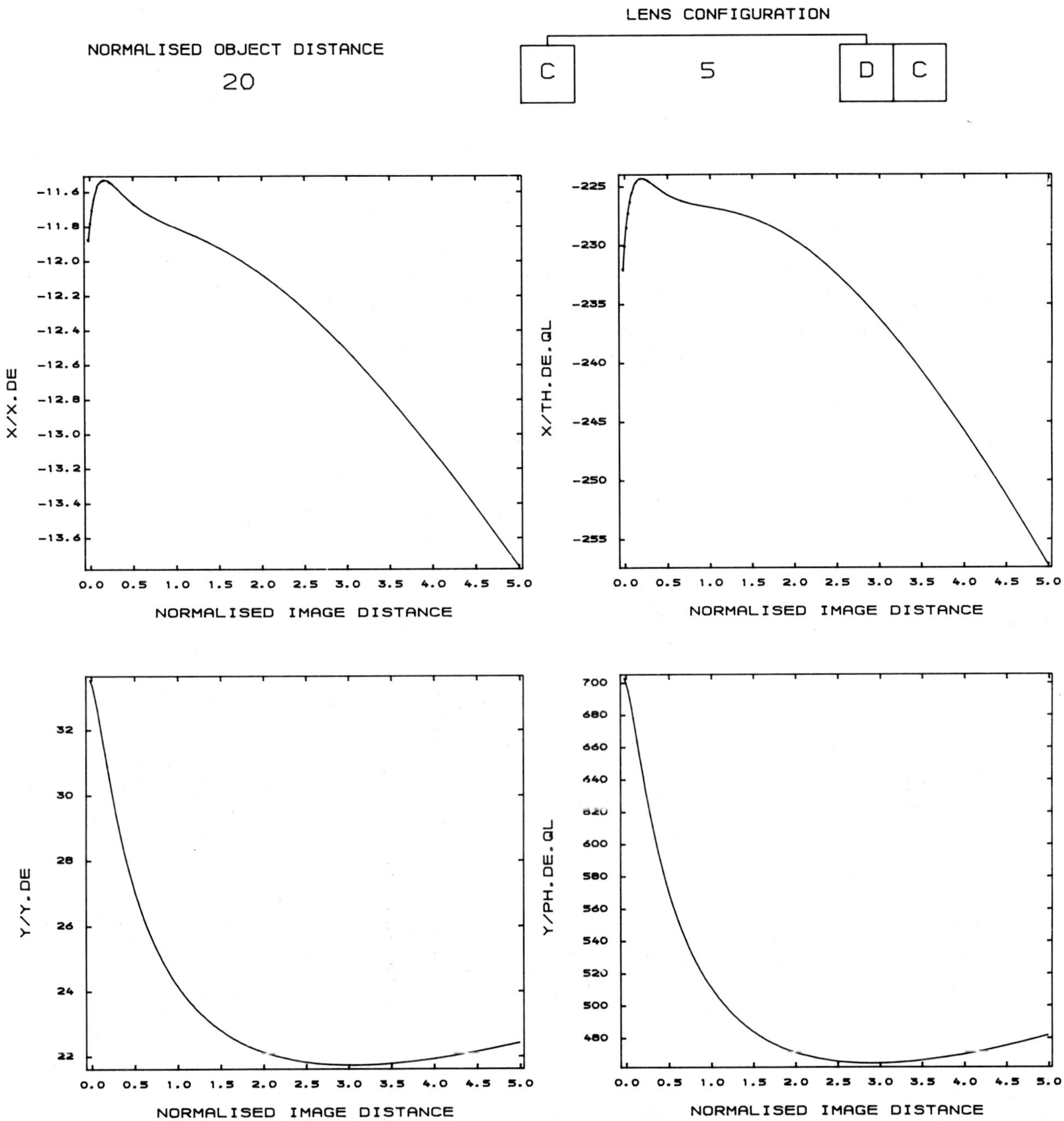
NORMALISED OBJECT DISTANCE
20
LENS CONFIGURATION
C
5
D
C
X/X.DE
X/TH.DE.QL
Y/Y.DE
Y/PH.DE.QL
NORMALISED IMAGE DISTANCE
NORMALISED IMAGE DISTANCE
NORMALISED IMAGE DISTANCE
NORMALISED IMAGE DISTANCE

Lens Configuration 20C5D0C

KL (QUAD 1)	KL (QUAD 3)	Normalised image distance	Normalised spherical aberration coefficients			
			x/θ^3QL	$x/\theta\phi^2$QL	y/ϕ^3QL	$y/\theta^2\phi$QL
1.091	1.489	0.5	8.65 (6)	6.68 (6)	−1.07 (7)	−3.56 (7)
1.030	1.267	1.0	4.81 (6)	5.24 (6)	−7.11 (6)	−2.06 (7)
0.989	1.149	1.5	3.28 (6)	4.55 (6)	−5.60 (6)	−1.45 (7)
0.959	1.074	2.0	2.48 (6)	4.12 (6)	−4.73 (6)	−1.14 (7)
0.935	1.021	2.5	2.01 (6)	3.85 (6)	−4.22 (6)	−9.59 (6)
0.916	0.982	3.0	1.70 (6)	3.66 (6)	−3.87 (6)	−8.37 (6)
0.900	0.952	3.5	1.49 (6)	3.53 (6)	−3.64 (6)	−7.51 (6)
0.887	0.928	4.0	1.34 (6)	3.42 (6)	−3.45 (6)	−6.91 (6)

Lens Configuration 20C0D0C

KL (QUAD 1)	KL (QUAD 2)	Normalised image distance	Normalised spherical aberration coefficients			
			x/θ^3QL	$x/\theta\phi^2QL$	y/ϕ^3QL	$y/\theta^2\phi QL$
0.883	1.272	0.5	−3.2 (3)	−4.12 (4)	−8.89 (4)	−4.30 (4)
0.759	1.185	1.0	−6.4 (3)	−3.48 (4)	−6.32 (4)	−6.44 (4)
0.633	1.096	1.5	−1.25 (4)	−3.20 (4)	−4.19 (4)	−8.19 (4)
0.537	1.024	2.0	−1.71 (4)	−3.08 (4)	−2.93 (4)	−8.65 (4)
0.467	0.969	2.5	−2.06 (4)	−2.90 (4)	−2.19 (4)	−8.73 (4)
0.413	0.924	3.0	−2.17 (4)	−2.82 (4)	−1.98 (4)	−8.39 (4)
0.371	0.889	3.5	−2.30 (4)	−2.77 (4)	−1.83 (4)	−8.22 (4)
0.337	0.860	4.0	−2.58 (4)	−2.65 (4)	−1.66 (4)	−8.14 (4)
2.132	1.679	0.5	3.04 (6)	4.75 (6)	−2.41 (7)	−2.22 (7)
1.371	1.467	1.0	1.15 (5)	2.82 (5)	−1.21 (6)	−9.09 (5)
1.211	1.397	1.5	2.8 (4)	1.58 (5)	−7.26 (5)	−3.21 (5)
1.155	1.367	2.0	1.9 (4)	1.54 (5)	−6.71 (5)	−2.39 (5)
1.128	1.351	2.5	1.7 (4)	1.69 (5)	−6.80 (5)	−2.26 (5)
1.112	1.341	3.0	2.0 (4)	1.91 (5)	−7.73 (5)	−2.22 (5)
1.102	1.334	3.5	2.0 (4)	2.15 (5)	−7.76 (5)	−2.34 (5)
1.095	1.329	4.0	2.1 (4)	2.42 (5)	−8.41 (5)	−2.39 (5)

NORMALISED OBJECT DISTANCE
20

LENS CONFIGURATION

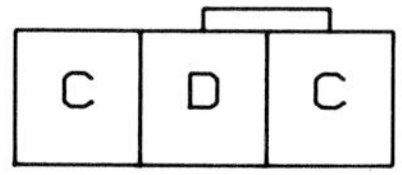

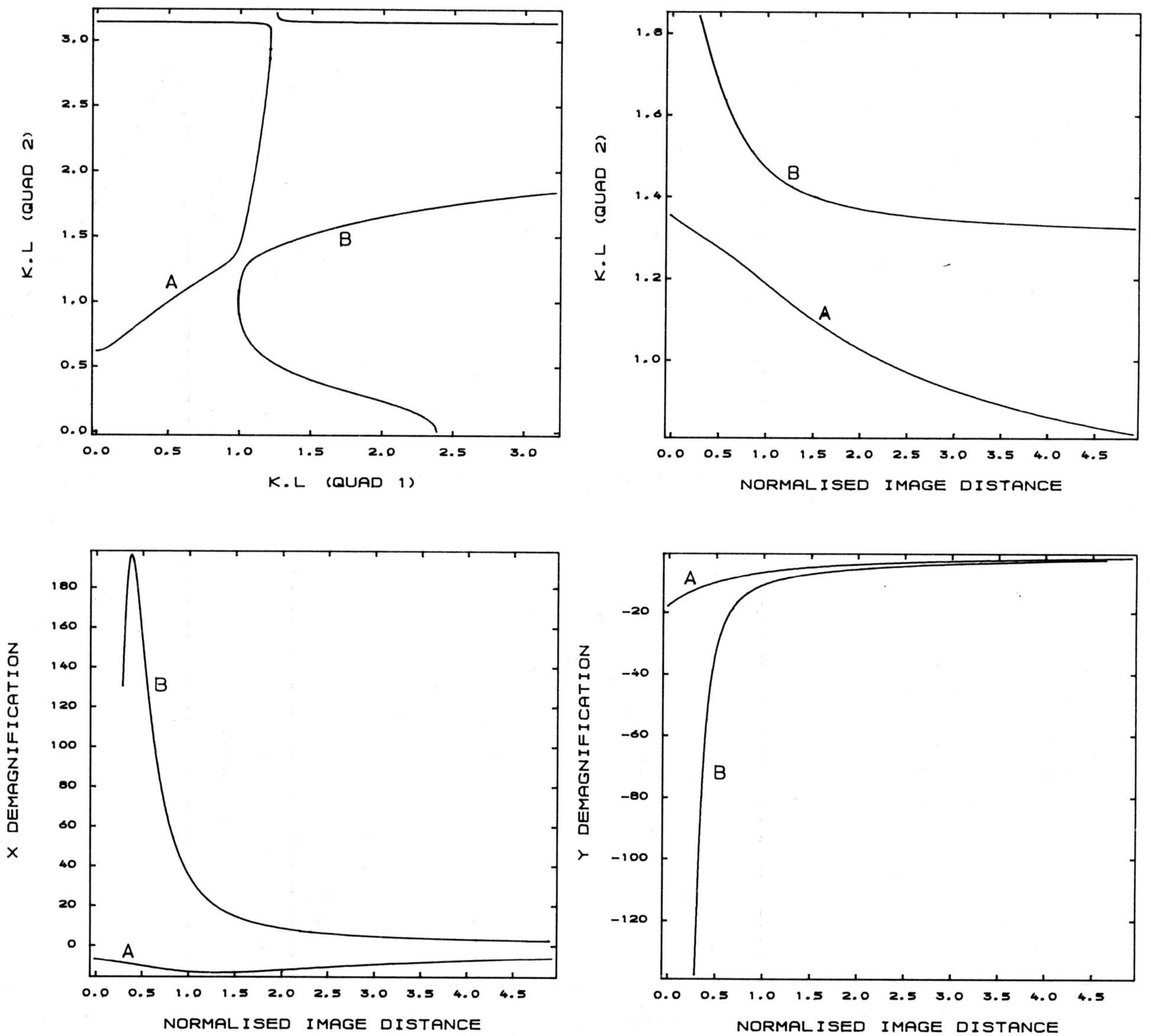

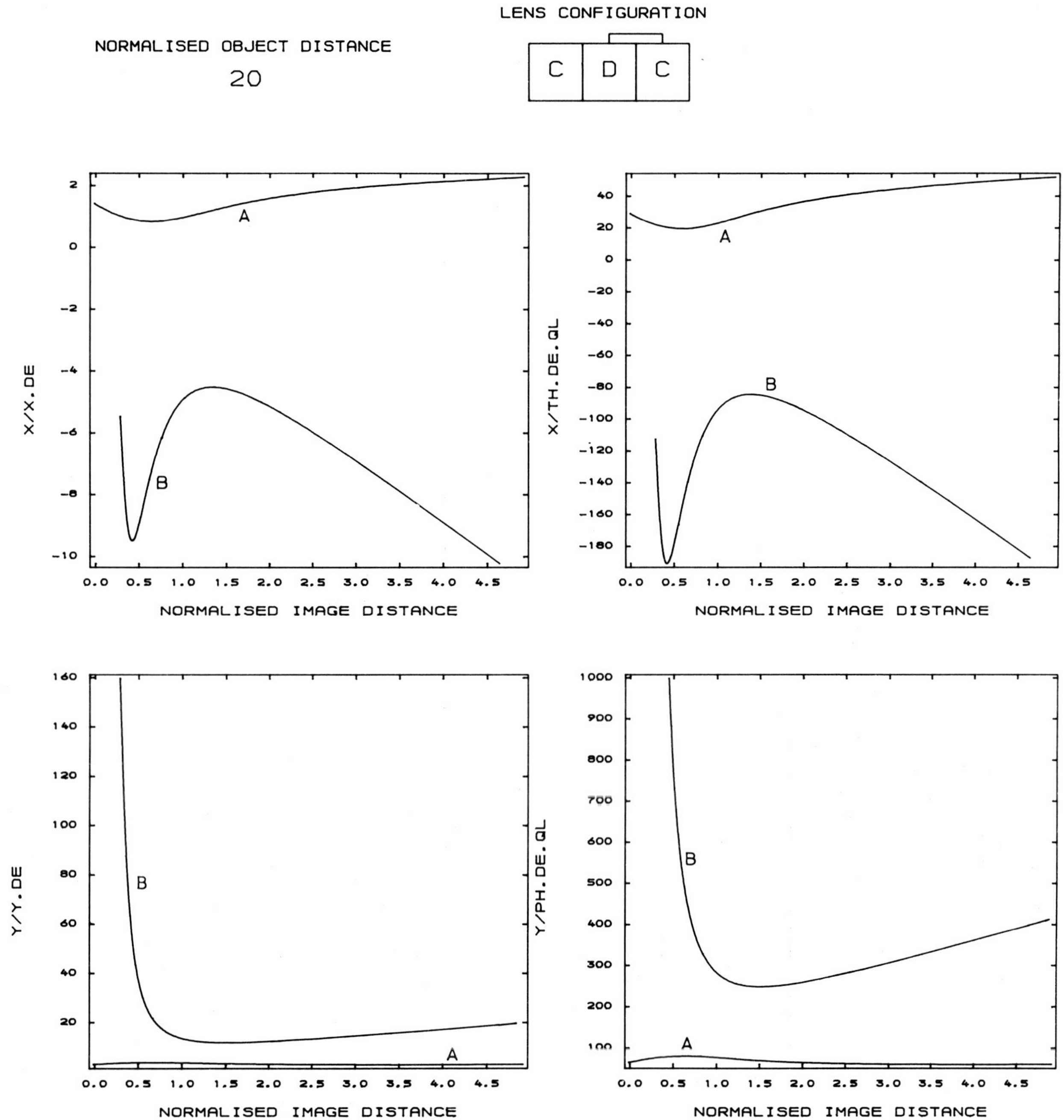
LENS CONFIGURATION
NORMALISED OBJECT DISTANCE
20
C D C
X/X.DE
X/TH.DE.QL
Y/Y.DE
Y/PH.DE.QL
A
B
NORMALISED IMAGE DISTANCE

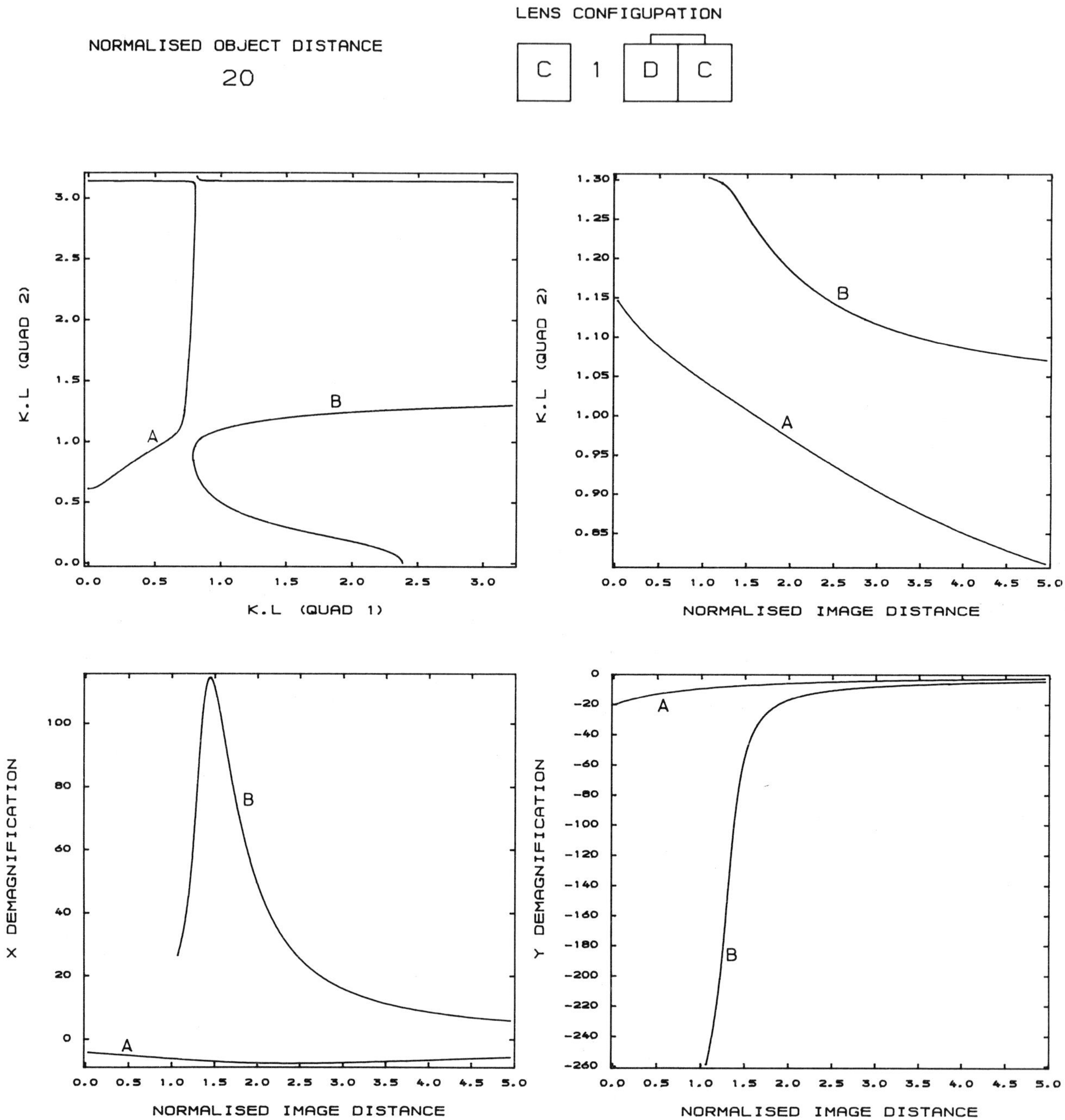
NORMALISED OBJECT DISTANCE
20
LENS CONFIGUPATION
C
1
D
C
K.L (QUAD 2)
K.L (QUAD 1)
A
B
NORMALISED IMAGE DISTANCE
X DEMAGNIFICATION
Y DEMAGNIFICATION

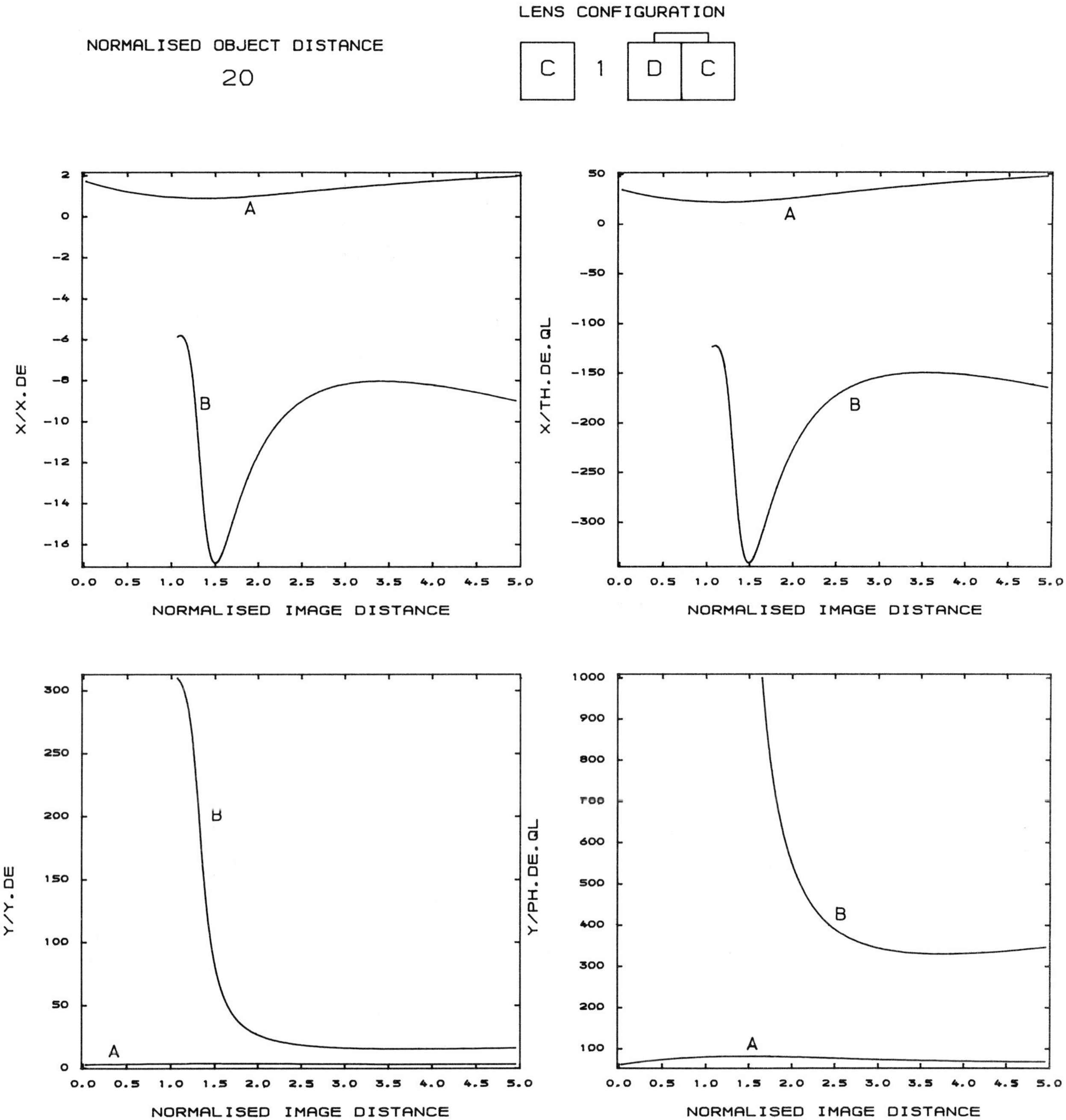
NORMALISED OBJECT DISTANCE
20
LENS CONFIGURATION
C 1 D C
X/X.DE
X/TH.DE.QL
Y/Y.DE
Y/PH.DE.QL
A
B
NORMALISED IMAGE DISTANCE

Lens Configuration 20C1$\overline{\mathrm{D0C}}$

KL (QUAD 1)	KL (QUAD 2)	Normalised image distance	Normalised spherical aberration coefficients			
			x/θ^3QL	$x/\theta\phi^2$QL	y/ϕ^3QL	$y/\theta^2\phi$QL
0.672	1.085	0.5	−1.7 (3)	−2.71 (4)	−6.25 (4)	−1.03 (4)
0.635	1.042	1.0	−2.0 (3)	−2.58 (4)	−5.74 (4)	−1.85 (4)
0.587	1.005	1.5	−1.5 (3)	−2.78 (4)	−5.45 (4)	−2.83 (4)
0.553	0.969	2.0	−4.7 (3)	−3.01 (4)	−4.69 (4)	−3.91 (4)
0.480	0.934	2.5	−7.2 (3)	−3.15 (4)	−3.63 (4)	−5.10 (4)
0.432	0.902	3.0	−1.10 (4)	−3.24 (4)	−3.10 (4)	−6.10 (4)
0.392	0.873	3.5	−1.53 (4)	−3.30 (4)	−2.92 (4)	−6.57 (4)
0.357	0.848	4.0	−1.37 (4)	−3.41 (4)	−2.40 (4)	−7.07 (4)
3.291	1.322	0.5	3.98 (5)	1.22 (8)	−7.66 (9)	2.56 (7)
3.227	1.302	1.0	5.47 (5)	1.06 (8)	−7.94 (9)	1.74 (7)
2.118	1.251	1.5	1.20 (7)	2.85 (7)	−1.16 (8)	−6.41 (7)
1.384	1.181	2.0	1.38 (6)	2.42 (6)	−4.70 (6)	−7.25 (6)
1.154	1.139	2.5	3.57 (5)	8.46 (5)	−1.68 (6)	−2.09 (6)
1.048	1.113	3.0	1.60 (5)	5.01 (5)	−1.09 (6)	−1.04 (6)
0.991	1.096	3.5	9.1 (4)	3.84 (5)	−9.00 (5)	−6.80 (5)
0.956	1.083	4.0	6.4 (4)	3.31 (5)	−8.21 (5)	−5.24 (5)

Lens Configuration 20C2$\overline{\mathrm{D0C}}$

KL (QUAD 1)	KL (QUAD 2)	Normalised image distance	Normalised spherical aberration coefficients			
			x/θ^3QL	$x/\theta\phi^2$QL	y/ϕ^3QL	$y/\theta^2\phi$QL
0.576	1.018	0.5	−1.4 (3)	−2.01 (4)	−5.83 (4)	−6.1 (3)
0.555	0.979	1.0	−1.4 (3)	−2.07 (4)	−6.06 (4)	−8.0 (3)
0.530	0.949	1.5	−0.6 (3)	−2.37 (4)	−5.87 (4)	−1.32 (4)
0.501	0.923	2.0	−1.1 (3)	−2.68 (4)	−5.80 (4)	−1.99 (4)
0.467	0.899	2.5	−2.4 (3)	−3.05 (4)	−5.03 (4)	−2.97 (4)
0.433	0.876	3.0	−5.4 (3)	−3.30 (4)	−4.25 (4)	−3.94 (4)
0.400	0.854	3.5	−6.7 (3)	−3.60 (4)	−3.69 (4)	−4.95 (4)
0.369	0.834	4.0	−9.6 (3)	−3.83 (4)	−3.18 (4)	−5.80 (4)
*	*	0.5	*	*	*	*
*	*	1.0	*	*	*	*
*	*	1.5	*	*	*	*
*	*	2.0	*	*	*	*
2.154	1.086	2.5	1.56 (7)	1.45 (8)	−3.86 (8)	−2.10 (8)
1.425	1.049	3.0	4.47 (6)	9.99 (6)	−1.42 (7)	−2.40 (7)
1.168	1.023	3.5	1.36 (6)	3.06 (6)	−4.16 (6)	−7.14 (6)
1.042	1.003	4.0	6.01 (5)	1.60 (6)	−2.30 (6)	−3.35 (6)

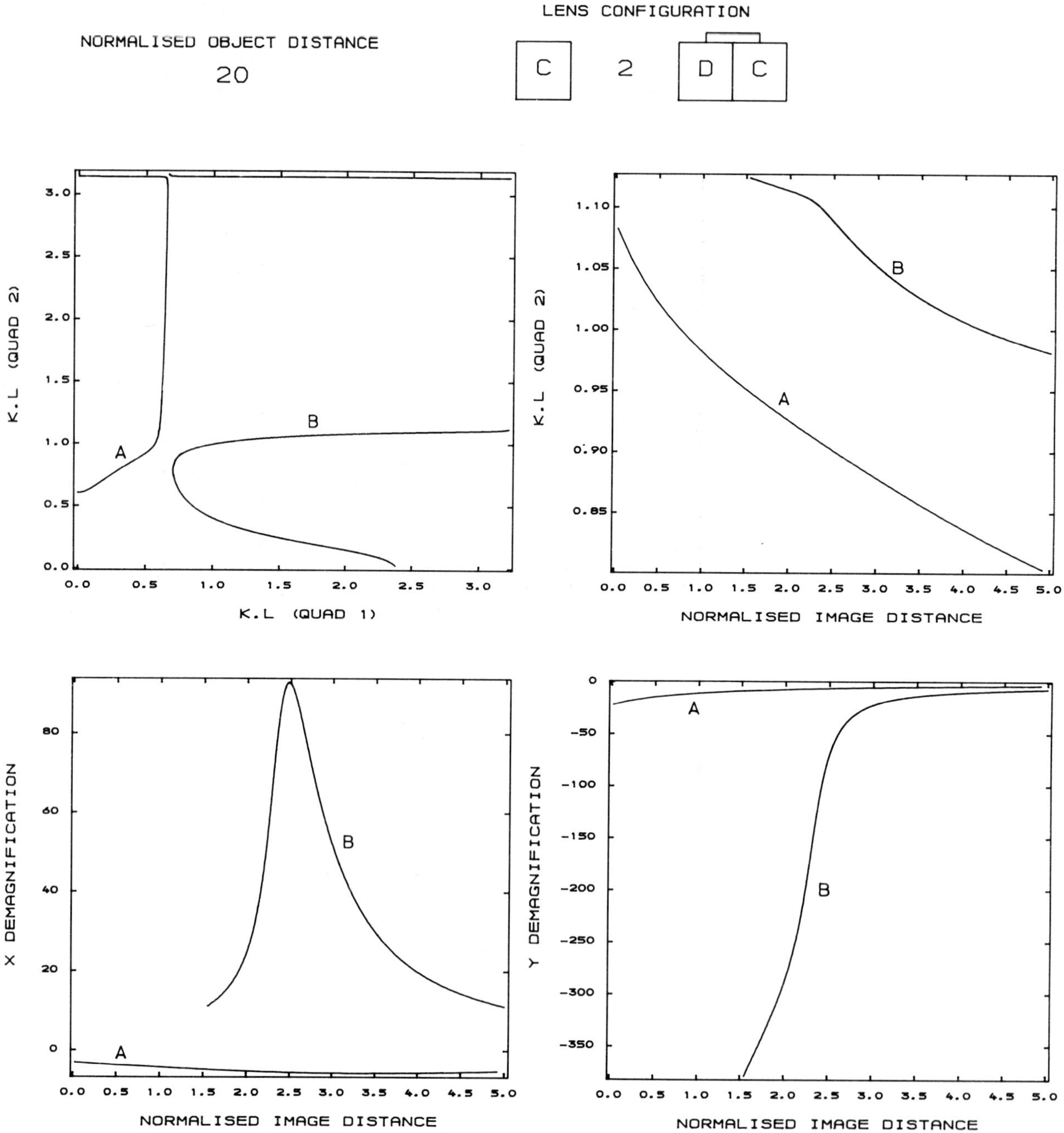
LENS CONFIGURATION
NORMALISED OBJECT DISTANCE
20
C
2
D
C
K.L (QUAD 2)
K.L (QUAD 1)
A
B
NORMALISED IMAGE DISTANCE
X DEMAGNIFICATION
Y DEMAGNIFICATION

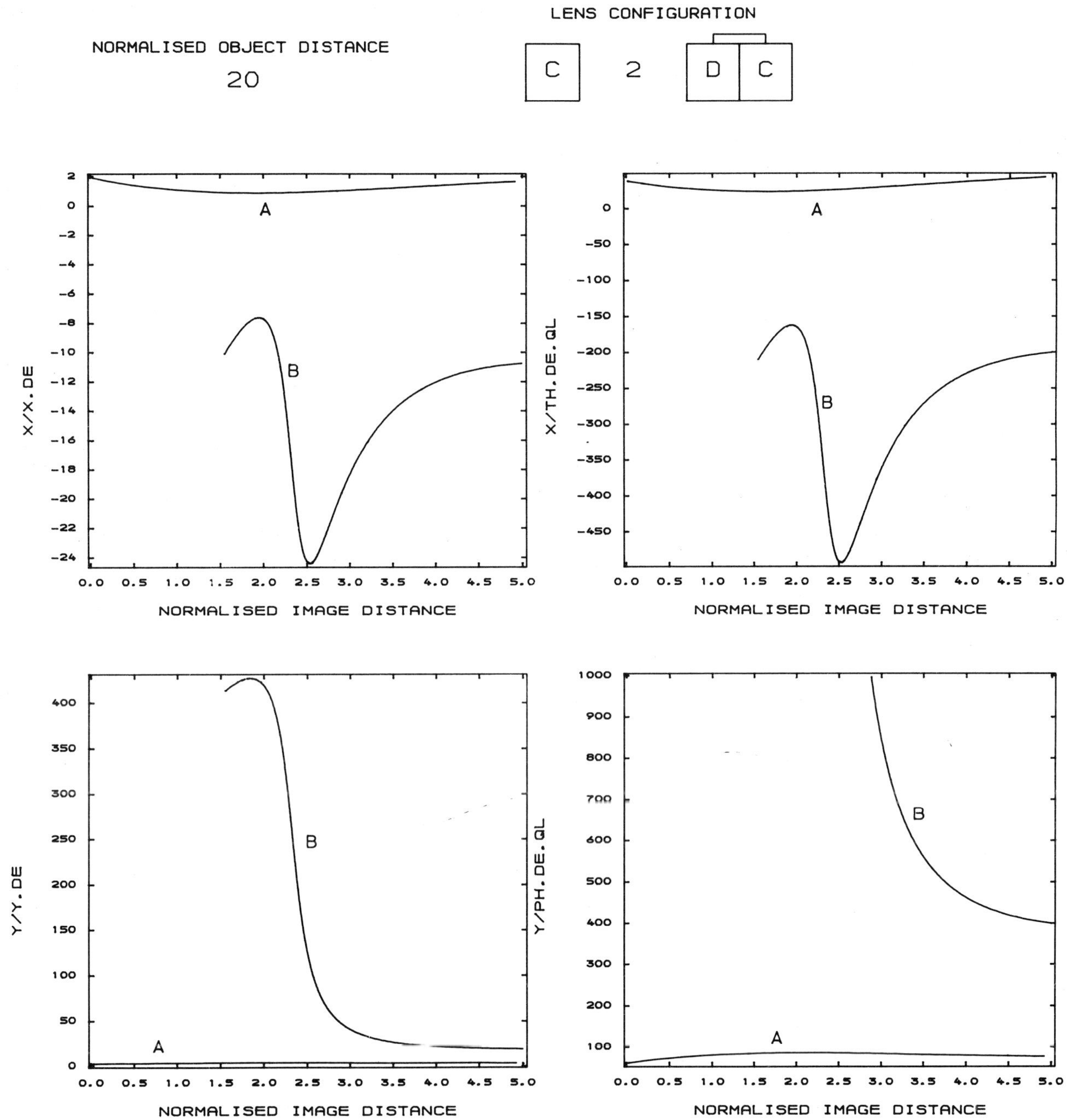

LENS CONFIGURATION
NORMALISED OBJECT DISTANCE
20
C
2
D
C
X/X.DE
X/TH.DE.QL
Y/Y.DE
Y/PH.DE.QL
A
B
NORMALISED IMAGE DISTANCE

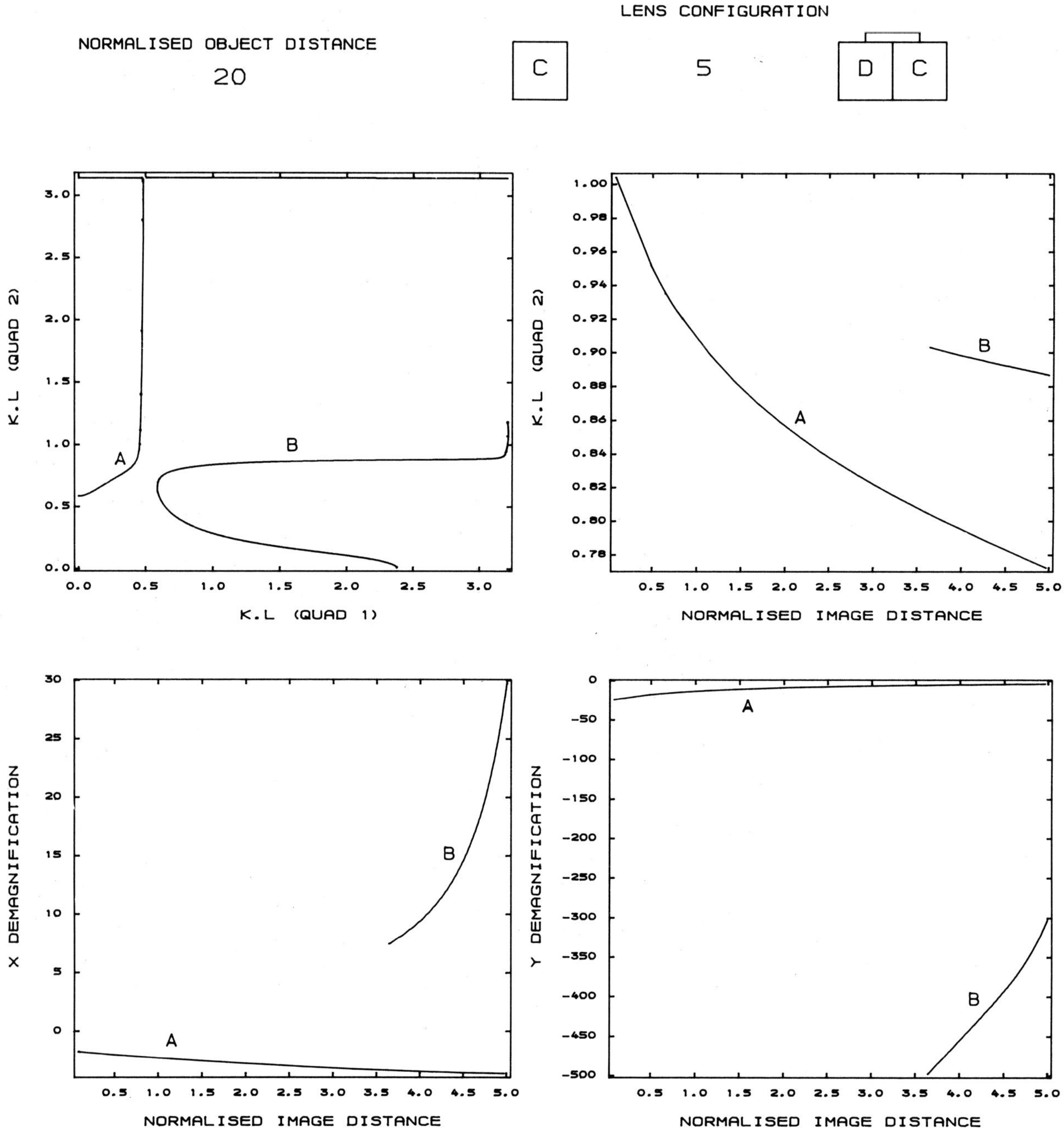
LENS CONFIGURATION
NORMALISED OBJECT DISTANCE
20
C
5
D
C
K.L (QUAD 2)
K.L (QUAD 1)
A
B
NORMALISED IMAGE DISTANCE
X DEMAGNIFICATION
Y DEMAGNIFICATION

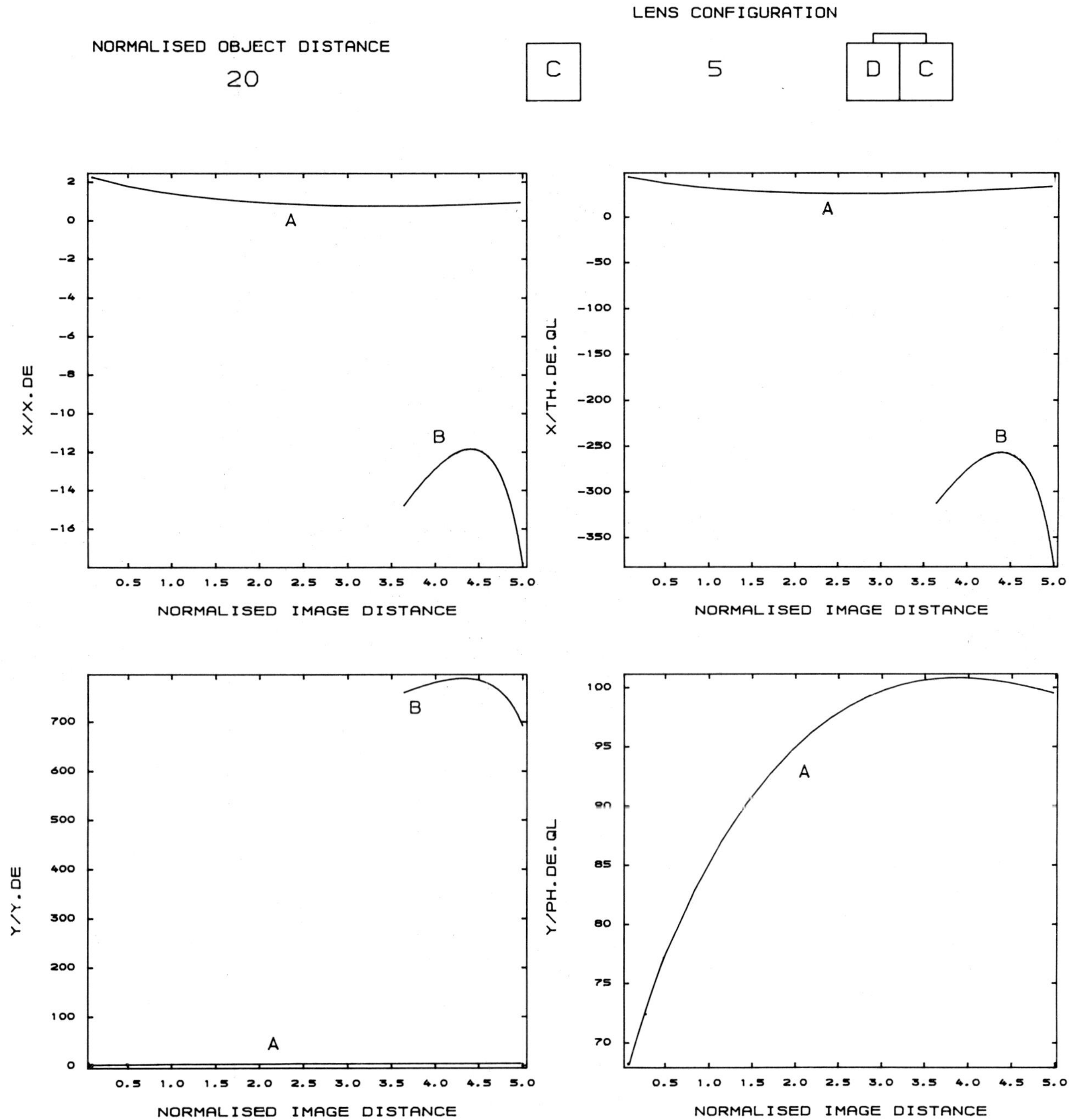

LENS CONFIGURATION
NORMALISED OBJECT DISTANCE
20
C
5
D
C
X/X.DE
X/TH.DE.QL
Y/Y.DE
Y/PH.DE.QL
A
B
NORMALISED IMAGE DISTANCE

Lens Configuration 20C5D0C

KL (QUAD 1)	KL (QUAD 2)	Normalised image distance	Normalised spherical aberration coefficients			
			x/θ^3QL	$x/\theta\phi^2QL$	y/ϕ^3QL	$y/\theta^2\phi QL$
0.448	0.947	0.5	−0.1 (3)	−1.34 (4)	−6.97 (4)	−1.7 (3)
0.440	0.906	1.0	−0.7 (3)	−1.39 (4)	−7.39 (4)	−2.1 (3)
0.431	0.876	1.5	−0.0 (3)	−1.67 (4)	−7.52 (4)	−3.6 (3)
0.421	0.854	2.0	−0.3 (3)	−2.00 (4)	−7.90 (4)	−6.2 (3)
0.409	0.835	2.5	−0.1 (3)	−2.48 (4)	−8.07 (4)	−7.1 (3)
0.396	0.819	3.0	−0.9 (3)	−2.93 (4)	−7.70 (4)	−1.36 (4)
0.382	0.805	3.5	−2.0 (3)	−3.38 (4)	−7.06 (4)	−2.01 (4)
0.367	0.792	4.0	−2.2 (3)	−3.89 (4)	−6.37 (4)	−2.60 (4)
*	*	0.5	*	*	*	*
*	*	1.0	*	*	*	*
*	*	1.5	*	*	*	*
*	*	2.0	*	*	*	*
*	*	2.5	*	*	*	*
*	*	3.0	*	*	*	*
*	*	3.5	*	*	*	*
*	*	4.0	*	*	*	*

Lens Configuration 20C0D0C

KL (QUAD 1)	KL (QUAD 2)	Normalised image distance	Normalised spherical aberration coefficients			
			x/θ^3QL	$x/\theta\phi^2$QL	y/ϕ^3QL	$y/\theta^2\phi$QL
0.916	1.215	0.5	−6.2 (3)	−6.41 (4)	−6.64 (4)	−2.48 (4)
0.877	1.161	1.0	−4.8 (3)	−5.96 (4)	−6.42 (4)	−3.00 (4)
0.843	1.119	1.5	−5.1 (3)	−5.84 (4)	−6.20 (4)	−3.60 (4)
0.814	1.084	2.0	−5.6 (3)	−5.74 (4)	−6.40 (4)	−3.94 (4)
0.788	1.054	2.5	−5.1 (3)	−5.83 (4)	−6.41 (4)	−4.12 (4)
0.766	1.029	3.0	−6.2 (3)	−5.92 (4)	−5.57 (4)	−4.66 (4)
0.747	1.007	3.5	−5.3 (3)	−6.10 (4)	−6.13 (4)	−4.78 (4)
0.731	0.987	4.0	−7.2 (3)	−6.23 (4)	−5.89 (4)	−5.11 (4)
1.667	1.589	0.5	7.22 (5)	8.61 (5)	−3.70 (6)	−4.47 (6)
1.458	1.489	1.0	2.25 (5)	4.33 (5)	−1.73 (6)	−1.55 (6)
1.358	1.433	1.5	1.15 (5)	3.32 (5)	−1.27 (6)	−9.02 (5)
1.302	1.398	2.0	7.9 (4)	3.00 (5)	−1.13 (6)	−6.76 (5)
1.266	1.375	2.5	6.3 (4)	2.95 (5)	−1.07 (6)	−5.75 (5)
1.242	1.357	3.0	5.6 (4)	3.00 (5)	−1.09 (6)	−5.32 (5)
1.225	1.345	3.5	5.4 (4)	3.12 (5)	−1.11 (6)	−5.25 (5)
1.212	1.335	4.0	5.1 (4)	3.28 (5)	−1.14 (6)	−5.14 (5)

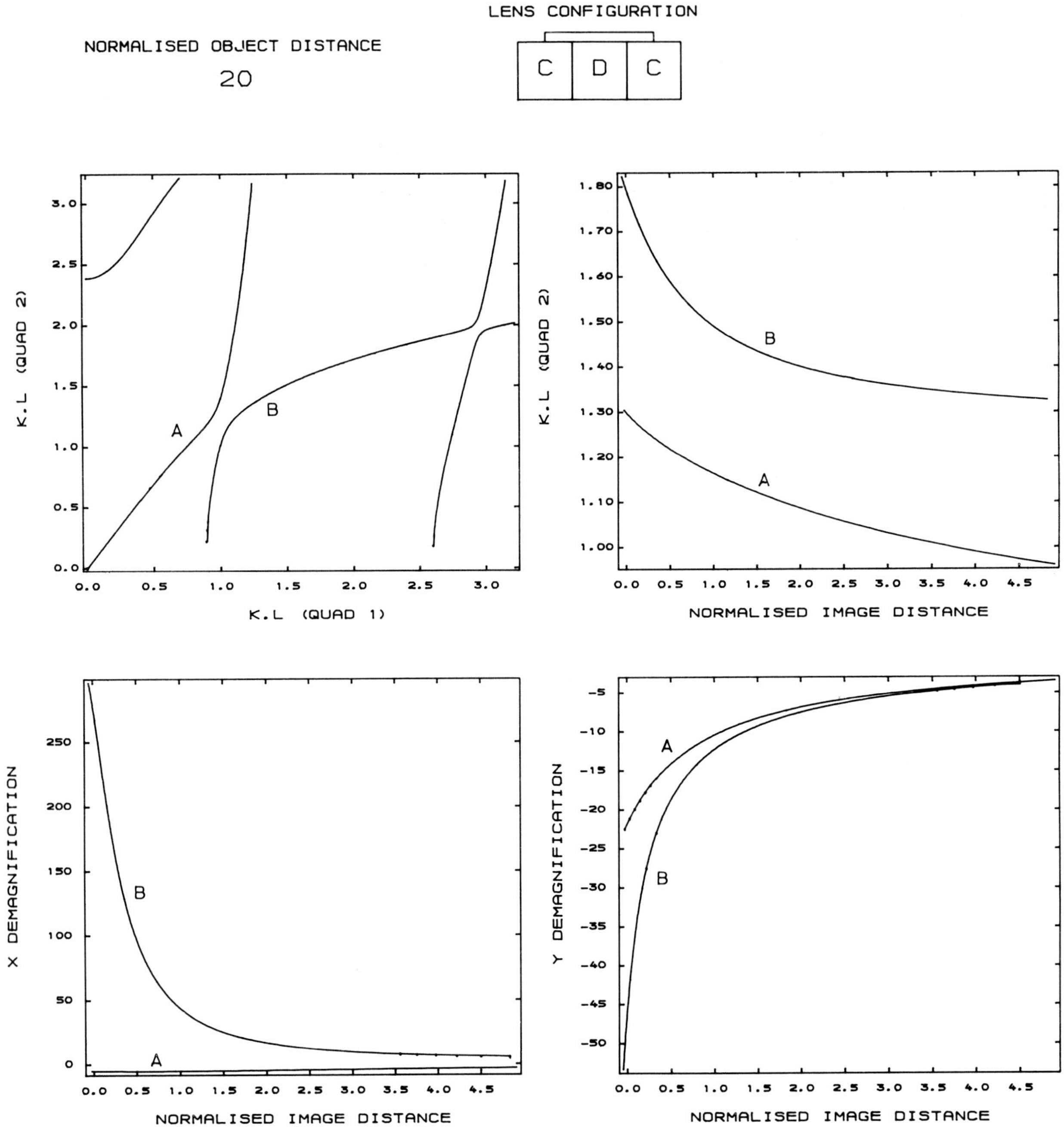
NORMALISED OBJECT DISTANCE
20
LENS CONFIGURATION
C
D
C
K.L (QUAD 2)
K.L (QUAD 1)
0.0
0.5
1.0
1.5
2.0
2.5
3.0
A
B
1.80
1.70
1.60
1.50
1.40
1.30
1.20
1.10
1.00
0.0
0.5
1.0
1.5
2.0
2.5
3.0
3.5
4.0
4.5
NORMALISED IMAGE DISTANCE
X DEMAGNIFICATION
250
200
150
100
50
0
Y DEMAGNIFICATION
-5
-10
-15
-20
-25
-30
-35
-40
-45
-50

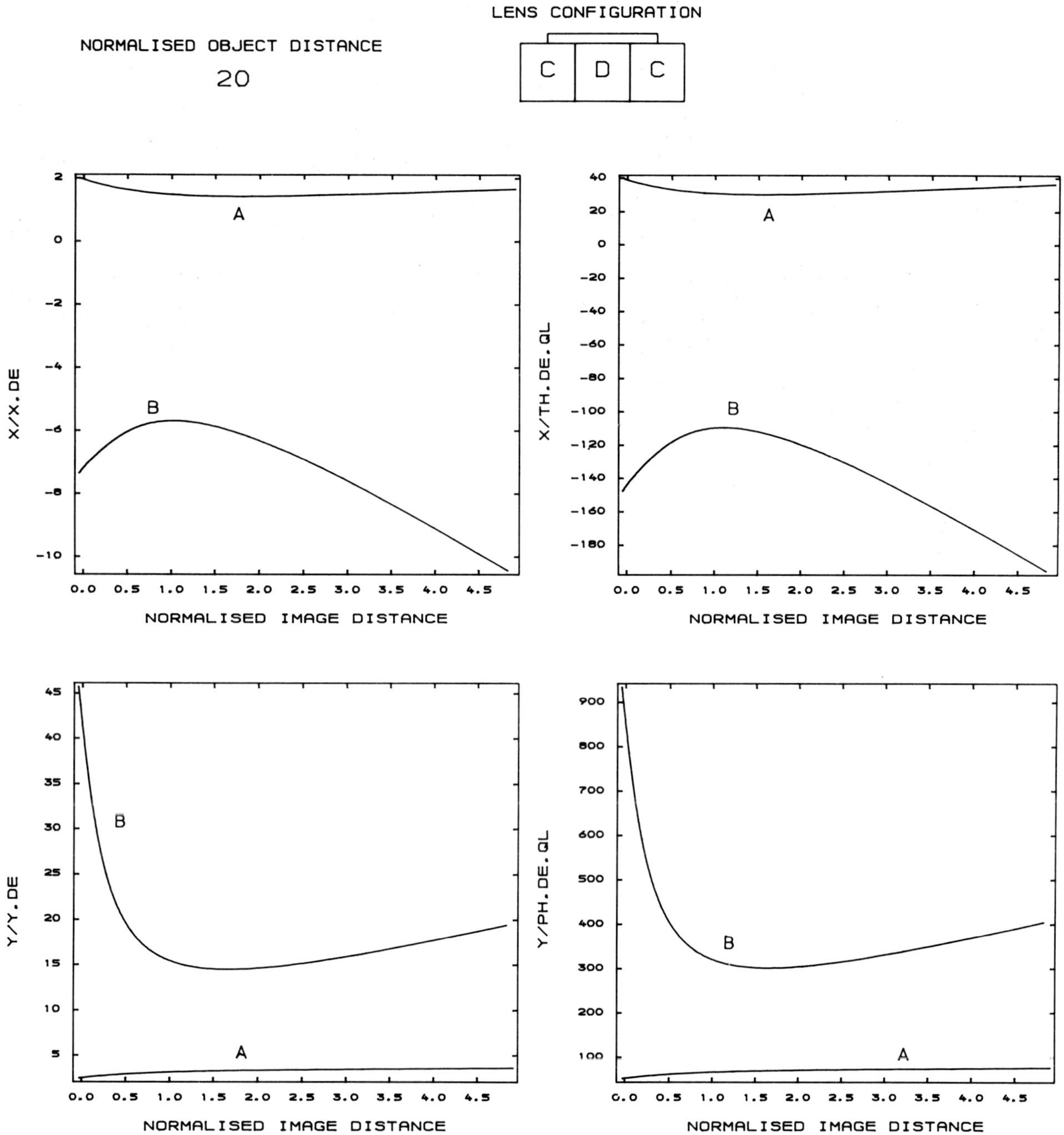

NORMALISED OBJECT DISTANCE
20
LENS CONFIGURATION
C D C
X/X.DE
X/TH.DE.QL
Y/Y.DE
Y/PH.DE.QL
A
B
NORMALISED IMAGE DISTANCE

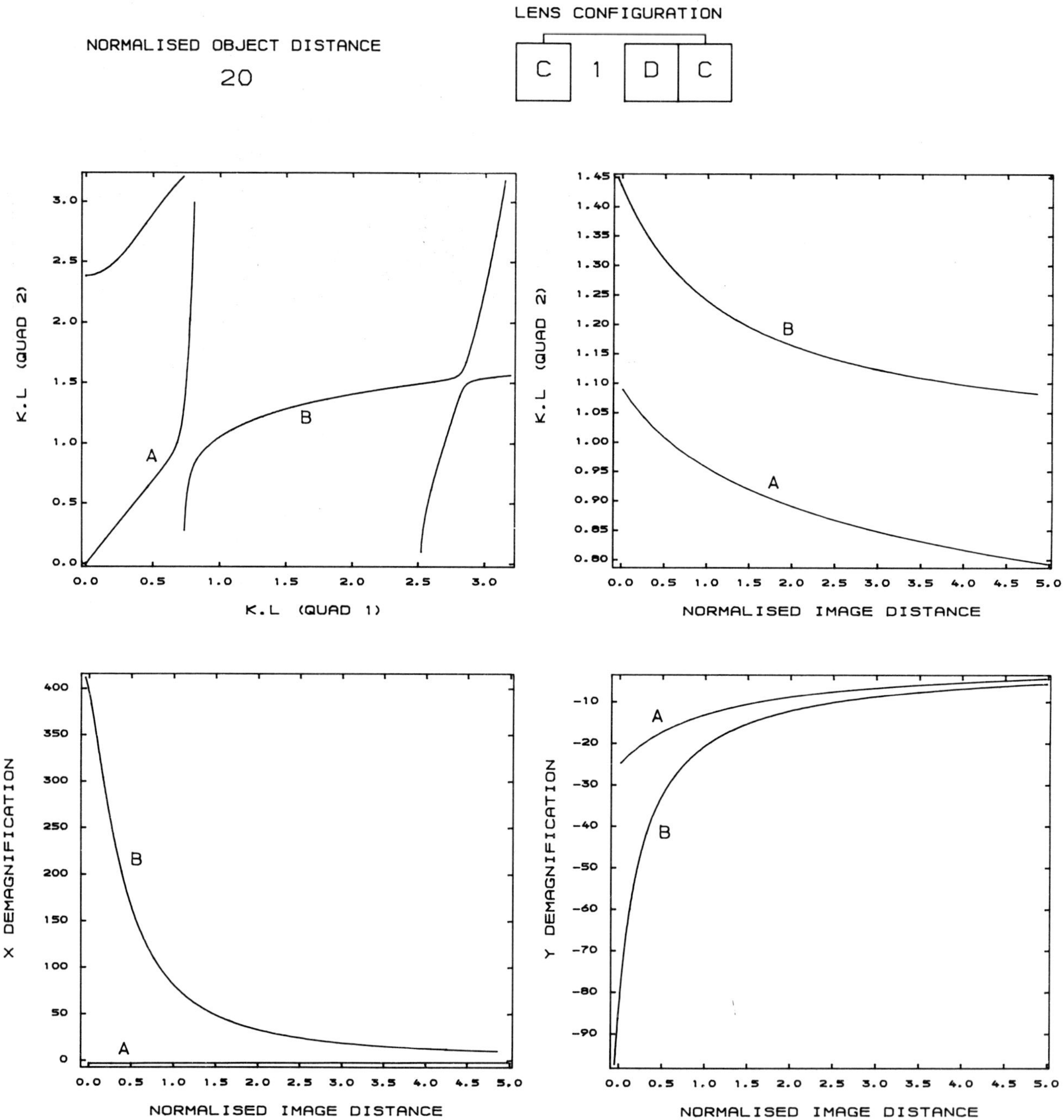
NORMALISED OBJECT DISTANCE
20
LENS CONFIGURATION
C
1
D
C
K.L (QUAD 2)
K.L (QUAD 1)
A
B
NORMALISED IMAGE DISTANCE
X DEMAGNIFICATION
Y DEMAGNIFICATION

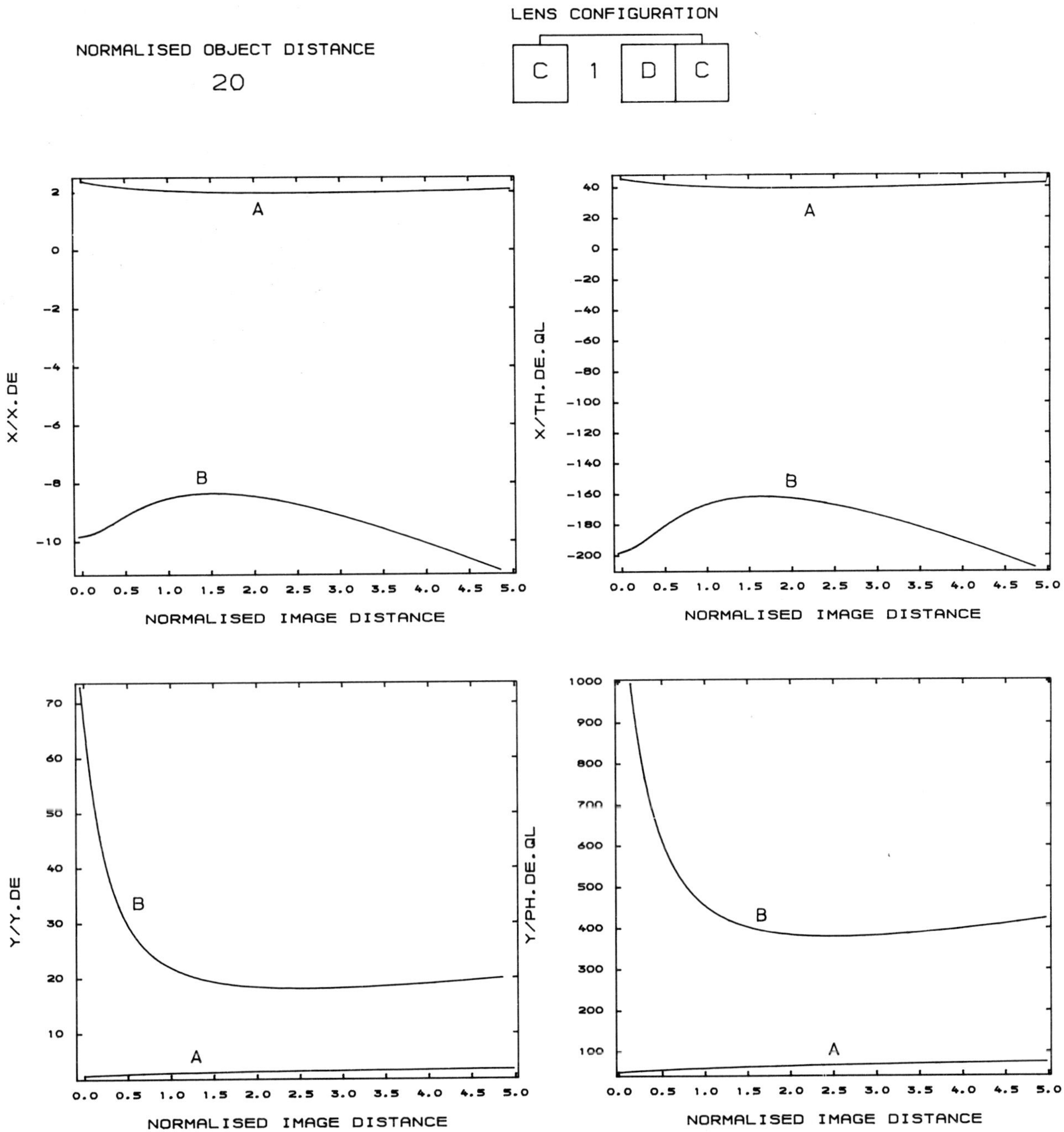
NORMALISED OBJECT DISTANCE
20
LENS CONFIGURATION
C 1 D C
X/X.DE
X/TH.DE.QL
Y/Y.DE
Y/PH.DE.QL
A
B
NORMALISED IMAGE DISTANCE

Lens Configuration 20C1D0C

KL (QUAD 1)	KL (QUAD 2)	Normalised image distance	Normalised spherical aberration coefficients			
			x/θ^3QL	$x/\theta\phi^2$QL	y/ϕ^3QL	$y/\theta^2\phi$QL
0.679	1.005	0.5	−4.1 (3)	−3.37 (4)	−3.53 (4)	−7.6 (3)
0.661	0.953	1.0	−3.7 (3)	−3.25 (4)	−3.75 (4)	−7.2 (3)
0.645	0.916	1.5	−0.8 (3)	−3.36 (4)	−3.69 (4)	−9.1 (3)
0.631	0.888	2.0	−3.0 (3)	−3.95 (4)	−3.95 (4)	−1.01 (4)
0.618	0.864	2.5	−2.6 (3)	−3.41 (4)	−3.54 (4)	−1.29 (4)
0.607	0.845	3.0	−2.2 (3)	−3.54 (4)	−3.52 (4)	−1.40 (4)
0.596	0.829	3.5	−2.0 (3)	−3.52 (4)	−3.33 (4)	−1.67 (4)
0.587	0.814	4.0	−3.5 (3)	−3.75 (4)	−3.77 (4)	−1.52 (4)
1.579	1.307	0.5	3.35 (6)	3.47 (6)	−9.78 (6)	−1.76 (7)
1.362	1.237	1.0	1.25 (6)	1.60 (6)	−3.73 (6)	−6.38 (6)
1.253	1.192	1.5	6.72 (5)	1.10 (6)	−2.37 (6)	−3.57 (6)
1.185	1.161	2.0	4.36 (5)	8.93 (5)	−1.83 (6)	−2.46 (6)
1.140	1.138	2.5	3.26 (5)	7.84 (5)	−1.60 (6)	−1.92 (6)
1.108	1.120	3.0	2.65 (5)	7.24 (5)	−1.47 (6)	−1.63 (6)
1.084	1.106	3.5	2.26 (5)	6.97 (5)	−1.40 (6)	−1.43 (6)
1.065	1.095	4.0	2.05 (5)	6.80 (5)	−1.36 (6)	−1.31 (6)

Lens Configuration 20C2D0C

KL (QUAD 1)	KL (QUAD 2)	Normalised image distance	Normalised spherical aberration coefficients			
			x/θ^3QL	$x/\theta\phi^2QL$	y/ϕ^3QL	$y/\theta^2\phi QL$
0.579	0.931	0.5	−1.3 (3)	−2.36 (4)	−3.31 (4)	−3.3 (3)
0.568	0.876	1.0	−2.0 (3)	−2.33 (4)	−3.13 (4)	−3.8 (3)
0.557	0.838	1.5	−1.4 (3)	−2.30 (4)	−3.10 (4)	−5.5 (3)
0.548	0.809	2.0	−1.2 (3)	−2.41 (4)	−3.04 (4)	−5.2 (3)
0.539	0.787	2.5	−0.5 (3)	−2.41 (4)	−3.12 (4)	−5.2 (3)
0.532	0.768	3.0	−2.7 (3)	−2.53 (4)	−2.94 (4)	−6.6 (3)
0.525	0.752	3.5	−2.5 (3)	−2.63 (4)	−3.09 (4)	−7.2 (3)
0.518	0.738	4.0	−0.8 (3)	−2.75 (4)	−3.28 (4)	−6.6 (3)
1.539	1.206	0.5	8.20 (6)	8.70 (6)	−2.02 (7)	−4.21 (7)
1.321	1.142	1.0	3.13 (6)	3.80 (6)	−7.03 (6)	−1.49 (7)
1.207	1.099	1.5	1.65 (6)	2.46 (6)	−4.09 (6)	−8.06 (6)
1.136	1.069	2.0	1.07 (6)	1.90 (6)	−2.99 (6)	−5.41 (6)
1.088	1.046	2.5	7.70 (5)	1.61 (6)	−2.45 (6)	−4.09 (6)
1.053	1.027	3.0	6.19 (5)	1.44 (6)	−2.16 (6)	−3.34 (6)
1.026	1.013	3.5	5.14 (5)	1.34 (6)	−1.97 (6)	−2.89 (6)
1.005	1.001	4.0	4.55 (5)	1.26 (6)	−1.87 (6)	−2.58 (6)

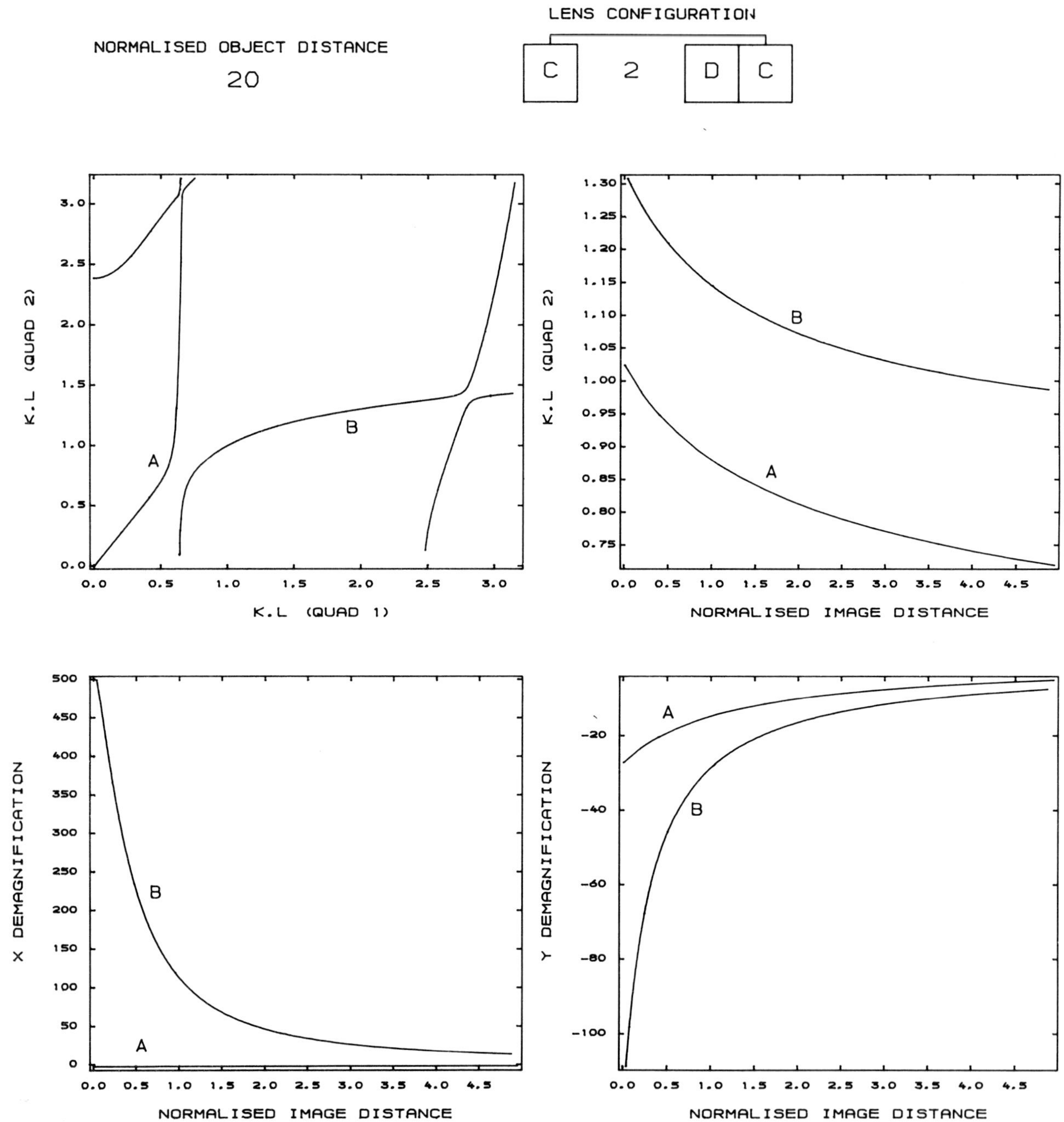

NORMALISED OBJECT DISTANCE
20
LENS CONFIGURATION
C
2
D
C
K.L (QUAD 2)
K.L (QUAD 1)
A
B
NORMALISED IMAGE DISTANCE
X DEMAGNIFICATION
Y DEMAGNIFICATION

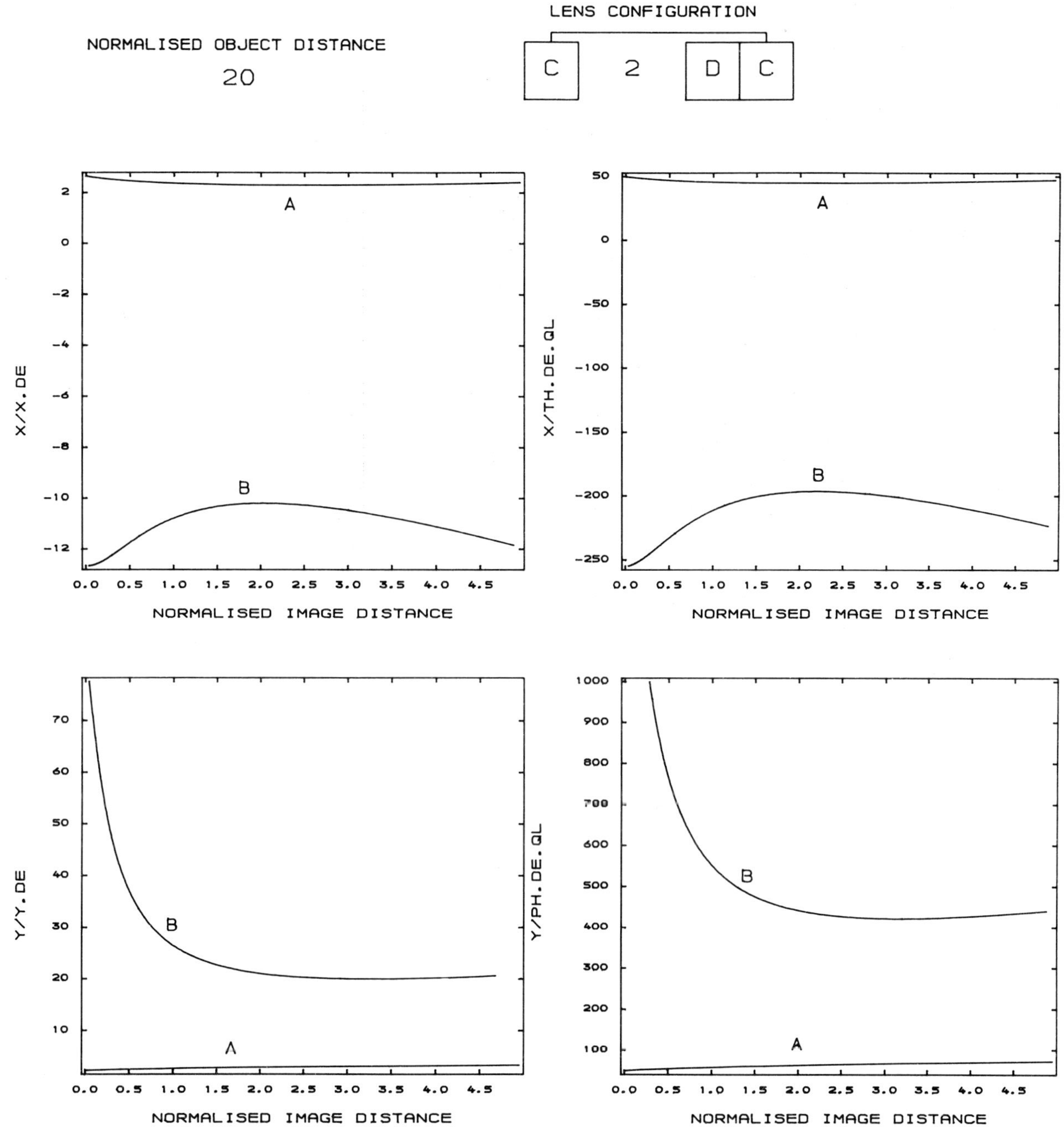
NORMALISED OBJECT DISTANCE
20
LENS CONFIGURATION
C
2
D
C
X/X.DE
X/TH.DE.QL
Y/Y.DE
Y/PH.DE.QL
A
B
NORMALISED IMAGE DISTANCE

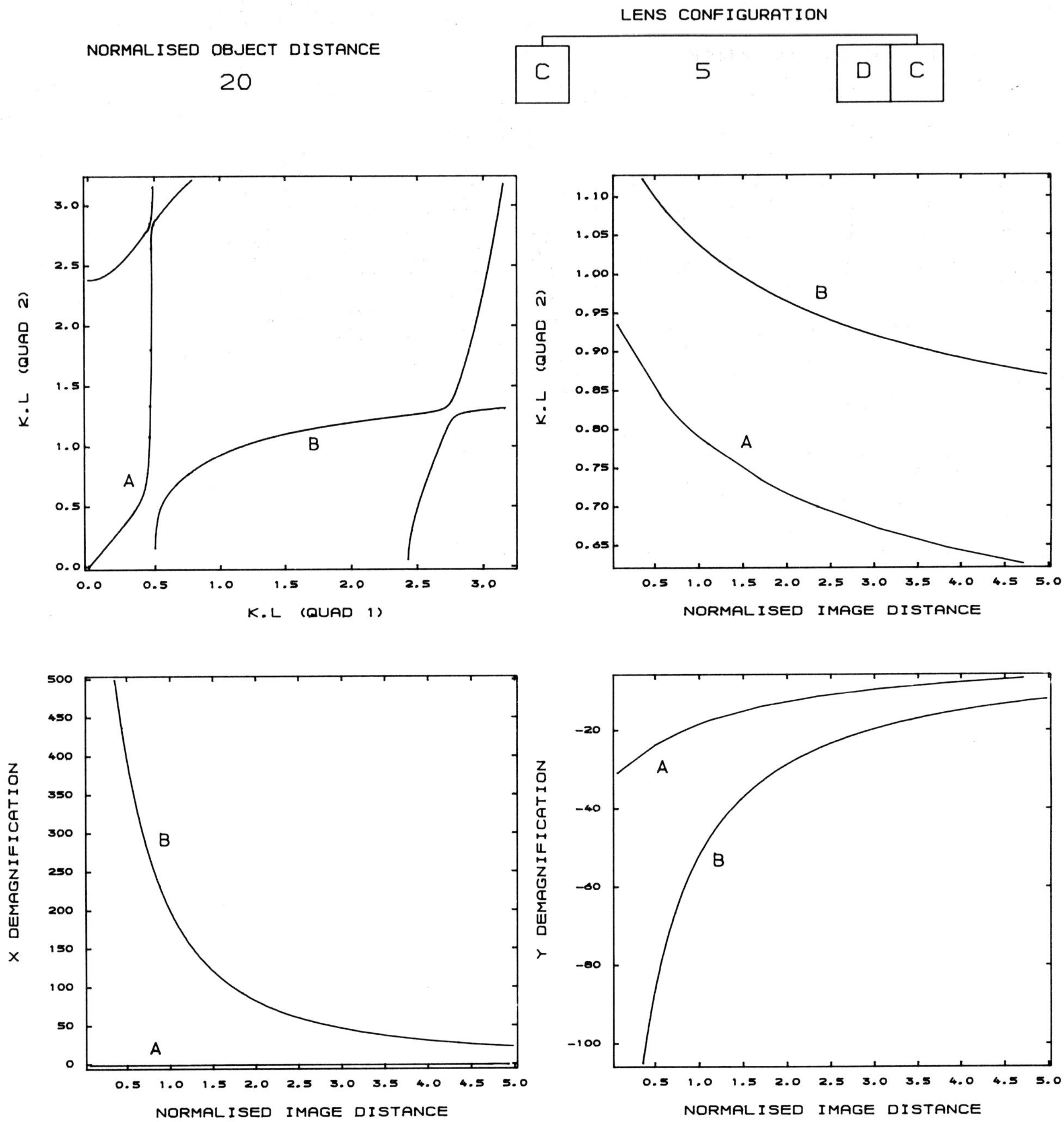

NORMALISED OBJECT DISTANCE
20
LENS CONFIGURATION
C
5
D
C
K.L (QUAD 2)
K.L (QUAD 1)
A
B
K.L (QUAD 2)
NORMALISED IMAGE DISTANCE
B
A
X DEMAGNIFICATION
NORMALISED IMAGE DISTANCE
B
A
Y DEMAGNIFICATION
NORMALISED IMAGE DISTANCE
A
B

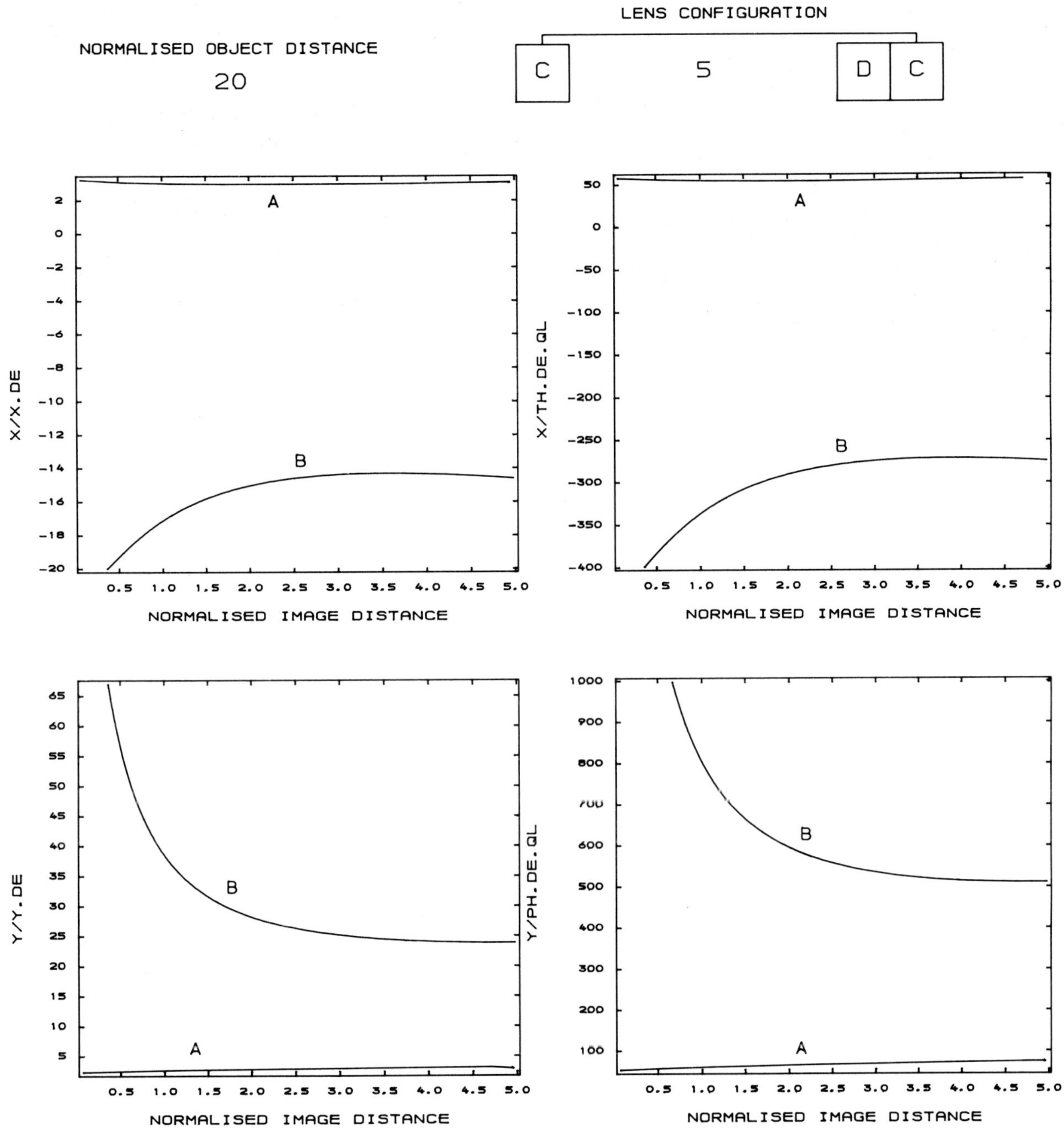
NORMALISED OBJECT DISTANCE
20
LENS CONFIGURATION
C
5
D
C
X/X.DE
A
B
NORMALISED IMAGE DISTANCE
X/TH.DE.QL
A
B
NORMALISED IMAGE DISTANCE
Y/Y.DE
B
A
NORMALISED IMAGE DISTANCE
Y/PH.DE.QL
B
A
NORMALISED IMAGE DISTANCE

Lens Configuration 20C5D0C

KL (QUAD 1)	KL (QUAD 2)	Normalised image distance	Normalised spherical aberration coefficients			
			x/θ^3QL	$x/\theta\phi^2QL$	y/ϕ^3QL	$y/\theta^2\phi QL$
0.449	0.850	0.5	−0.7 (3)	−1.14 (4)	−3.61 (4)	−0.4 (3)
0.444	0.788	1.0	−2.1 (3)	−1.28 (4)	−3.34 (4)	−0.5 (3)
0.439	0.745	1.5	−1.1 (3)	−1.31 (4)	−3.08 (4)	−0.2 (3)
0.435	0.714	2.0	−0.6 (3)	−1.39 (4)	−2.94 (4)	−2.1 (3)
0.431	0.690	2.5	−0.6 (3)	−1.36 (4)	−2.65 (4)	−3.1 (3)
0.427	0.670	3.0	−0.8 (3)	−1.47 (4)	−3.00 (4)	−1.8 (3)
0.424	0.654	3.5	−0.8 (3)	−1.57 (4)	−2.78 (4)	−2.2 (3)
0.421	0.640	4.0	−1.3 (3)	−1.63 (4)	−2.93 (4)	−2.3 (3)
1.489	1.097	0.5	4.40 (7)	5.52 (7)	−9.17 (7)	−2.45 (8)
1.268	1.035	1.0	1.63 (7)	2.00 (7)	−2.55 (7)	−7.69 (7)
1.149	0.993	1.5	8.35 (6)	1.17 (7)	−1.40 (7)	−3.78 (7)
1.073	0.962	2.0	5.21 (6)	8.41 (6)	−9.33 (6)	−2.38 (7)
1.021	0.937	2.5	3.64 (6)	6.69 (6)	−7.06 (6)	−1.70 (7)
0.982	0.918	3.0	2.77 (6)	5.66 (6)	−5.81 (6)	−1.33 (7)
0.952	0.901	3.5	2.22 (6)	4.99 (6)	−4.95 (6)	−1.09 (7)
0.928	0.888	4.0	1.87 (6)	4.54 (6)	−4.46 (6)	−9.38 (6)

Lens Configuration 20C0D0D0C

KL (QUAD 1)	KL (QUAD 2)	Normalised image distance	Normalised spherical aberration coefficients			
			x/θ^3QL	$x/\theta\phi^2QL$	y/ϕ^3QL	$y/\theta^2\phi QL$
0.801	0.762	0.5	−5.2 (3)	−3.92 (4)	−2.26 (4)	−1.13 (4)
0.777	0.733	1.0	−4.7 (3)	−3.51 (4)	−2.31 (4)	−1.33 (4)
0.756	0.711	1.5	−3.9 (3)	−3.21 (4)	−2.23 (4)	−1.65 (4)
0.736	0.692	2.0	−3.6 (3)	−3.17 (4)	−2.49 (4)	−1.63 (4)
0.718	0.676	2.5	−3.1 (3)	−3.17 (4)	−2.18 (4)	−2.01 (4)
0.701	0.662	3.0	−3.8 (3)	−3.10 (4)	−2.17 (4)	−2.28 (4)
0.687	0.650	3.5	−4.8 (3)	−3.13 (4)	−2.19 (4)	−2.41 (4)
0.673	0.639	4.0	−2.3 (3)	−3.26 (4)	−2.27 (4)	−2.50 (4)
1.512	0.977	0.5	2.18 (6)	6.83 (5)	−1.34 (6)	−5.77 (6)
1.301	0.917	1.0	5.36 (5)	3.04 (5)	−5.49 (5)	−1.70 (6)
1.194	0.880	1.5	2.13 (5)	2.03 (5)	−3.80 (5)	−8.37 (5)
1.131	0.855	2.0	1.15 (5)	1.67 (5)	−3.06 (5)	−5.44 (5)
1.090	0.837	2.5	7.6 (4)	1.53 (5)	−2.83 (5)	−4.13 (5)
1.062	0.824	3.0	5.9 (4)	1.46 (5)	−2.65 (5)	−3.44 (5)
1.041	0.814	3.5	5.0 (4)	1.45 (5)	−2.66 (5)	−3.08 (5)
1.026	0.806	4.0	4.3 (4)	1.48 (5)	−2.58 (5)	−2.90 (5)
1.751	2.220	0.5	*	*	*	*
1.529	2.157	1.0	1.37 (9)	6.03 (7)	9.29 (6)	1.09 (9)
1.418	2.120	1.5	4.36 (8)	1.30 (7)	7.05 (6)	3.38 (8)
1.350	2.096	2.0	1.93 (8)	7.78 (6)	6.23 (6)	1.74 (8)
1.305	2.079	2.5	1.02 (8)	5.76 (6)	5.94 (6)	1.08 (8)
1.273	2.066	3.0	6.01 (7)	4.72 (6)	5.84 (6)	7.45 (7)
1.248	2.056	3.5	3.80 (7)	4.03 (6)	5.89 (6)	5.45 (7)
1.229	2.048	4.0	2.53 (7)	3.59 (6)	5.98 (6)	4.17 (7)
1.113	2.038	0.5	−1.60 (5)	−2.87 (6)	9.86 (5)	5.4 (4)
1.111	2.025	1.0	−1.32 (5)	−2.70 (6)	1.32 (6)	8.0 (4)
1.110	2.017	1.5	−1.52 (5)	−2.56 (6)	1.65 (6)	1.00 (5)
1.109	2.012	2.0	−1.55 (5)	−2.45 (6)	1.99 (6)	1.21 (5)
1.108	2.008	2.5	−1.39 (5)	−2.37 (6)	2.33 (6)	1.39 (5)
1.108	2.006	3.0	−1.41 (5)	−2.28 (6)	2.67 (6)	1.64 (5)
1.107	2.003	3.5	−1.24 (5)	−2.21 (6)	3.00 (6)	1.84 (5)
1.107	2.002	4.0	−1.49 (5)	−2.14 (6)	3.34 (6)	2.03 (5)

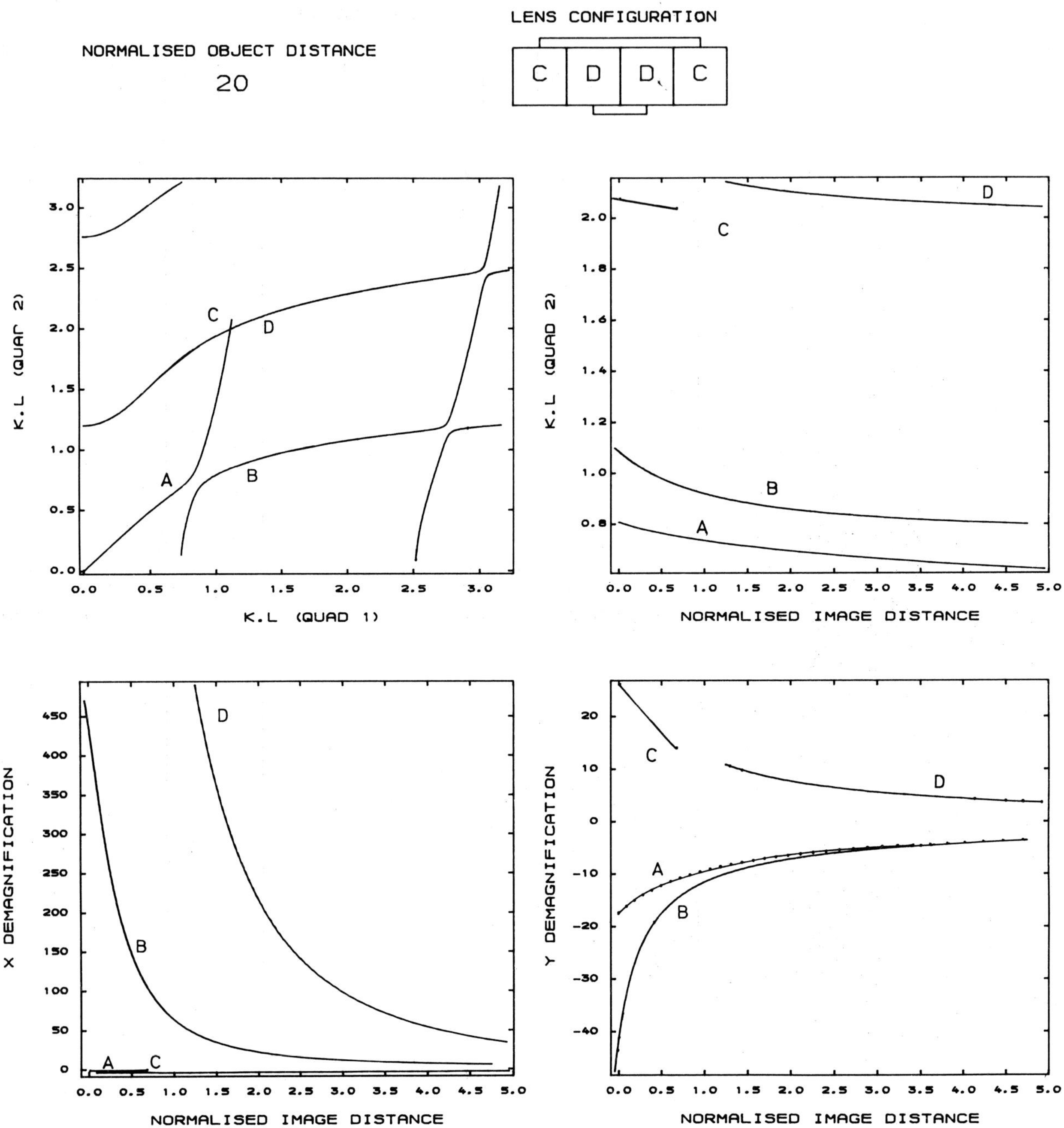
NORMALISED OBJECT DISTANCE
20
LENS CONFIGURATION
C D D C
K.L (QUAD 2)
K.L (QUAD 1)
NORMALISED IMAGE DISTANCE
X DEMAGNIFICATION
Y DEMAGNIFICATION
A
B
C
D

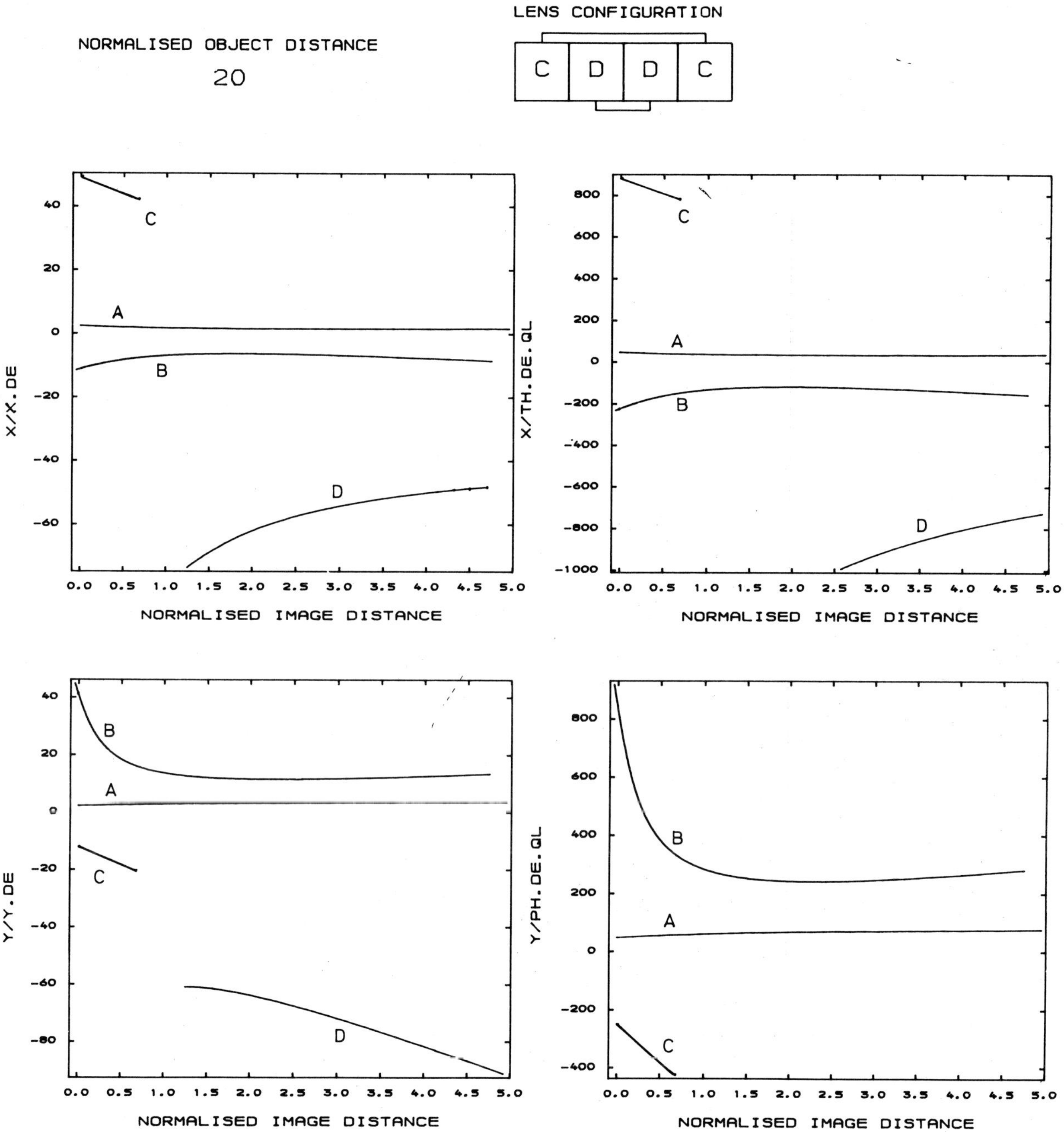
LENS CONFIGURATION
NORMALISED OBJECT DISTANCE
20
C
D
D
C
X/X.DE
X/TH.DE.QL
Y/Y.DE
Y/PH.DE.QL
A
B
C
D
NORMALISED IMAGE DISTANCE
0.0
0.5
1.0
1.5
2.0
2.5
3.0
3.5
4.0
4.5
5.0

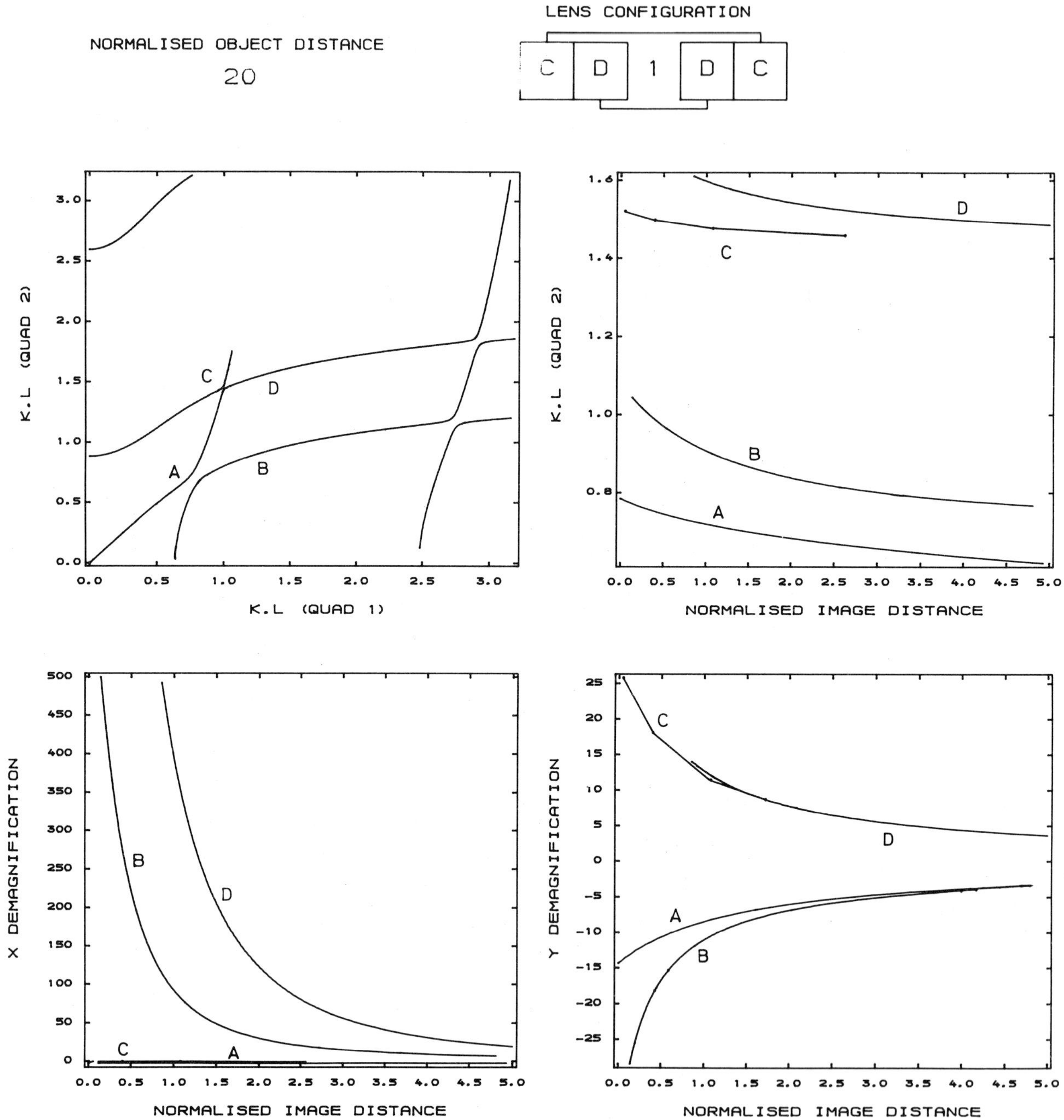
NORMALISED OBJECT DISTANCE
20
LENS CONFIGURATION
C D 1 D C
K.L (QUAD 2)
K.L (QUAD 1)
NORMALISED IMAGE DISTANCE
X DEMAGNIFICATION
Y DEMAGNIFICATION
A
B
C
D

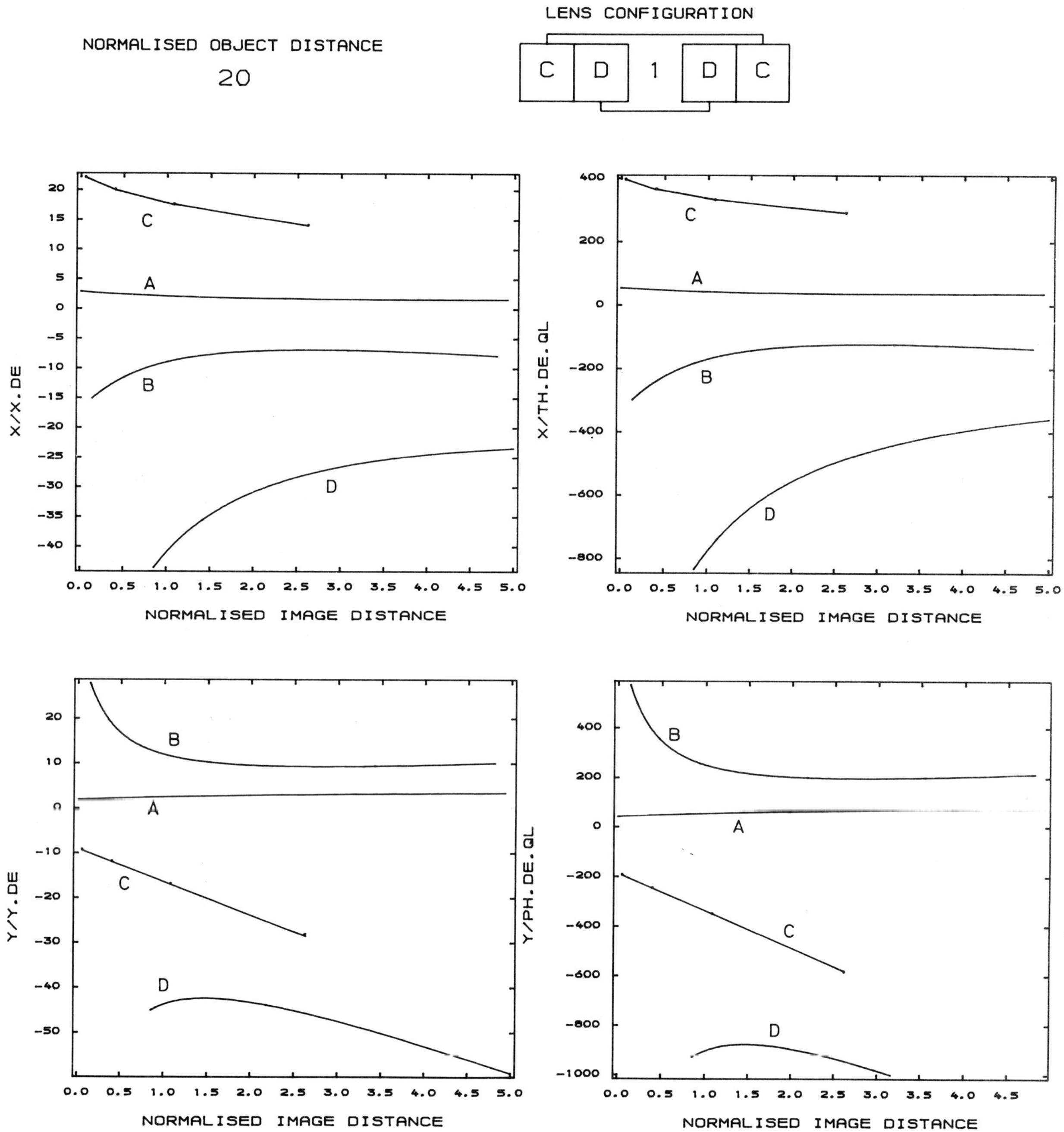

NORMALISED OBJECT DISTANCE
20
LENS CONFIGURATION
C D 1 D C
X/X.DE
X/TH.DE.QL
Y/Y.DE
Y/PH.DE.QL
A
B
C
D
NORMALISED IMAGE DISTANCE

Lens Configuration 20C0D1D0C

KL (QUAD 1)	KL (QUAD 2)	Normalised image distance	Normalised spherical aberration coefficients			
			x/θ^3QL	$x/\theta\phi^2QL$	y/ϕ^3QL	$y/\theta^2\phi QL$
0.764	0.743	0.5	−5.3 (3)	−4.43 (4)	−1.69 (4)	−9.9 (3)
0.746	0.717	1.0	−4.5 (3)	−3.91 (4)	−1.70 (4)	−1.26 (4)
0.730	0.697	1.5	−6.1 (3)	−3.48 (4)	−1.83 (4)	−1.39 (4)
0.715	0.681	2.0	−3.8 (3)	−3.36 (4)	−2.19 (4)	−1.34 (4)
0.701	0.667	2.5	−4.1 (3)	−3.20 (4)	−2.06 (4)	−1.61 (4)
0.688	0.655	3.0	−3.9 (3)	−3.17 (4)	−2.10 (4)	−1.89 (4)
0.676	0.643	3.5	−2.9 (3)	−3.21 (4)	−2.26 (4)	−1.91 (4)
0.665	0.634	4.0	−3.1 (3)	−3.25 (4)	−2.12 (4)	−2.18 (4)
1.477	0.968	0.5	7.12 (6)	8.68 (5)	−1.03 (6)	−1.11 (7)
1.259	0.905	1.0	1.50 (6)	3.43 (5)	−4.15 (5)	−2.84 (6)
1.145	0.863	1.5	5.16 (5)	2.14 (5)	−2.64 (5)	−1.23 (6)
1.076	0.835	2.0	2.43 (5)	1.58 (5)	−2.05 (5)	−7.11 (5)
1.030	0.814	2.5	1.38 (5)	1.35 (5)	−1.80 (5)	−4.84 (5)
0.998	0.798	3.0	9.3 (4)	1.21 (5)	−1.67 (5)	−3.74 (5)
0.974	0.786	3.5	7.0 (4)	1.14 (5)	−1.60 (5)	−3.14 (5)
0.956	0.777	4.0	5.2 (4)	1.13 (5)	−1.60 (3)	−2.74 (5)
1.647	1.647	0.5	8.79 (8)	3.58 (7)	7.64 (6)	8.36 (8)
1.429	1.593	1.0	1.73 (8)	5.57 (6)	3.65 (6)	1.48 (8)
1.317	1.561	1.5	5.98 (7)	3.06 (6)	2.64 (6)	6.11 (7)
1.248	1.539	2.0	2.70 (7)	2.19 (6)	2.27 (6)	3.33 (7)
1.202	1.523	2.5	1.43 (7)	1.72 (6)	2.09 (6)	2.10 (7)
1.169	1.511	3.0	8.44 (6)	1.48 (6)	2.02 (6)	1.45 (7)
1.144	1.501	3.5	5.37 (6)	1.29 (6)	2.00 (6)	1.06 (7)
1.124	1.493	4.0	3.58 (6)	1.16 (6)	2.00 (6)	8.18 (6)
1.004	1.489	0.5	−7.6 (4)	−9.20 (5)	3.68 (5)	3.1 (4)
1.000	1.474	1.0	−5.4 (4)	−8.58 (5)	4.64 (5)	3.8 (4)
0.998	1.465	1.5	−5.9 (4)	−8.14 (5)	5.55 (5)	5.2 (4)
0.996	1.459	2.0	−6.3 (4)	−7.74 (5)	6.42 (5)	6.2 (4)
0.995	1.455	2.5	−5.4 (4)	−7.49 (5)	7.41 (5)	7.8 (4)
0.993	1.452	3.0	−5.1 (4)	−7.25 (5)	8.33 (5)	8.9 (4)
0.992	1.449	3.5	−5.5 (4)	−7.01 (5)	9.38 (5)	9.6 (4)
0.992	1.448	4.0	−5.0 (4)	−6.85 (5)	1.02 (6)	1.1 (5)

Lens Configuration 20C0D2D0C

KL (QUAD 1)	KL (QUAD 2)	Normalised image distance	Normalised spherical aberration coefficients			
			x/θ^3QL	$x/\theta\phi^2QL$	y/ϕ^3QL	$y/\theta^2\phi QL$
0.739	0.728	0.5	−7.8 (3)	−5.00 (4)	−1.38 (4)	−1.05 (4)
0.724	0.704	1.0	−6.9 (3)	−4.34 (4)	−1.48 (4)	−1.11 (4)
0.711	0.686	1.5	−4.0 (3)	−3.87 (4)	−1.50 (4)	−1.20 (4)
0.699	0.671	2.0	−5.3 (3)	−3.61 (4)	−1.68 (4)	−1.48 (4)
0.687	0.659	2.5	−5.7 (3)	−3.44 (4)	−1.92 (4)	−1.47 (4)
0.677	0.648	3.0	−3.2 (3)	−3.43 (4)	−1.51 (4)	−1.99 (4)
0.667	0.638	3.5	−3.2 (3)	−3.36 (4)	−1.77 (4)	−1.91 (4)
0.657	0.629	4.0	−5.0 (3)	−3.18 (4)	−1.91 (4)	−2.15 (4)
1.460	0.964	0.5	1.67 (7)	1.08 (6)	−9.29 (5)	−1.86 (7)
1.239	0.898	1.0	3.24 (6)	4.13 (5)	−3.57 (5)	−4.39 (6)
1.122	0.855	1.5	1.05 (6)	2.28 (5)	−2.21 (5)	−1.78 (6)
1.048	0.824	2.0	4.56 (5)	1.70 (5)	−1.69 (5)	−9.59 (5)
0.999	0.802	2.5	2.44 (5)	1.35 (5)	−1.43 (5)	−6.18 (5)
0.963	0.784	3.0	1.44 (5)	1.19 (5)	−1.40 (5)	−4.46 (5)
0.937	0.770	3.5	9.7 (4)	1.09 (5)	−1.28 (5)	−3.53 (5)
0.917	0.759	4.0	7.3 (4)	1.04 (5)	−1.22 (5)	−2.94 (5)
1.592	1.429	0.5	5.08 (8)	1.59 (7)	4.63 (6)	4.42 (8)
1.374	1.375	1.0	1.00 (8)	3.67 (6)	2.12 (6)	8.81 (7)
1.261	1.342	1.5	3.40 (7)	2.07 (6)	1.49 (6)	3.60 (7)
1.191	1.319	2.0	1.50 (7)	1.47 (6)	1.25 (6)	1.93 (7)
1.144	1.302	2.5	7.84 (6)	1.17 (6)	1.12 (6)	1.20 (7)
1.110	1.289	3.0	4.57 (6)	9.77 (5)	1.07 (6)	8.20 (6)
1.084	1.279	3.5	2.88 (6)	8.42 (5)	1.04 (6)	5.97 (6)
1.064	1.271	4.0	1.91 (6)	7.56 (5)	1.04 (6)	4.59 (6)
0.941	1.272	0.5	−4.5 (4)	−5.75 (5)	2.23 (5)	2.1 (4)
0.935	1.253	1.0	−4.5 (4)	−5.23 (5)	2.55 (5)	2.7 (4)
0.932	1.243	1.5	−3.3 (4)	−5.04 (5)	2.90 (5)	3.7 (4)
0.929	1.236	2.0	−4.1 (4)	−4.67 (5)	3.36 (5)	4.3 (4)
0.927	1.231	2.5	−4.1 (4)	−4.45 (5)	3.78 (5)	4.8 (4)
0.925	1.227	3.0	−3.9 (4)	−4.35 (5)	4.14 (5)	6.2 (4)
0.924	1.224	3.5	−3.6 (4)	−4.22 (5)	4.61 (5)	6.8 (4)
0.922	1.221	4.0	−3.5 (4)	−4.15 (5)	5.07 (5)	7.6 (4)

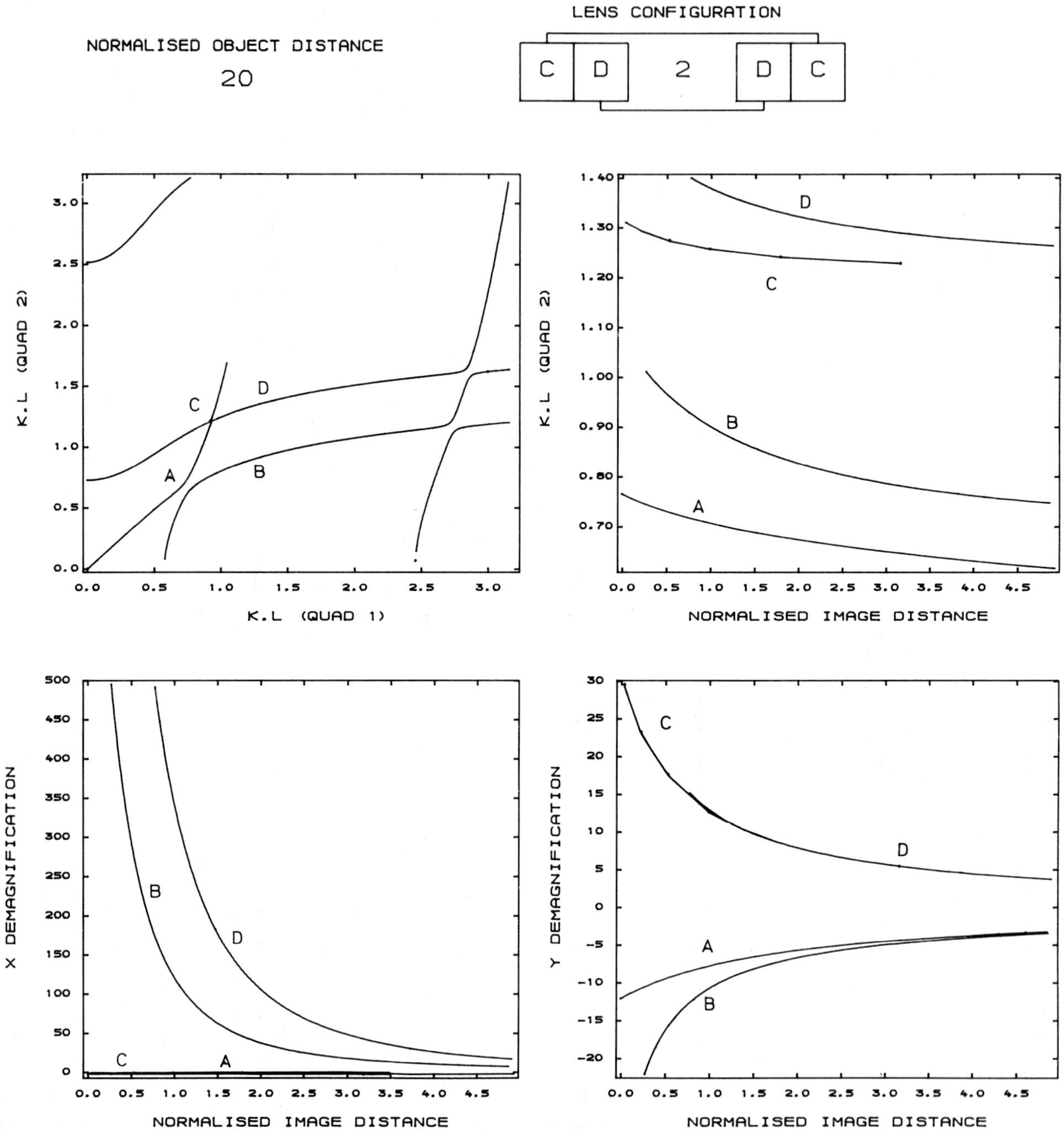
NORMALISED OBJECT DISTANCE
20
LENS CONFIGURATION
C D 2 D C
K.L (QUAD 2)
K.L (QUAD 1)
A
B
C
D
NORMALISED IMAGE DISTANCE
X DEMAGNIFICATION
Y DEMAGNIFICATION

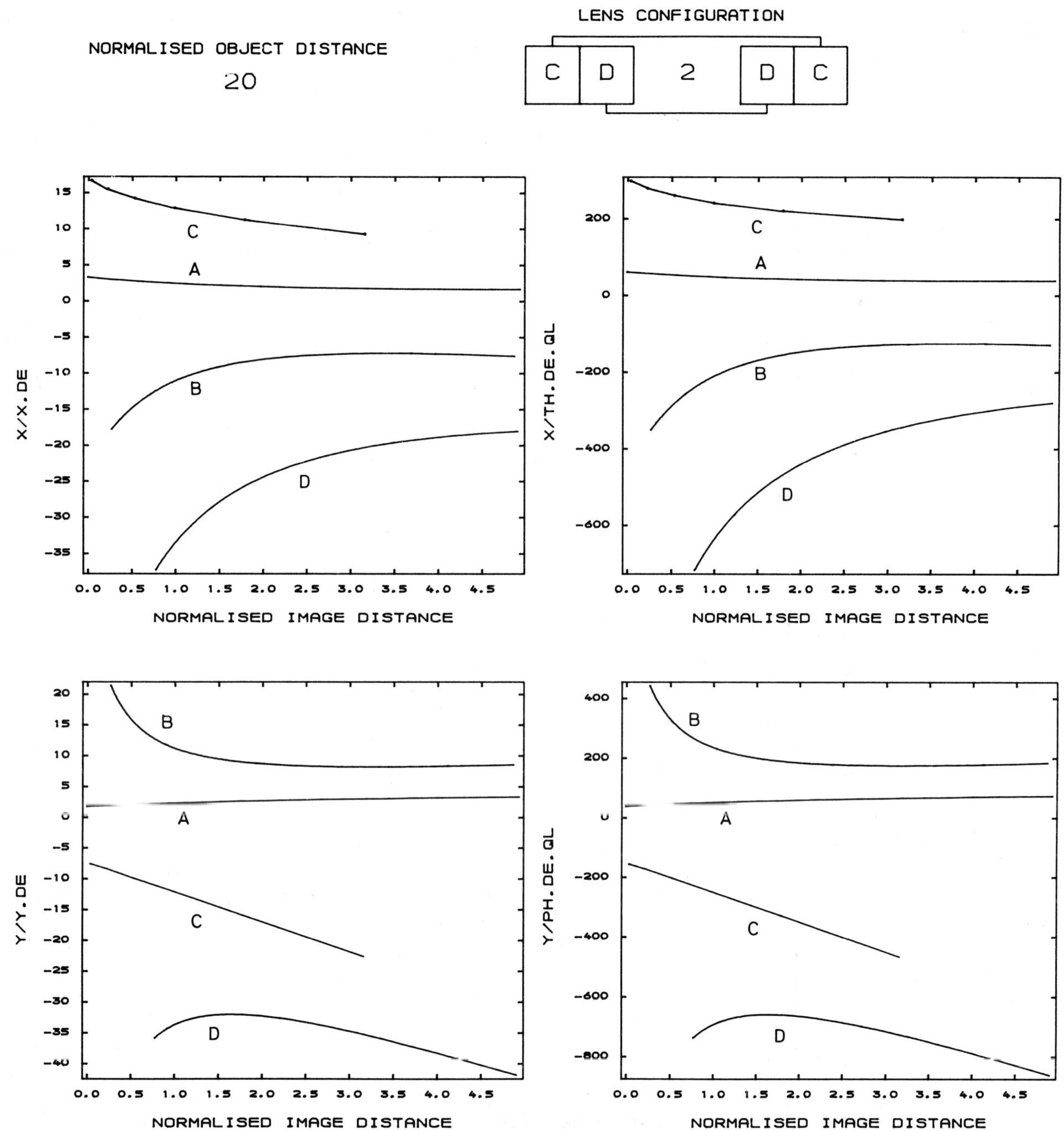

NORMALISED OBJECT DISTANCE
20
LENS CONFIGURATION
C D 2 D C
X/X.DE
X/TH.DE.QL
Y/Y.DE
Y/PH.DE.QL
NORMALISED IMAGE DISTANCE
A
B
C
D

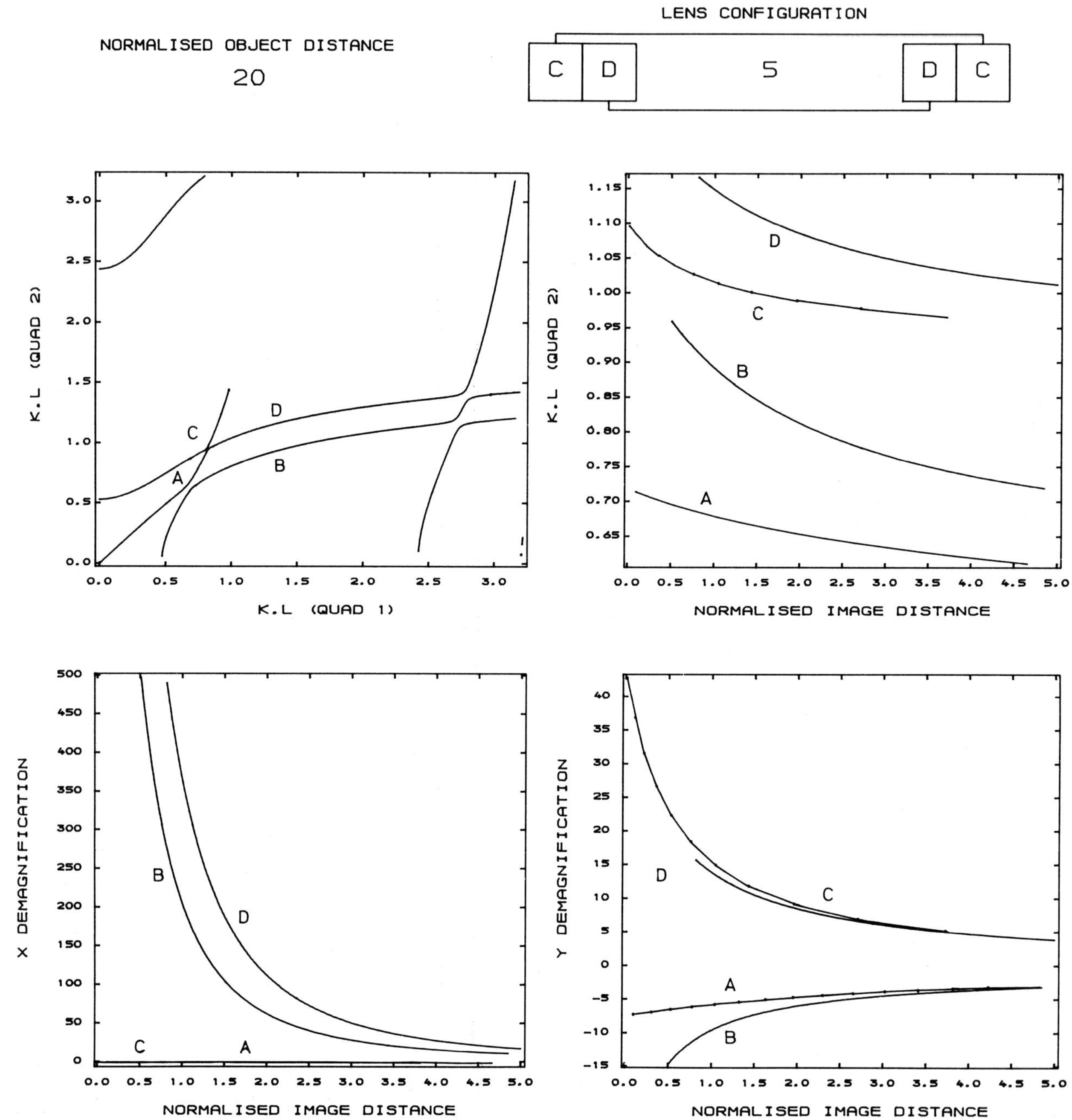
NORMALISED OBJECT DISTANCE
20
LENS CONFIGURATION
C D 5 D C
K.L (QUAD 2)
K.L (QUAD 1)
NORMALISED IMAGE DISTANCE
X DEMAGNIFICATION
Y DEMAGNIFICATION
A
B
C
D

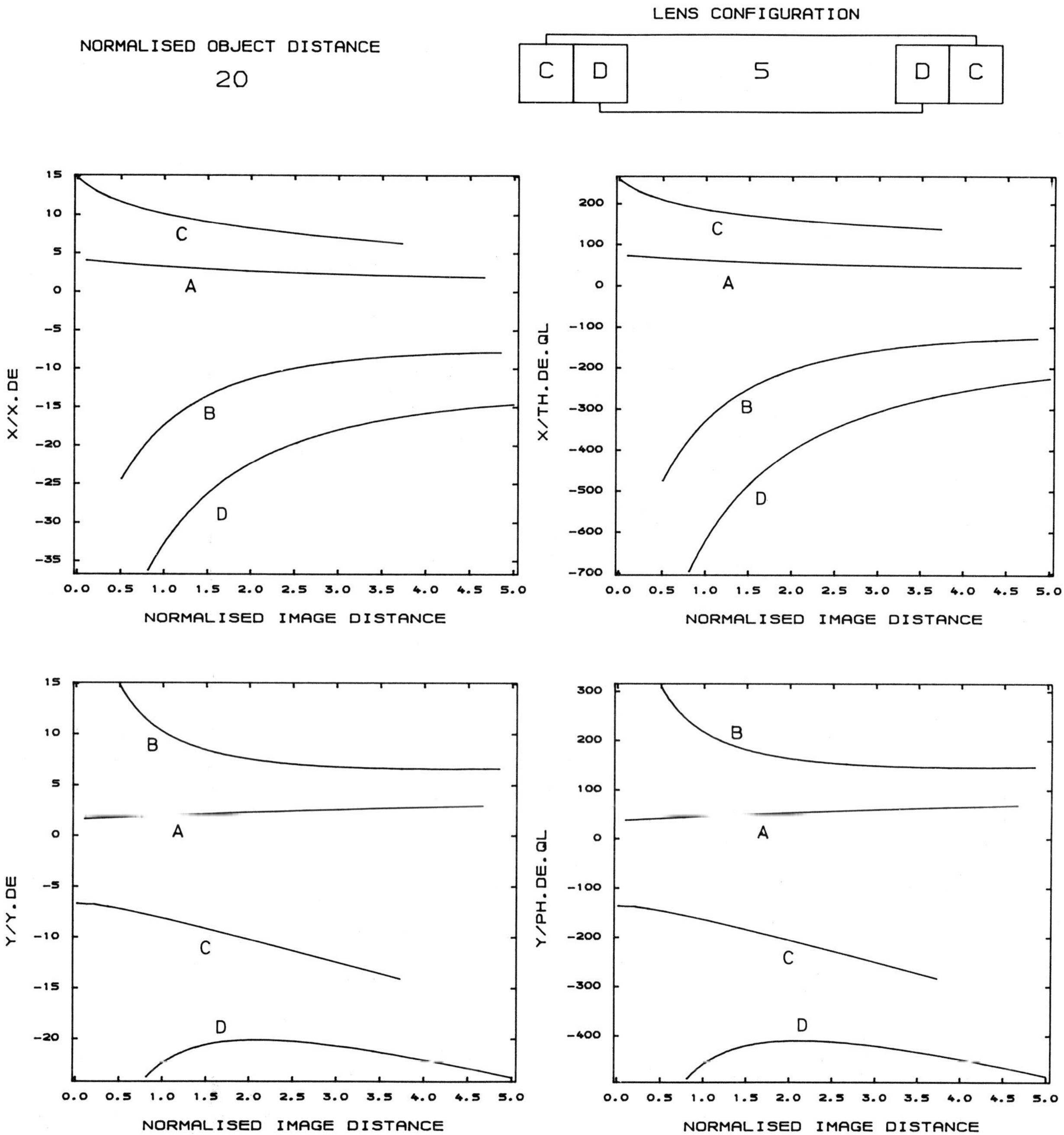
NORMALISED OBJECT DISTANCE
20
LENS CONFIGURATION
C D 5 D C
X/X.DE
X/TH.DE.QL
Y/Y.DE
Y/PH.DE.QL
A
B
C
D
NORMALISED IMAGE DISTANCE
0.0 0.5 1.0 1.5 2.0 2.5 3.0 3.5 4.0 4.5 5.0

Lens Configuration 20C0D5D0C

KL (QUAD 1)	KL (QUAD 2)	Normalised image distance	Normalised spherical aberration coefficients			
			x/θ^3QL	$x/\theta\phi^2QL$	y/ϕ^3QL	$y/\theta^2\phi QL$
0.695	0.694	0.5	−6.0 (3)	−6.46 (4)	−9.6 (3)	−1.31 (4)
0.684	0.676	1.0	−8.5 (3)	−5.48 (4)	−1.27 (4)	−1.25 (4)
0.675	0.662	1.5	−4.5 (3)	−5.22 (4)	−1.20 (4)	−1.33 (4)
0.667	0.651	2.0	−6.0 (3)	−4.62 (4)	−1.44 (4)	−1.48 (4)
0.659	0.641	2.5	−5.5 (3)	−4.44 (4)	−1.48 (4)	−1.39 (4)
0.652	0.632	3.0	−6.6 (3)	−3.92 (4)	−1.39 (4)	−1.66 (4)
0.645	0.624	3.5	−6.1 (3)	−3.96 (4)	−1.71 (4)	−1.51 (4)
0.638	0.617	4.0	−3.2 (3)	−3.80 (4)	−1.59 (4)	−2.00 (4)
1.439	0.959	0.5	9.19 (7)	1.95 (6)	−7.96 (5)	−5.54 (7)
1.214	0.889	1.0	1.59 (7)	5.50 (5)	−2.92 (5)	−1.09 (7)
1.091	0.843	1.5	4.67 (6)	2.95 (5)	−1.69 (5)	−4.04 (6)
1.013	0.810	2.0	1.84 (6)	1.95 (5)	−1.30 (5)	−1.99 (6)
0.959	0.784	2.5	8.88 (5)	1.46 (5)	−1.04 (5)	−1.17 (6)
0.919	0.764	3.0	4.84 (5)	1.26 (5)	−9.5 (4)	−7.75 (5)
0.888	0.748	3.5	2.89 (5)	1.12 (5)	−8.6 (4)	−5.63 (5)
0.864	0.734	4.0	1.88 (5)	1.01 (5)	−7.7 (4)	−4.31 (5)
1.521	1.202	0.5	6.33 (8)	2.06 (7)	2.56 (6)	5.13 (8)
1.300	1.145	1.0	1.11 (8)	3.64 (6)	1.11 (6)	8.43 (7)
1.183	1.107	1.5	3.48 (7)	1.87 (6)	7.37 (5)	3.16 (7)
1.110	1.081	2.0	1.45 (7)	1.25 (6)	5.78 (5)	1.59 (7)
1.060	1.062	2.5	7.23 (6)	9.45 (5)	4.93 (5)	9.46 (6)
1.023	1.046	3.0	4.04 (6)	7.58 (5)	4.61 (5)	6.19 (6)
0.995	1.034	3.5	2.45 (6)	6.36 (5)	4.25 (5)	4.35 (6)
0.973	1.024	4.0	1.59 (6)	5.51 (5)	4.17 (5)	3.21 (6)
0.851	1.039	0.5	−3.8 (4)	−3.97 (5)	1.73 (5)	1.0 (4)
0.840	1.012	1.0	−3.1 (4)	−3.34 (5)	1.49 (5)	1.3 (4)
0.832	0.996	1.5	−2.7 (4)	−3.03 (3)	1.46 (5)	1.8 (4)
0.828	0.985	2.0	−3.2 (4)	−2.75 (5)	1.47 (5)	2.5 (4)
0.824	0.977	2.5	−2.7 (4)	−2.65 (5)	1.50 (5)	3.0 (4)
0.821	0.971	3.0	−2.8 (4)	−2.55 (5)	1.60 (5)	3.0 (4)
0.819	0.967	3.5	−2.4 (4)	−2.47 (5)	1.66 (5)	3.3 (4)
0.816	0.963	4.0	−2.4 (4)	−2.37 (5)	1.79 (5)	4.6 (4)

Lens Configuration 20C0D0C0D

KL (QUAD 1)	KL (QUAD 2)	Normalised image distance	Normalised spherical aberration coefficients x/θ^3QL	$x/\theta\phi^2QL$	y/ϕ^3QL	$y/\theta^2\phi QL$
0.842	1.044	0.5	−9.2 (3)	−9.88 (4)	−2.86 (4)	−1.29 (4)
0.803	0.958	1.0	−8.5 (3)	−9.37 (4)	−2.54 (4)	−1.17 (4)
0.769	0.890	1.5	−9.3 (3)	−8.68 (4)	−2.61 (4)	−1.15 (4)
0.742	0.838	2.0	−5.4 (3)	−8.35 (4)	−2.57 (4)	−1.29 (4)
0.719	0.798	2.5	−6.8 (3)	−7.89 (4)	−2.60 (4)	−1.34 (4)
0.700	0.766	3.0	−8.9 (3)	−7.82 (4)	−2.66 (4)	−1.39 (4)
0.684	0.740	3.5	−1.0 (4)	−7.68 (4)	−2.57 (4)	−1.46 (4)
0.670	0.718	4.0	−7.3 (3)	−7.54 (4)	−2.53 (4)	−1.56 (4)
1.564	1.457	0.5	1.28 (6)	3.32 (6)	−7.94 (5)	−9.79 (5)
1.513	1.335	1.0	9.30 (5)	6.08 (6)	−1.62 (6)	−8.96 (5)
1.475	1.243	1.5	7.24 (5)	8.39 (6)	−3.26 (6)	−9.62 (5)
1.449	1.182	2.0	6.41 (5)	1.00 (7)	−4.86 (6)	−1.03 (6)
1.430	1.139	2.5	6.05 (5)	1.13 (7)	−6.30 (6)	−1.10 (6)
1.415	1.108	3.0	5.70 (5)	1.24 (7)	−7.56 (6)	−1.18 (6)
1.404	1.084	3.5	5.64 (5)	1.35 (7)	−8.76 (6)	−1.26 (6)
1.395	1.064	4.0	5.72 (5)	1.45 (7)	−9.85 (6)	−1.33 (6)
1.025	1.643	0.5	−5.2 (4)	−8.91 (5)	4.20 (5)	9.0 (3)
0.988	1.495	1.0	−5.1 (4)	−7.38 (5)	1.79 (5)	1.6 (4)
0.971	1.436	1.5	−5.7 (4)	−7.49 (5)	1.61 (5)	2.4 (4)
0.962	1.406	2.0	−6.0 (4)	−8.08 (5)	1.77 (5)	3.3 (4)
0.956	1.388	2.5	−7.6 (4)	−8.84 (5)	1.96 (5)	4.6 (4)
0.952	1.376	3.0	−7.6 (4)	−9.78 (5)	2.29 (5)	5.1 (4)
0.950	1.368	3.5	−8.3 (4)	−1.07 (6)	2.56 (5)	6.4 (4)
0.948	1.362	4.0	−8.8 (4)	−1.17 (6)	2.88 (5)	7.0 (4)
1.653	1.744	0.5	5.21 (6)	1.41 (7)	2.62 (6)	1.98 (6)
1.613	1.637	1.0	4.26 (6)	6.59 (6)	1.61 (6)	2.04 (6)
1.600	1.611	1.5	4.76 (6)	6.76 (6)	2.48 (6)	3.00 (6)
1.593	1.599	2.0	5.44 (6)	7.63 (6)	3.48 (6)	4.06 (6)
1.589	1.593	2.5	6.26 (6)	8.67 (6)	4.48 (6)	5.17 (6)
1.587	1.589	3.0	7.06 (6)	9.75 (6)	5.51 (6)	6.27 (6)
1.585	1.587	3.5	7.85 (6)	1.08 (7)	6.54 (6)	7.39 (6)
1.583	1.585	4.0	8.71 (6)	1.19 (7)	7.56 (6)	8.50 (6)

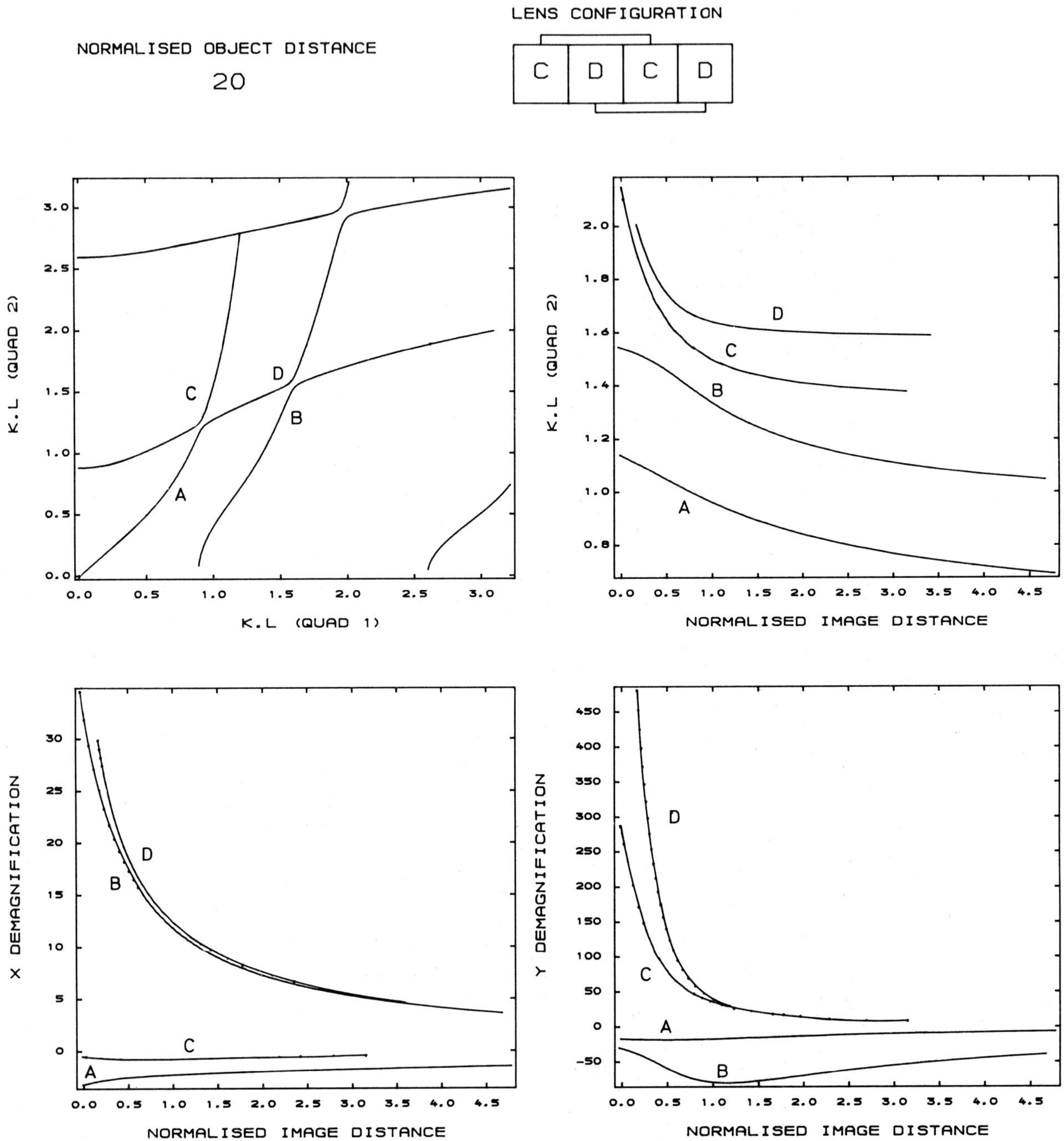
NORMALISED OBJECT DISTANCE
20
LENS CONFIGURATION
C D C D
K.L (QUAD 2)
K.L (QUAD 1)
A
B
C
D
NORMALISED IMAGE DISTANCE
X DEMAGNIFICATION
Y DEMAGNIFICATION

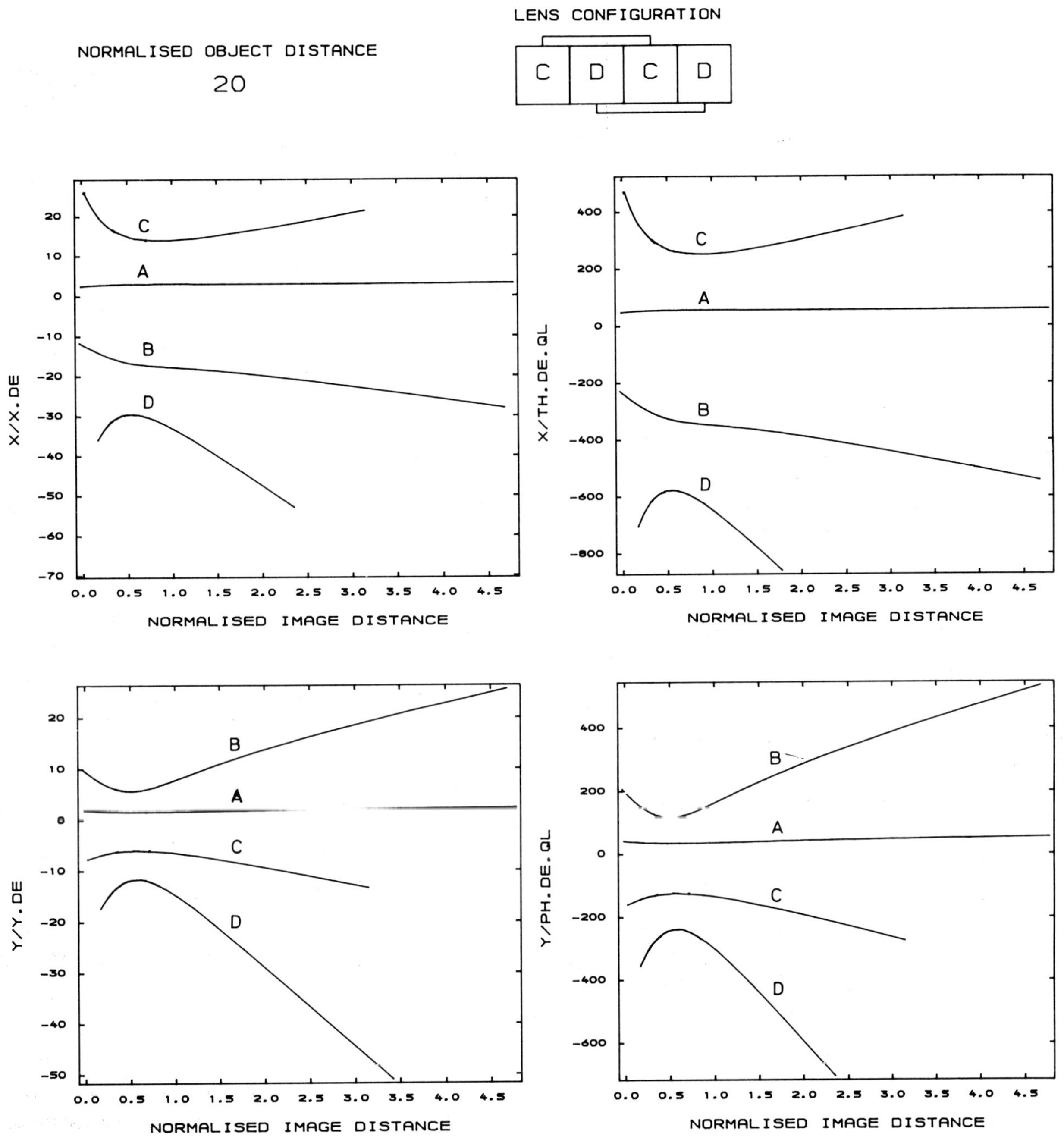
NORMALISED OBJECT DISTANCE
20
LENS CONFIGURATION
C D C D
X/X.DE
X/TH.DE.QL
Y/Y.DE
Y/PH.DE.QL
A
B
C
D
NORMALISED IMAGE DISTANCE

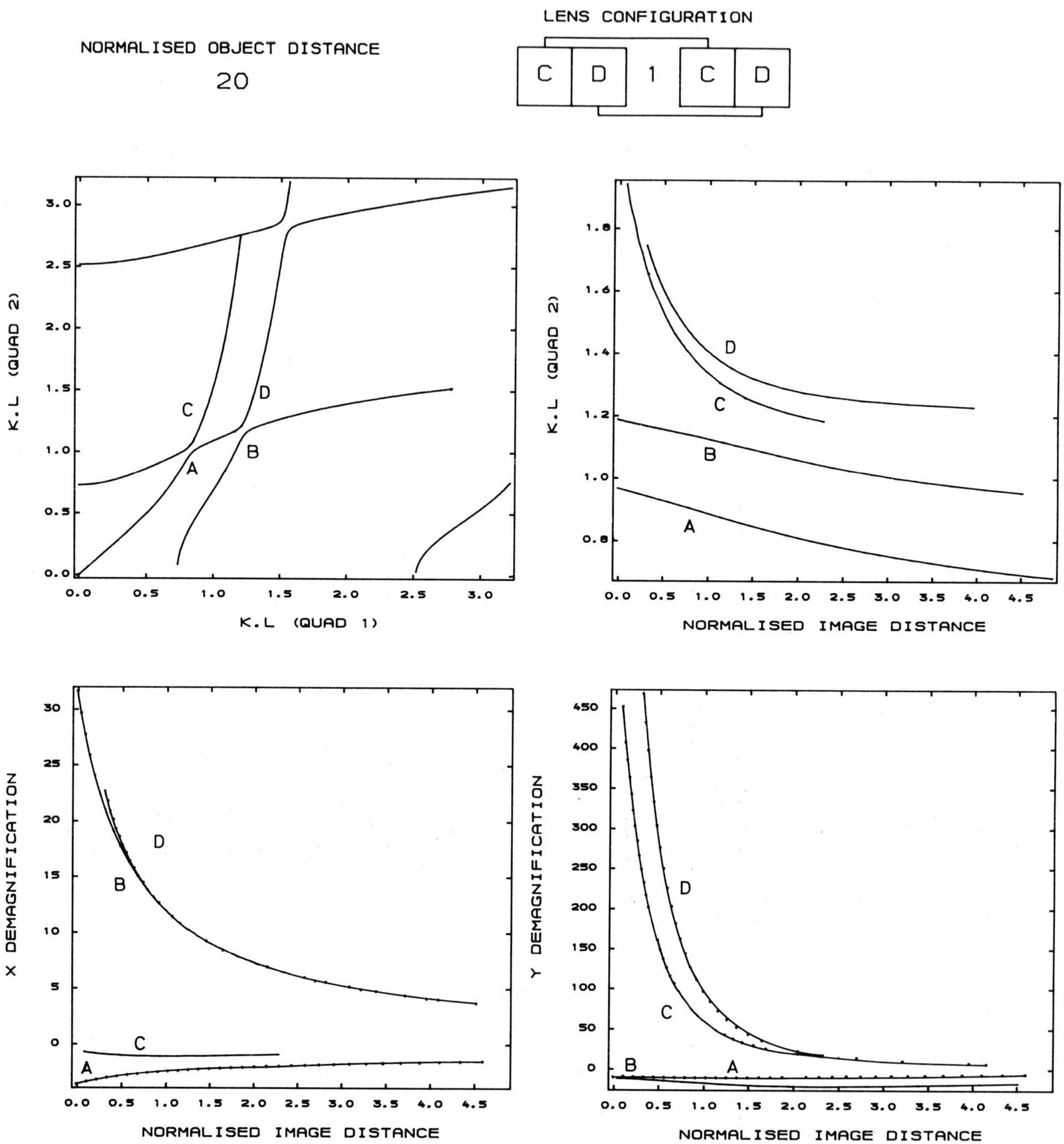
NORMALISED OBJECT DISTANCE
20
LENS CONFIGURATION
C D 1 C D
K.L (QUAD 2)
K.L (QUAD 1)
A
B
C
D
NORMALISED IMAGE DISTANCE
X DEMAGNIFICATION
Y DEMAGNIFICATION

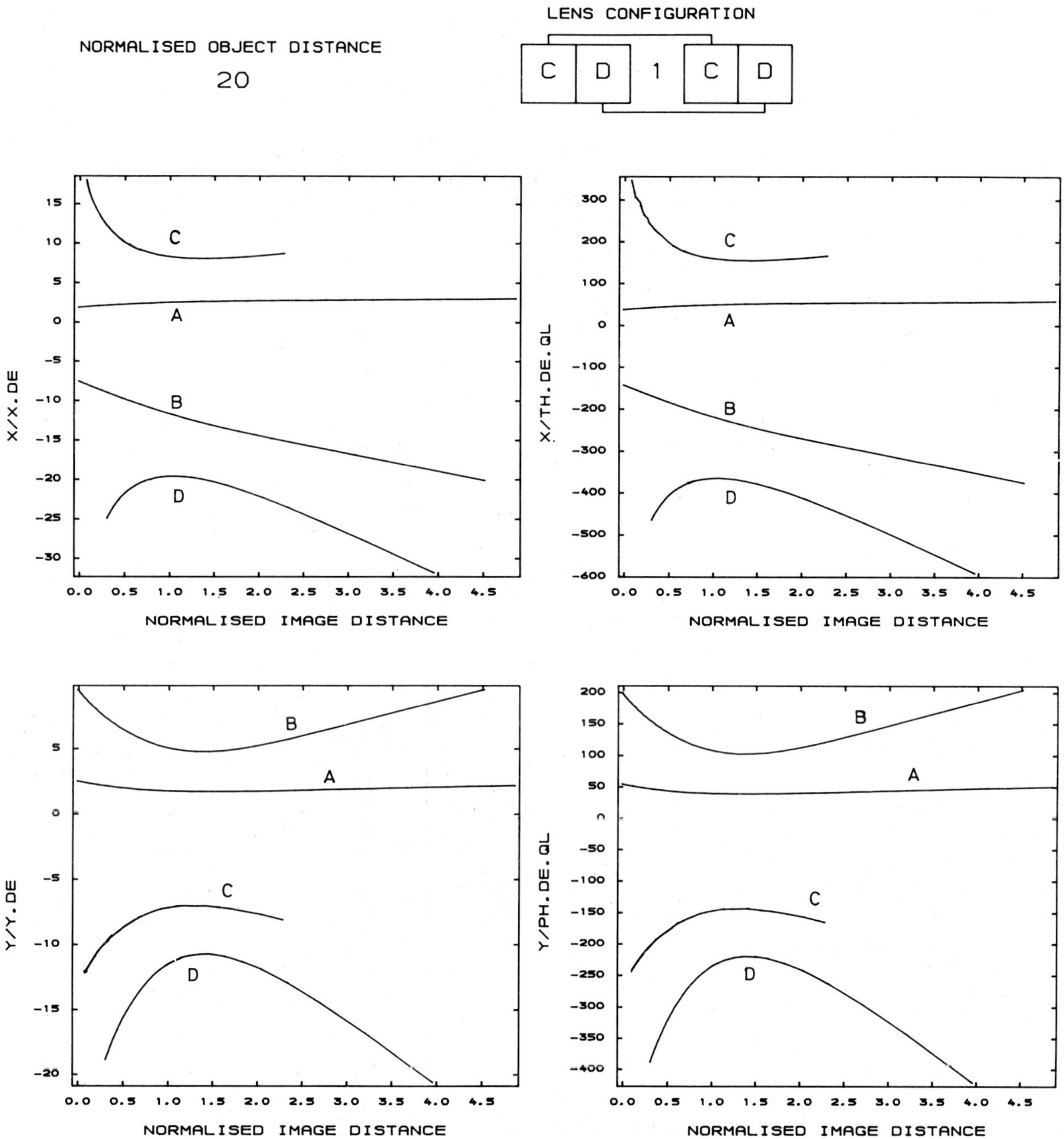
NORMALISED OBJECT DISTANCE
20
LENS CONFIGURATION
C D 1 C D
X/X.DE
X/TH.DE.QL
Y/Y.DE
Y/PH.DE.QL
A
B
C
D
NORMALISED IMAGE DISTANCE

Lens Configuration 20C0D1C0D

KL (QUAD 1)	KL (QUAD 2)	Normalised image distance	Normalised spherical aberration coefficients			
			x/θ^3QL	$x/\theta\phi^2$QL	y/ϕ^3QL	$y/\theta^2\phi$QL
0.794	0.926	0.5	−5.3 (3)	−6.29 (4)	−2.55 (4)	−1.69 (4)
0.770	0.884	1.0	−6.8 (3)	−6.62 (4)	−2.14 (4)	−1.52 (4)
0.747	0.844	1.5	−7.6 (3)	−6.72 (4)	−1.88 (4)	−1.37 (4)
0.727	0.808	2.0	−7.5 (3)	−6.57 (4)	−1.75 (4)	−1.48 (4)
0.708	0.777	2.5	−7.2 (3)	−6.67 (4)	−1.79 (4)	−1.49 (4)
0.691	0.751	3.0	−8.6 (3)	−6.52 (4)	−1.97 (4)	−1.48 (4)
0.677	0.728	3.5	−7.8 (3)	−6.69 (4)	−2.07 (4)	−1.46 (4)
0.664	0.709	4.0	−6.2 (3)	−6.86 (4)	−2.06 (4)	−1.58 (4)
1.244	1.154	0.5	2.66 (5)	2.22 (5)	−2.97 (5)	−2.82 (5)
1.219	1.124	1.0	2.64 (5)	3.30 (5)	−2.17 (5)	−2.35 (5)
1.201	1.090	1.5	2.56 (5)	5.15 (5)	−1.92 (5)	−2.38 (5)
1.185	1.057	2.0	2.50 (5)	7.48 (5)	−2.29 (5)	−2.66 (5)
1.172	1.027	2.5	2.46 (5)	1.00 (6)	−3.01 (5)	−2.98 (5)
1.161	1.003	3.0	2.34 (5)	1.26 (6)	−4.04 (5)	−3.29 (5)
1.152	0.983	3.5	2.37 (5)	1.50 (6)	−5.24 (5)	−3.56 (5)
1.145	0.967	4.0	2.25 (5)	1.74 (6)	−6.29 (5)	−3.86 (5)
0.995	1.522	0.5	−4.0 (4)	−8.54 (5)	2.40 (6)	6.0 (3)
0.940	1.331	1.0	−3.0 (4)	−5.06 (5)	4.78 (5)	8.0 (3)
0.911	1.243	1.5	−2.4 (4)	−4.20 (5)	2.04 (5)	1.2 (4)
0.894	1.197	2.0	−3.8 (4)	−3.94 (5)	1.39 (5)	1.7 (4)
0.883	1.169	2.5	−3.0 (4)	−4.01 (5)	1.21 (5)	2.5 (4)
0.876	1.152	3.0	−3.7 (4)	−4.15 (5)	1.20 (5)	2.9 (4)
0.870	1.139	3.5	−3.1 (4)	−4.35 (5)	1.24 (5)	3.6 (4)
0.867	1.131	4.0	−4.0 (4)	−4.61 (5)	1.31 (3)	4.1 (4)
1.332	1.597	0.5	1.79 (6)	2.94 (7)	1.66 (7)	1.92 (6)
1.282	1.400	1.0	9.14 (5)	5.83 (6)	2.18 (6)	7.30 (5)
1.256	1.315	1.5	7.30 (5)	2.31 (6)	6.71 (5)	4.75 (5)
1.241	1.274	2.0	6.88 (5)	1.45 (6)	4.34 (5)	4.19 (5)
1.232	1.253	2.5	7.09 (5)	1.17 (6)	4.26 (5)	4.22 (5)
1.226	1.240	3.0	7.26 (5)	1.08 (6)	4.68 (5)	4.59 (5)
1.222	1.232	3.5	7.74 (5)	1.06 (6)	5.40 (5)	5.04 (5)
1.218	1.226	4.0	8.31 (5)	1.07 (6)	6.13 (5)	5.53 (5)

Lens Configuration 20C0D2C0D

KL (QUAD 1)	KL (QUAD 2)	Normalised image distance	Normalised spherical aberration coefficients			
			x/θ^3QL	$x/\theta\phi^2QL$	y/ϕ^3QL	$y/\theta^2\phi QL$
0.761	0.854	0.5	−6.0 (3)	−4.93 (4)	−2.36 (4)	−2.09 (4)
0.744	0.830	1.0	−6.4 (3)	−5.33 (4)	−2.17 (4)	−1.74 (4)
0.728	0.804	1.5	−6.7 (3)	−5.51 (4)	−1.83 (4)	−1.70 (4)
0.712	0.779	2.0	−7.9 (3)	−5.57 (4)	−1.96 (4)	−1.38 (4)
0.697	0.756	2.5	−5.9 (3)	−5.79 (4)	−1.93 (4)	−1.36 (4)
0.683	0.735	3.0	−8.8 (3)	−5.78 (4)	−1.44 (4)	−1.50 (4)
0.671	0.716	3.5	−6.3 (3)	−6.01 (4)	−1.68 (4)	−1.51 (4)
0.659	0.699	4.0	−7.9 (3)	−6.04 (4)	−1.67 (4)	−1.47 (4)
1.126	1.034	0.5	1.70 (5)	7.4 (4)	−2.10 (5)	−1.77 (5)
1.100	1.014	1.0	1.57 (5)	1.01 (5)	−1.63 (5)	−1.43 (5)
1.082	0.994	1.5	1.48 (5)	1.46 (5)	−1.27 (5)	−1.34 (5)
1.069	0.975	2.0	1.50 (5)	2.01 (5)	−1.19 (5)	−1.43 (5)
1.057	0.956	2.5	1.45 (5)	2.78 (5)	−1.16 (5)	−1.57 (5)
1.047	0.938	3.0	1.43 (5)	3.63 (5)	−1.26 (5)	−1.73 (5)
1.039	0.923	3.5	1.39 (5)	4.57 (5)	−1.51 (5)	−1.90 (5)
1.032	0.909	4.0	1.37 (5)	5.49 (5)	−1.77 (5)	−2.08 (5)
0.986	1.485	0.5	−3.2 (4)	−1.21 (6)	8.05 (6)	8.0 (3)
0.923	1.276	1.0	−2.7 (4)	−5.87 (5)	1.37 (6)	7.0 (3)
0.886	1.174	1.5	−2.1 (4)	−4.06 (5)	4.41 (5)	1.2 (4)
0.863	1.116	2.0	−2.6 (4)	−3.36 (5)	2.16 (5)	1.2 (4)
0.848	1.080	2.5	−2.0 (4)	−3.17 (5)	1.44 (5)	1.6 (4)
0.838	1.056	3.0	−1.9 (4)	−3.07 (5)	1.16 (5)	1.9 (4)
0.830	1.040	3.5	−1.9 (4)	−3.07 (5)	1.05 (5)	2.5 (4)
0.824	1.028	4.0	−3.0 (4)	−3.12 (5)	1.00 (5)	3.0 (4)
1.222	1.549	0.5	1.21 (6)	4.63 (7)	3.83 (7)	2.30 (6)
1.171	1.340	1.0	5.76 (5)	9.81 (6)	5.90 (6)	8.56 (5)
1.141	1.238	1.5	4.16 (5)	3.60 (6)	1.59 (6)	4.94 (5)
1.122	1.182	2.0	3.62 (5)	1.83 (6)	6.47 (5)	3.59 (5)
1.109	1.148	2.5	3.39 (5)	1.17 (6)	3.84 (5)	3.02 (5)
1.100	1.127	3.0	3.27 (5)	8.88 (5)	2.92 (5)	2.79 (5)
1.093	1.113	3.5	3.35 (5)	7.48 (5)	2.74 (5)	2.63 (5)
1.088	1.103	4.0	3.47 (5)	6.75 (5)	2.78 (5)	2.66 (5)

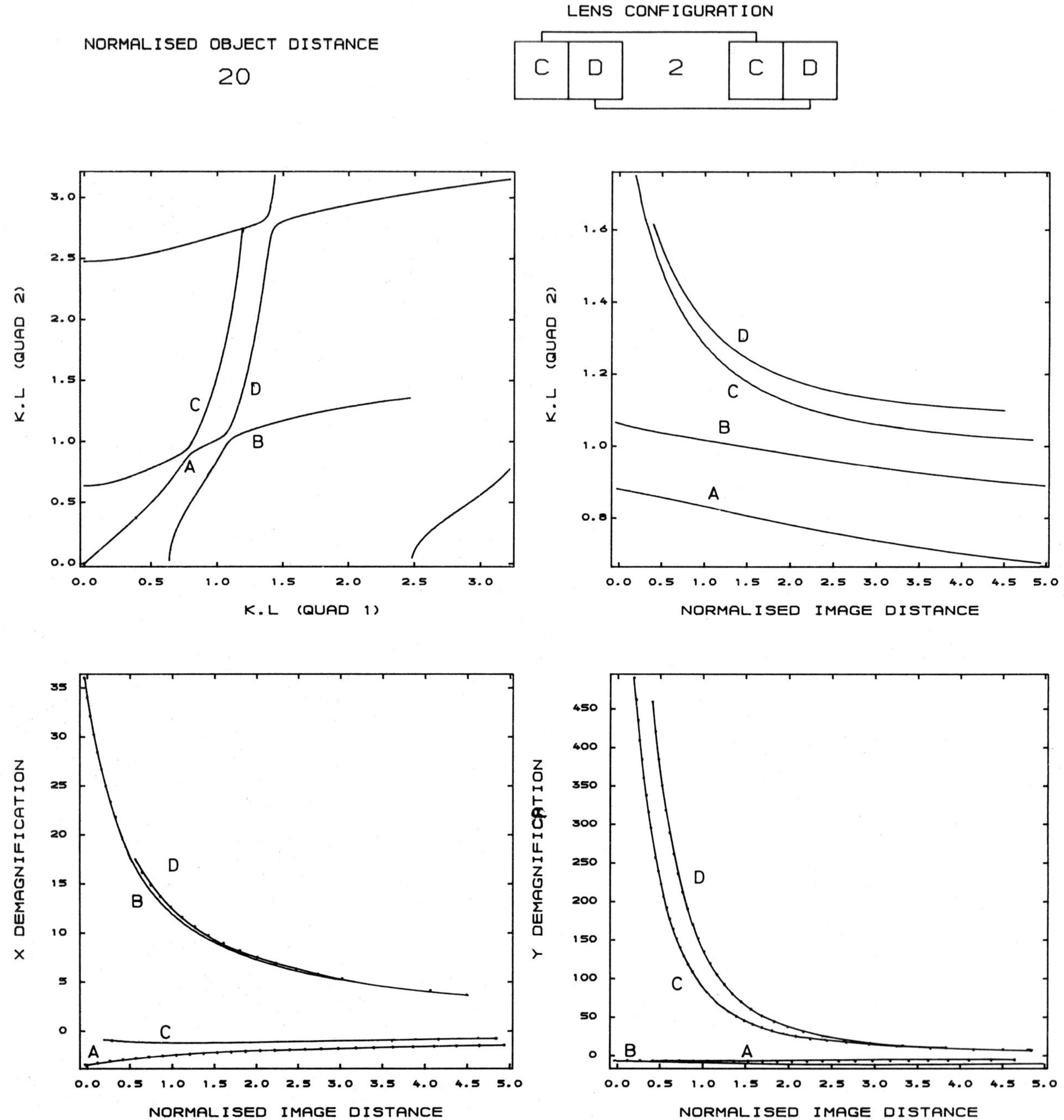
NORMALISED OBJECT DISTANCE
20
LENS CONFIGURATION
C D 2 C D
K.L (QUAD 2)
K.L (QUAD 1)
A
B
C
D
NORMALISED IMAGE DISTANCE
X DEMAGNIFICATION
Y DEMAGNIFICATION

NORMALISED OBJECT DISTANCE

20

LENS CONFIGURATION

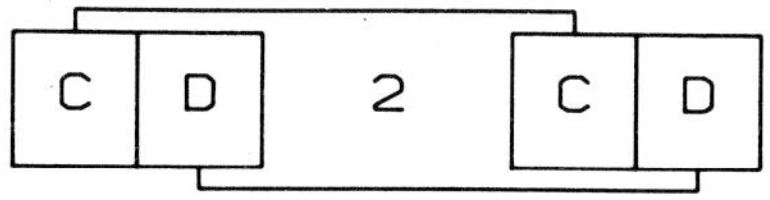

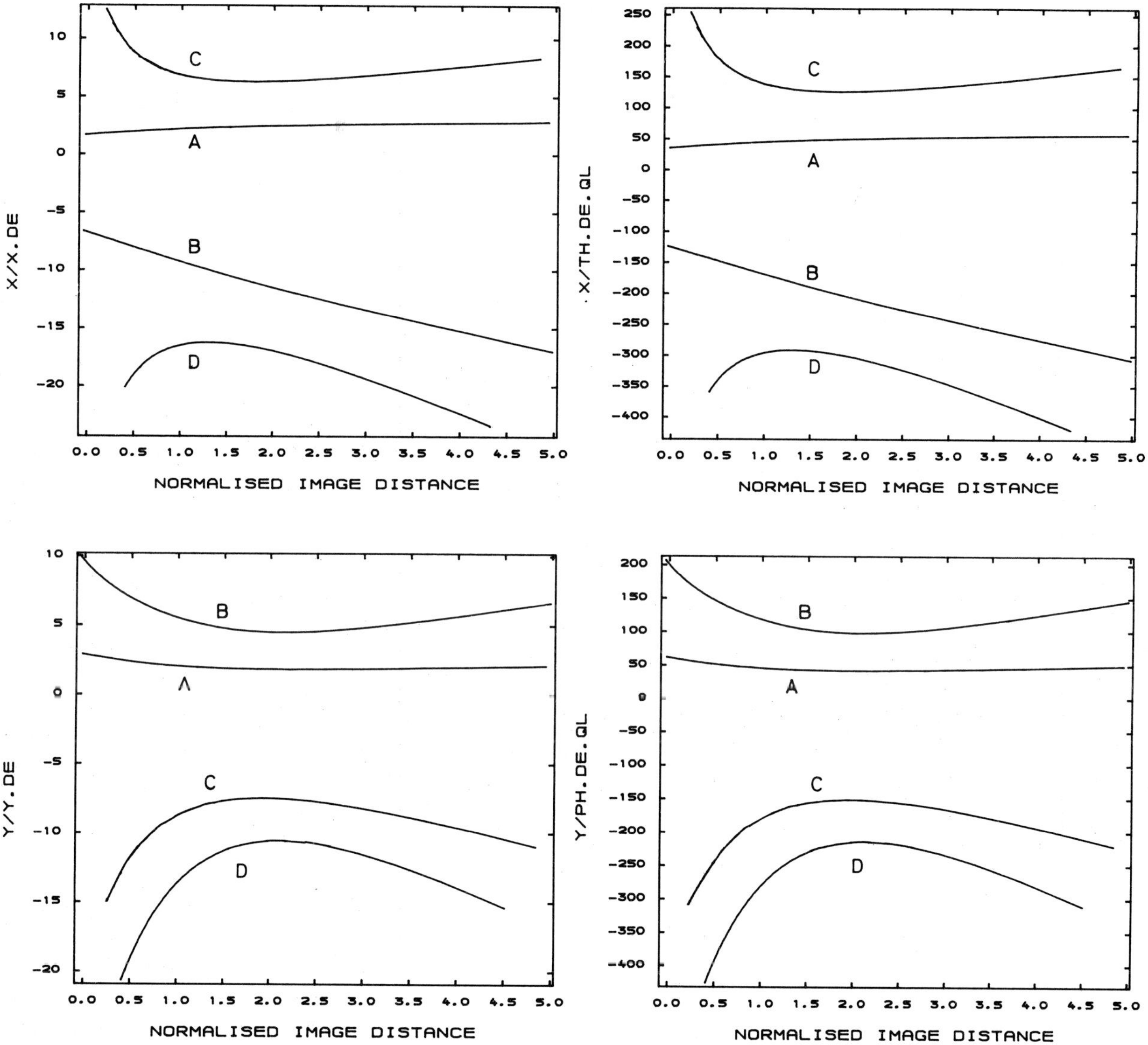

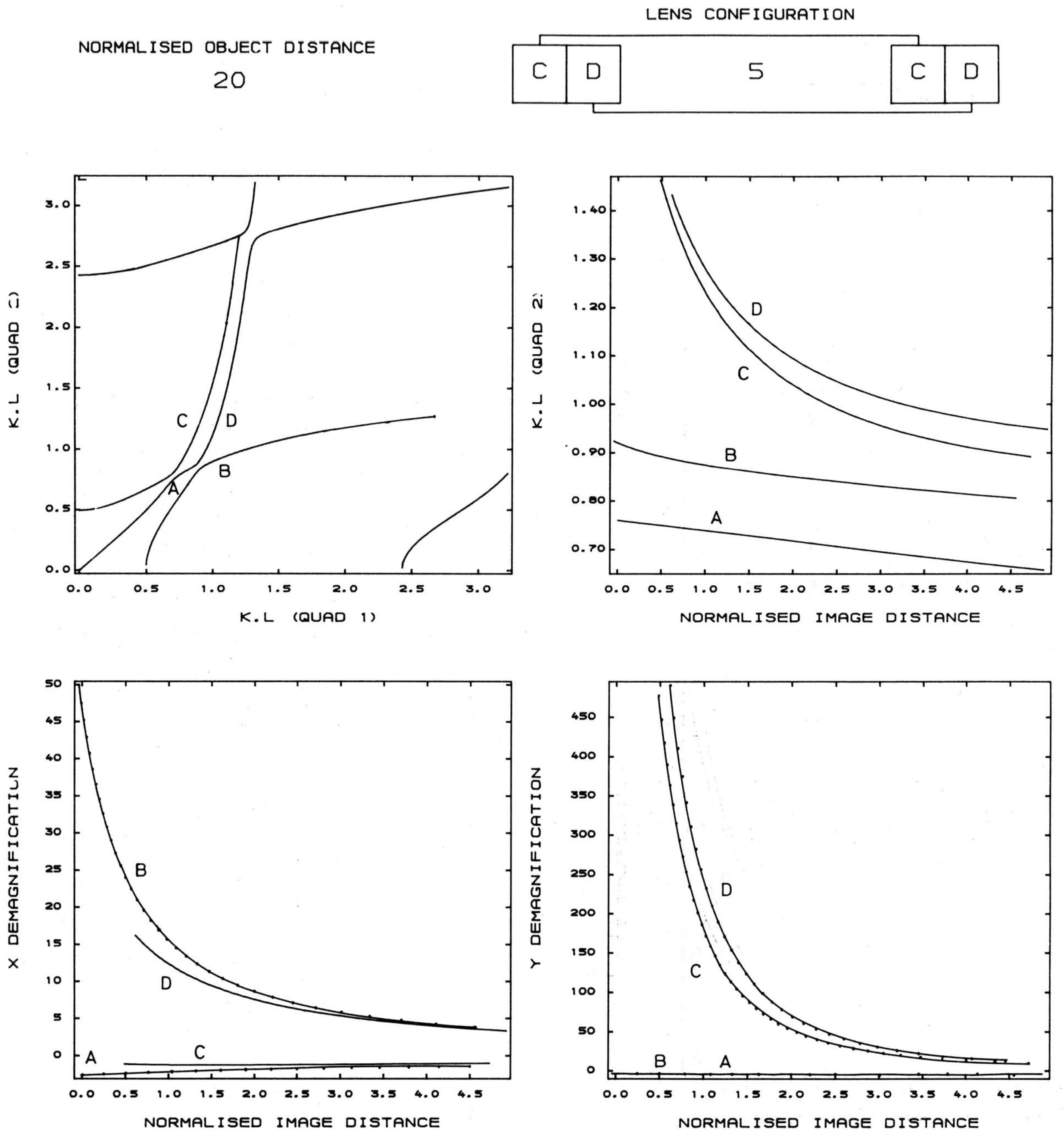

LENS CONFIGURATION
NORMALISED OBJECT DISTANCE
20
C
D
5
C
D
K.L (QUAD 1)
NORMALISED IMAGE DISTANCE
X DEMAGNIFICATION
Y DEMAGNIFICATION
NORMALISED IMAGE DISTANCE
NORMALISED IMAGE DISTANCE

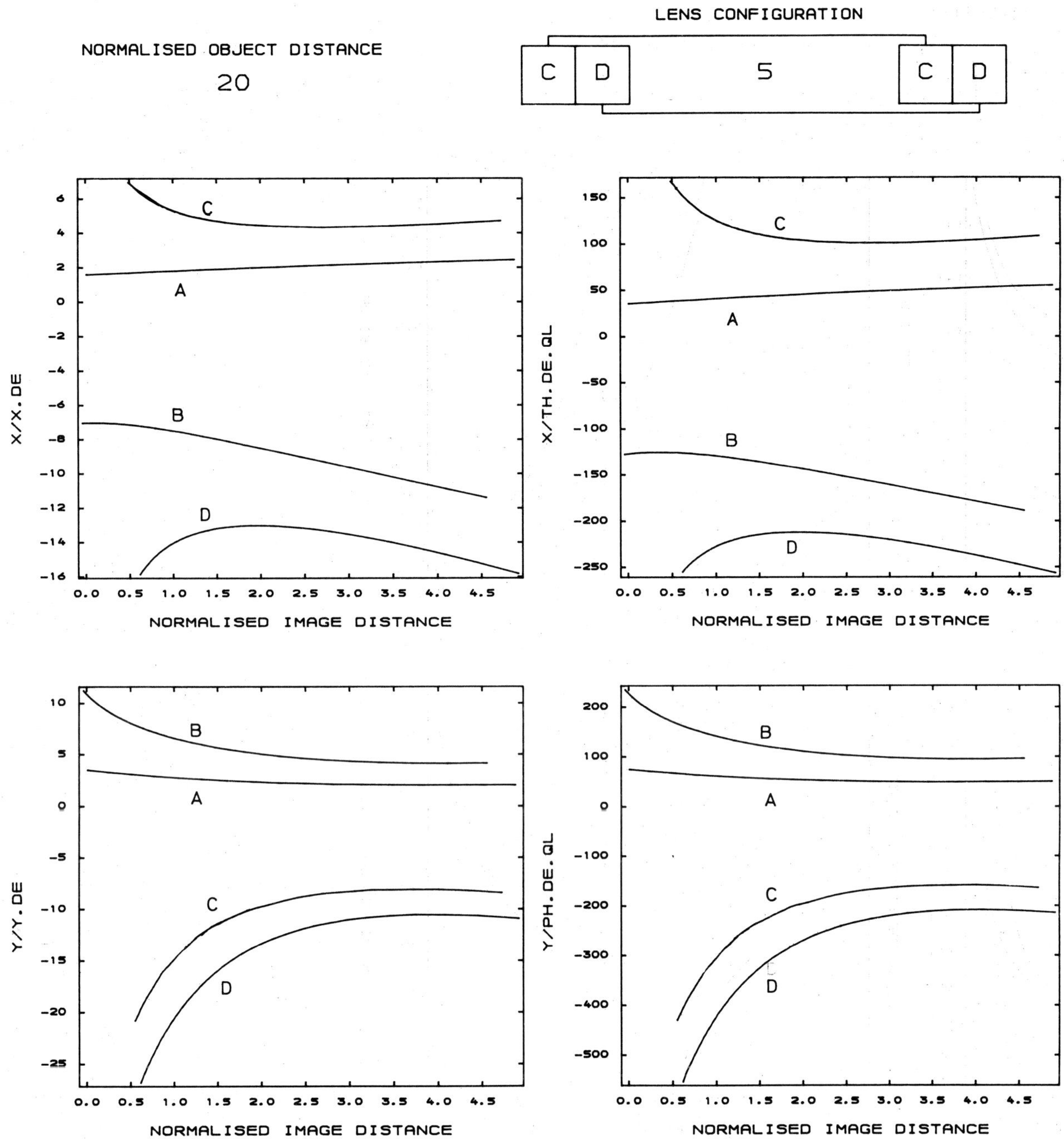
NORMALISED OBJECT DISTANCE
20
LENS CONFIGURATION
C
D
5
C
D
X/X.DE
X/TH.DE.QL
Y/Y.DE
Y/PH.DE.QL
A
B
C
D
NORMALISED IMAGE DISTANCE

Lens Configuration 20C0D5C0D

KL (QUAD 1)	KL (QUAD 2)	Normalised image distance	Normalised spherical aberration coefficients x/θ^3QL	$x/\theta\phi^2QL$	y/ϕ^3QL	$y/\theta^2\phi QL$
0.705	0.747	0.5	−5.5 (3)	−3.96 (4)	−2.60 (4)	−2.54 (4)
0.695	0.737	1.0	−6.2 (3)	−4.06 (4)	−2.46 (4)	−2.35 (4)
0.685	0.726	1.5	−4.8 (3)	−4.25 (4)	−1.87 (4)	−2.20 (4)
0.676	0.715	2.0	−5.6 (3)	−4.42 (4)	−1.75 (4)	−2.03 (4)
0.667	0.704	2.5	−5.2 (3)	−4.52 (4)	−1.79 (4)	−1.87 (4)
0.659	0.693	3.0	−7.0 (3)	−4.55 (4)	−1.52 (4)	−1.92 (4)
0.651	0.682	3.5	−6.4 (3)	−4.69 (4)	−1.46 (4)	−1.88 (4)
0.643	0.672	4.0	−6.8 (3)	−4.77 (4)	−1.47 (4)	−1.81 (4)
0.988	0.889	0.5	1.67 (5)	1.9 (4)	−1.60 (5)	−1.35 (5)
0.955	0.872	1.0	1.21 (5)	2.4 (4)	−1.12 (5)	−1.02 (5)
0.934	0.858	1.5	9.7 (4)	3.0 (4)	−9.7 (4)	−9.6 (4)
0.918	0.847	2.0	8.6 (4)	4.0 (4)	−7.9 (4)	−8.8 (4)
0.907	0.837	2.5	8.1 (4)	5.0 (4)	−7.4 (4)	−9.0 (4)
0.897	0.828	3.0	7.4 (4)	6.6 (4)	−6.8 (4)	−8.7 (4)
0.889	0.820	3.5	7.4 (4)	8.0 (4)	−6.5 (4)	−8.7 (4)
0.883	0.812	4.0	6.9 (4)	9.9 (4)	−6.2 (4)	−9.3 (4)
0.976	1.450	0.5	−3.0 (4)	−3.79 (6)	6.99 (7)	1.1 (4)
0.906	1.227	1.0	−1.8 (4)	−1.38 (6)	1.09 (7)	1.1 (4)
0.862	1.109	1.5	−2.3 (4)	−7.40 (5)	3.05 (6)	9.0 (3)
0.831	1.036	2.0	−1.4 (4)	−4.92 (5)	1.18 (6)	9.0 (3)
0.809	0.987	2.5	−1.9 (4)	−3.72 (5)	5.71 (5)	1.4 (4)
0.792	0.953	3.0	−1.5 (4)	−3.09 (5)	3.25 (5)	1.9 (4)
0.779	0.928	3.5	−1.7 (4)	−2.72 (5)	2.14 (5)	1.9 (4)
0.769	0.909	4.0	−1.9 (4)	−2.50 (5)	1.56 (5)	1.9 (4)
1.105	1.493	0.5	8.04 (5)	1.33 (8)	1.84 (8)	4.73 (6)
1.050	1.275	1.0	3.82 (5)	2.70 (7)	3.08 (7)	1.43 (6)
1.014	1.161	1.5	2.55 (5)	9.90 (6)	8.74 (6)	7.77 (5)
0.989	1.090	2.0	1.97 (5)	4.75 (6)	3.33 (6)	5.20 (5)
0.970	1.043	2.5	1.69 (5)	2.70 (6)	1.55 (6)	3.83 (5)
0.955	1.010	3.0	1.51 (5)	1.72 (6)	8.53 (5)	3.07 (5)
0.945	0.986	3.5	1.39 (5)	1.21 (6)	5.20 (5)	2.60 (5)
0.936	0.968	4.0	1.32 (5)	9.24 (5)	3.52 (5)	2.27 (5)

Lens Configuration 20C0D0D0C

KL (QUAD 1)	KL (QUAD 2)	Normalised image distance	Normalised spherical aberration coefficients			
			x/θ^3QL	$x/\theta\phi^2QL$	y/ϕ^3QL	$y/\theta^2\phi QL$
0.790	0.713	0.5	−5.5 (3)	−3.99 (4)	−2.02 (4)	−9.3 (3)
0.764	0.664	1.0	−5.0 (3)	−3.66 (4)	−1.83 (4)	−1.09 (4)
0.742	0.626	1.5	−4.2 (3)	−3.56 (4)	−2.02 (4)	−1.14 (4)
0.724	0.596	2.0	−4.8 (3)	−3.54 (4)	−1.79 (4)	−1.29 (4)
0.708	0.571	2.5	−3.6 (3)	−3.59 (4)	−2.07 (4)	−1.17 (4)
0.694	0.550	3.0	−3.9 (3)	−3.52 (4)	−1.83 (4)	−1.38 (4)
0.682	0.531	3.5	−4.9 (3)	−3.46 (4)	−2.01 (4)	−1.37 (4)
0.671	0.514	4.0	−5.5 (3)	−3.56 (4)	−1.87 (4)	−1.39 (4)

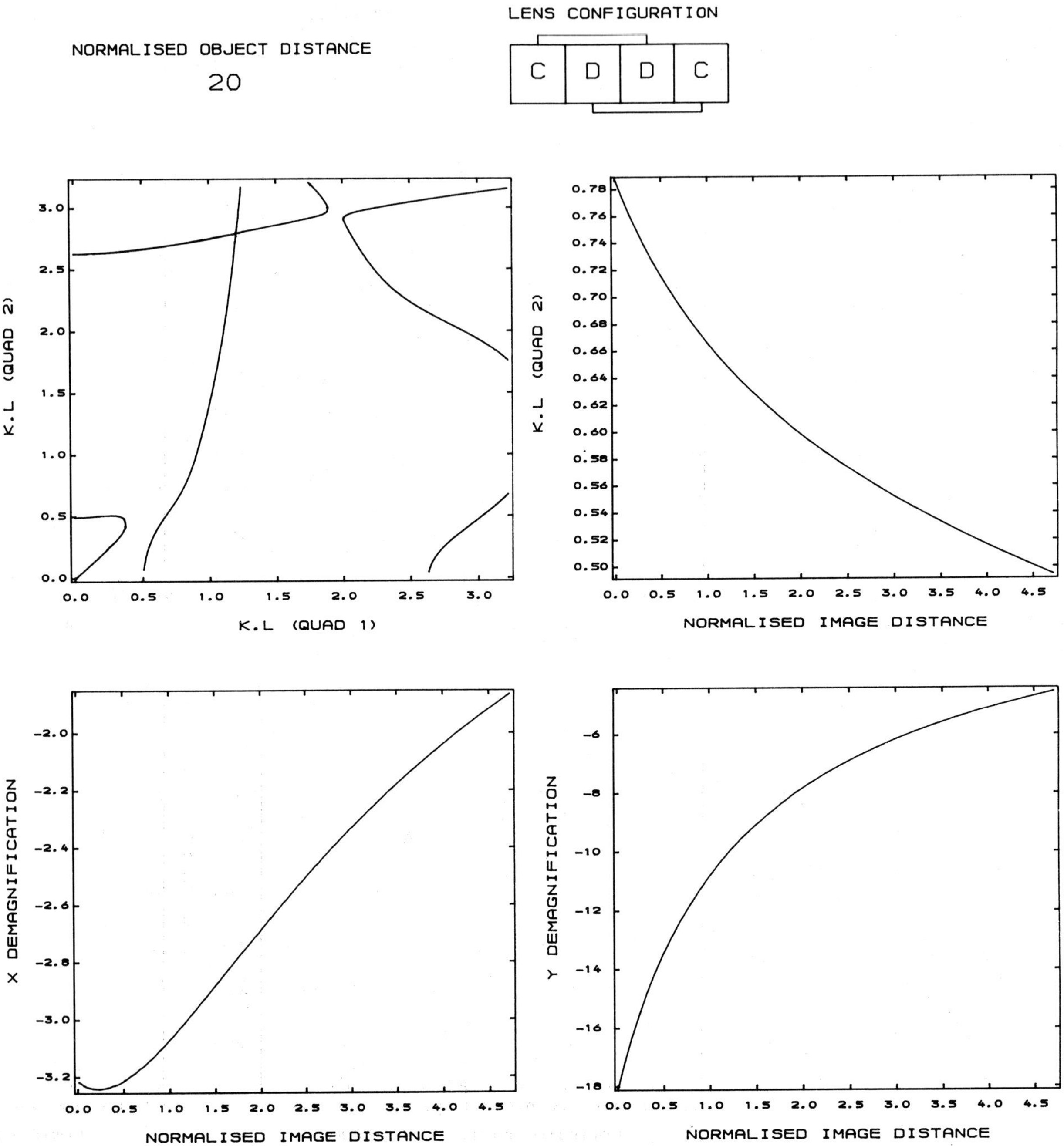

NORMALISED OBJECT DISTANCE
20
LENS CONFIGURATION
C
D
D
C
K.L (QUAD 2)
K.L (QUAD 1)
NORMALISED IMAGE DISTANCE
X DEMAGNIFICATION
Y DEMAGNIFICATION

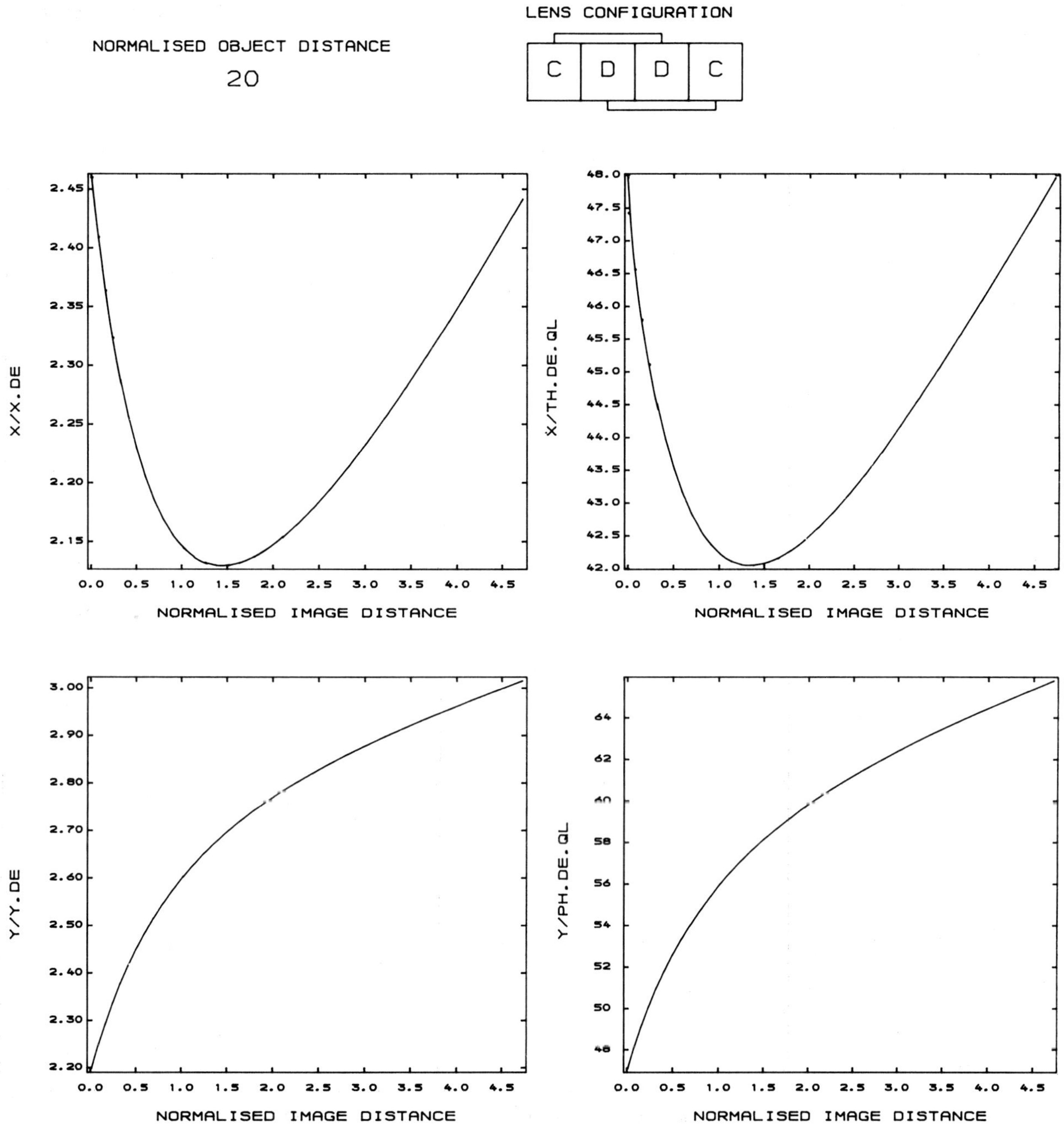

NORMALISED OBJECT DISTANCE
20
LENS CONFIGURATION
C D D C
X/X.DE
X/TH.DE.QL
Y/Y.DE
Y/PH.DE.QL
NORMALISED IMAGE DISTANCE
NORMALISED IMAGE DISTANCE
NORMALISED IMAGE DISTANCE
NORMALISED IMAGE DISTANCE

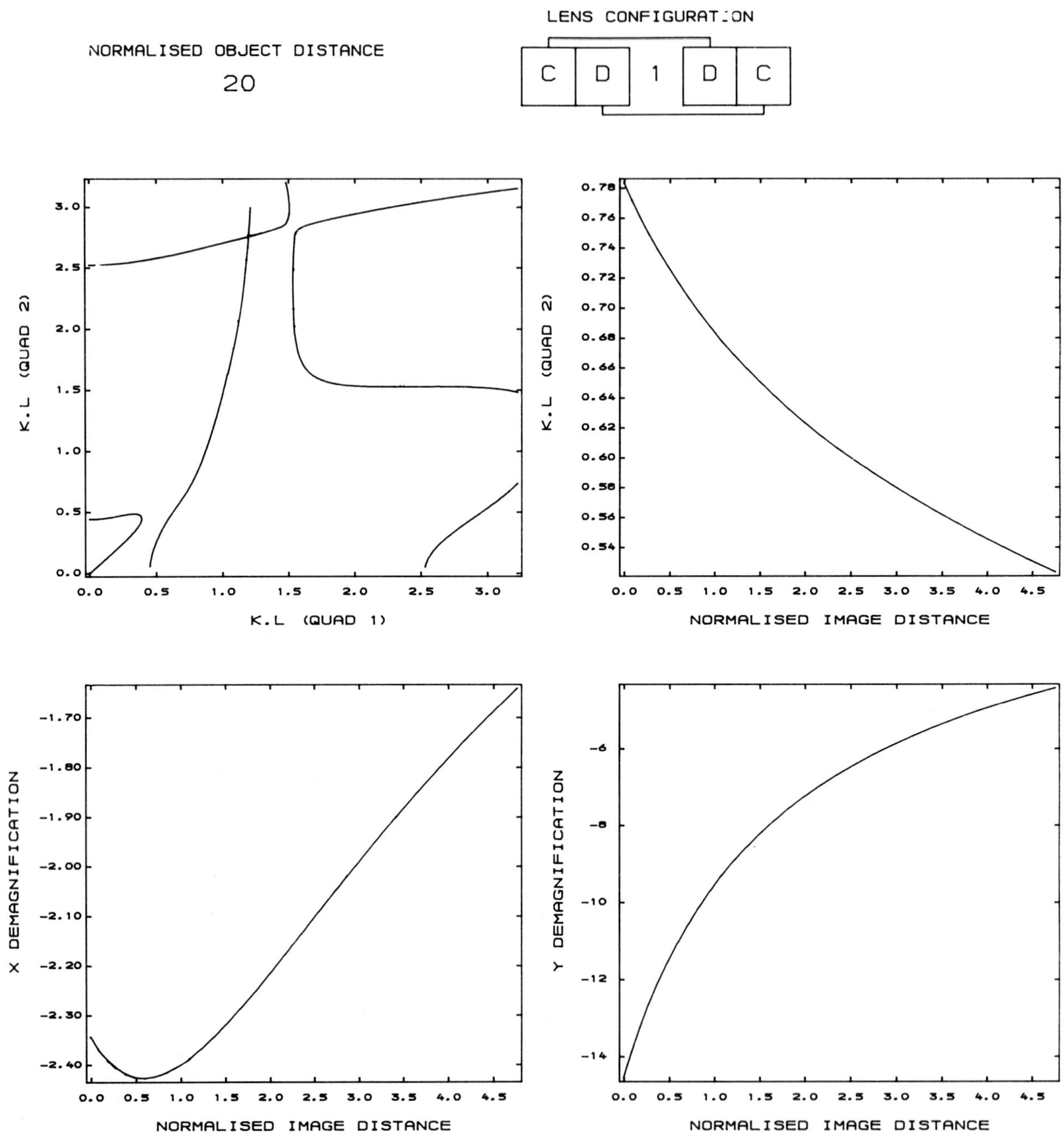
LENS CONFIGURATION
NORMALISED OBJECT DISTANCE
20
C
D
1
D
C
K.L (QUAD 2)
K.L (QUAD 1)
NORMALISED IMAGE DISTANCE
X DEMAGNIFICATION
Y DEMAGNIFICATION

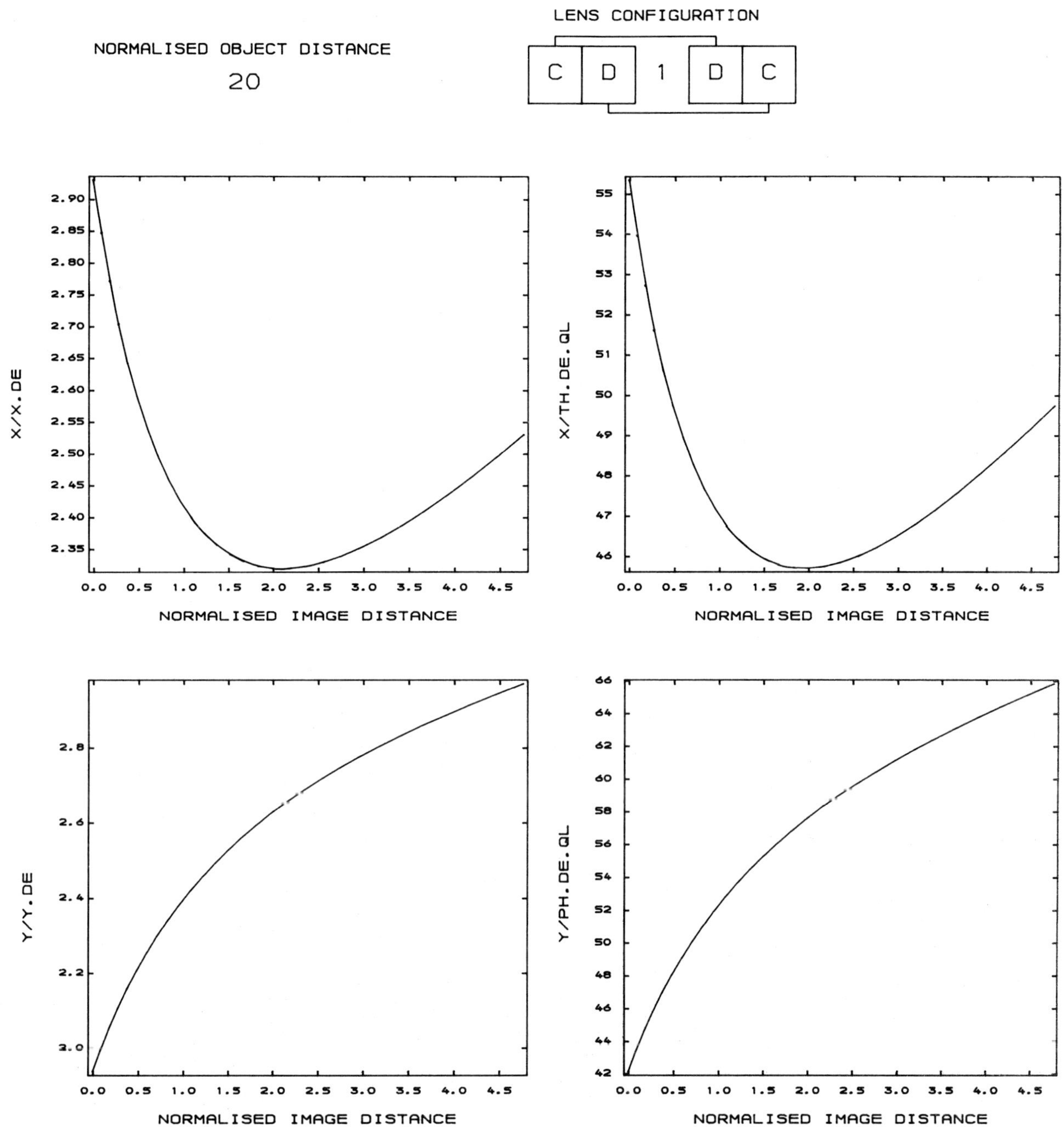
NORMALISED OBJECT DISTANCE
20
LENS CONFIGURATION
C D 1 D C
X/X.DE
X/TH.DE.QL
Y/Y.DE
Y/PH.DE.QL
NORMALISED IMAGE DISTANCE

Lens Configuration 20C0D1D0C

KL (QUAD 1)	KL (QUAD 2)	Normalised image distance	Normalised spherical aberration coefficients			
			x/θ^3QL	$x/\theta\phi^2QL$	y/ϕ^3QL	$y/\theta^2\phi QL$
0.757	0.723	0.5	−5.6 (3)	−4.35 (4)	−1.52 (4)	−9.3 (3)
0.735	0.681	1.0	−3.9 (3)	−3.91 (4)	−1.61 (4)	−1.06 (4)
0.716	0.648	1.5	−5.2 (3)	−3.59 (4)	−1.59 (4)	−1.08 (4)
0.699	0.621	2.0	−4.6 (3)	−3.49 (4)	−1.39 (4)	−1.32 (4)
0.684	0.598	2.5	−5.7 (3)	−3.36 (4)	−1.49 (4)	−1.29 (4)
0.672	0.577	3.0	−4.3 (3)	−3.50 (4)	−1.74 (4)	−1.12 (4)
0.660	0.559	3.5	−6.0 (3)	−3.33 (4)	−1.62 (4)	−1.31 (4)
0.650	0.543	4.0	−5.6 (3)	−3.45 (4)	−1.91 (4)	−1.37 (4)

Lens Configuration 20C0D2D0C

KL (QUAD 1)	KL (QUAD 2)	Normalised image distance	Normalised spherical aberration coefficients			
			x/θ^3QL	$x/\theta\phi^2$QL	y/ϕ^3QL	$y/\theta^2\phi$QL
0.736	0.715	0.5	−6.2 (3)	−5.00 (4)	−1.28 (4)	−8.8 (3)
0.717	0.684	1.0	−4.7 (3)	−4.30 (4)	−1.23 (4)	−1.10 (4)
0.700	0.655	1.5	−4.1 (3)	−3.99 (4)	−1.54 (4)	−1.07 (4)
0.685	0.631	2.0	−4.0 (3)	−3.79 (4)	−1.49 (4)	−1.10 (4)
0.672	0.601	2.5	−5.0 (3)	−3.69 (4)	−1.58 (4)	−1.09 (4)
0.660	0.591	3.0	−4.5 (3)	−3.61 (4)	−1.48 (4)	−1.22 (4)
0.649	0.574	3.5	−5.8 (3)	−3.46 (4)	−1.44 (4)	−1.26 (4)
0.639	0.559	4.0	−3.1 (3)	−3.78 (4)	−1.77 (4)	−1.08 (4)
1.327	1.567	0.5	7.11 (7)	7.35 (6)	5.49 (6)	1.16 (8)
1.376	1.375	1.0	1.02 (8)	3.69 (6)	2.12 (6)	8.92 (7)
1.540	1.307	1.5	5.51 (8)	5.45 (6)	4.12 (6)	2.34 (8)

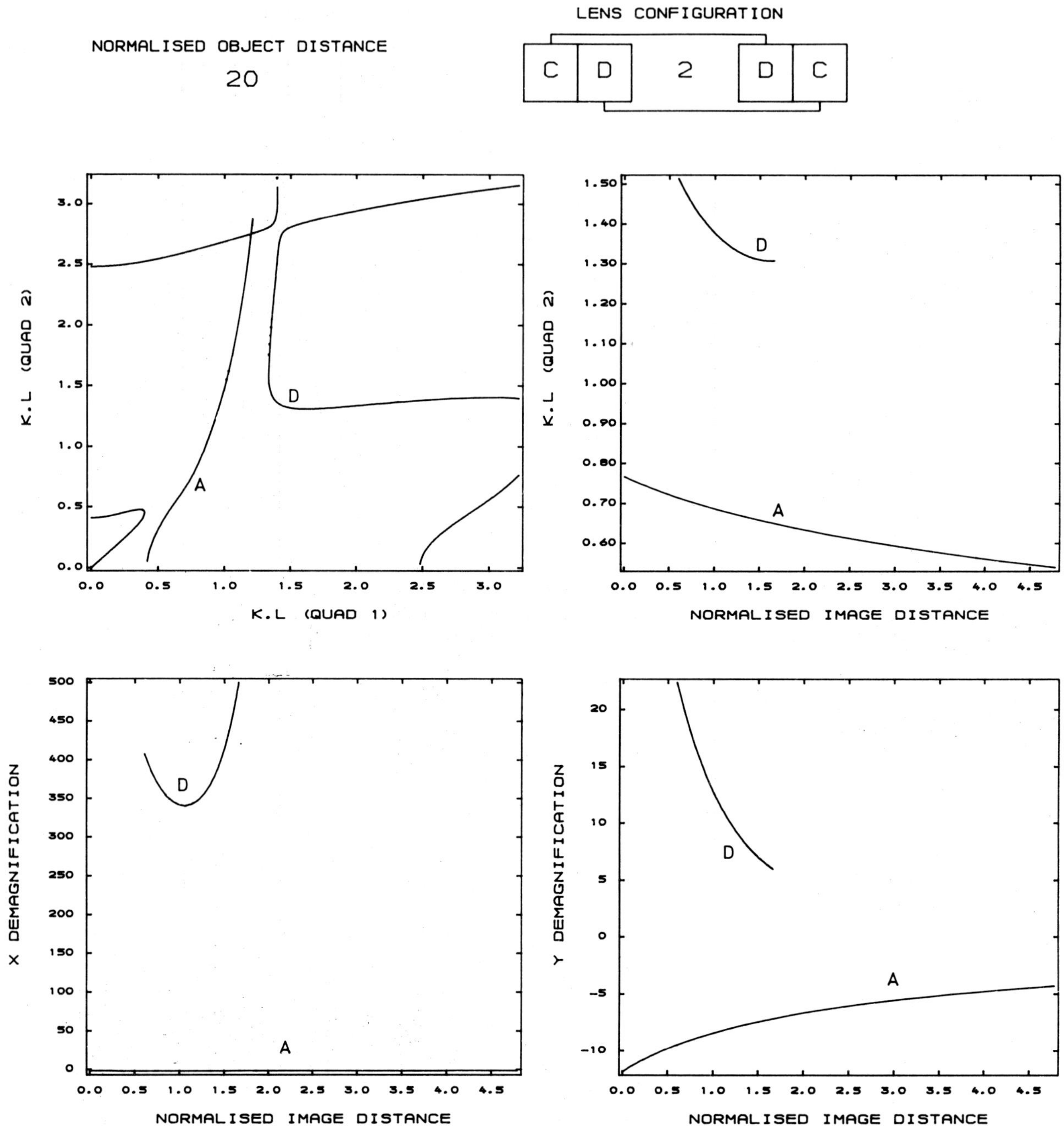
NORMALISED OBJECT DISTANCE
20
LENS CONFIGURATION
C
D
2
D
C
K.L (QUAD 2)
K.L (QUAD 1)
D
A
NORMALISED IMAGE DISTANCE
X DEMAGNIFICATION
Y DEMAGNIFICATION

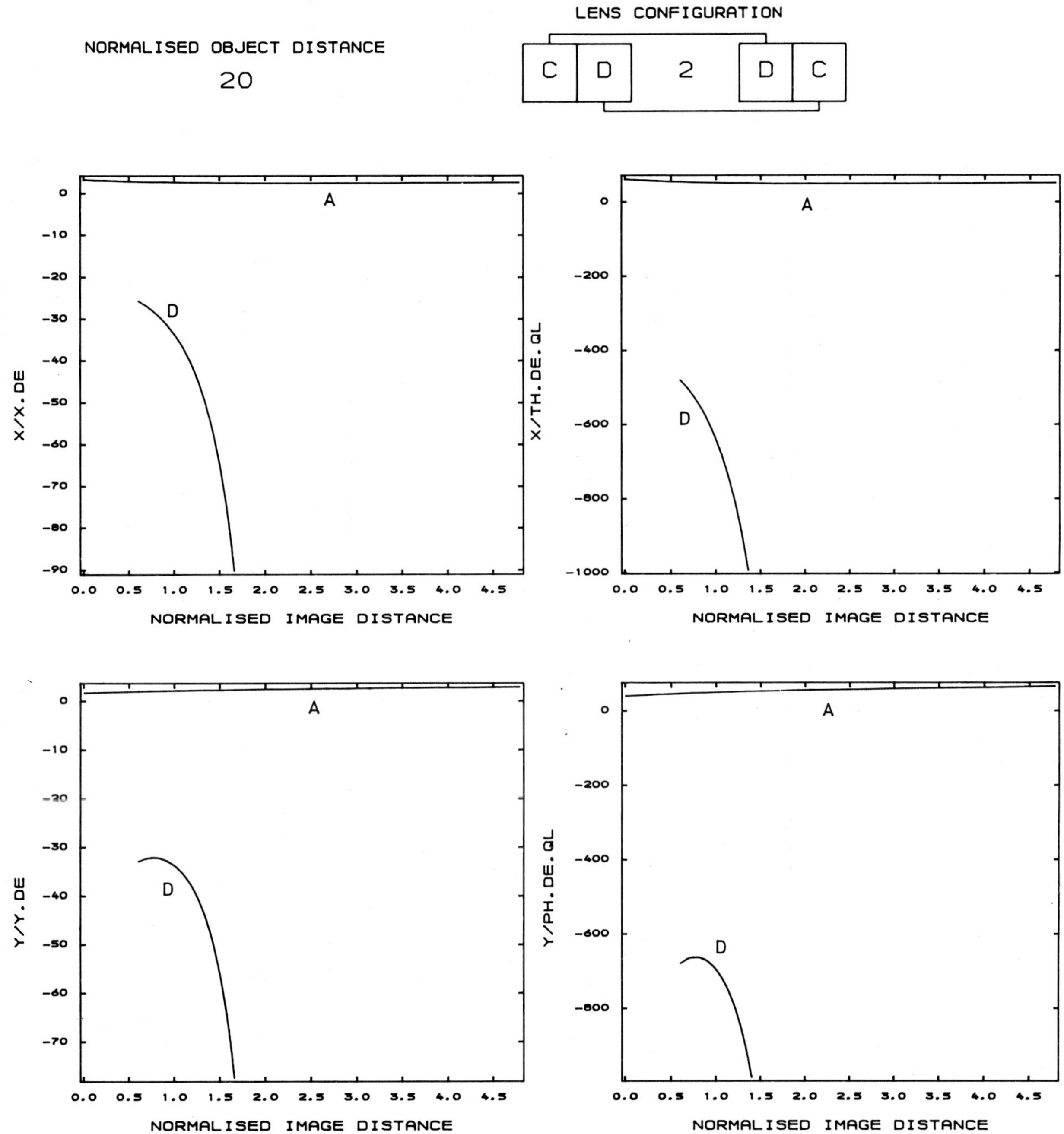
LENS CONFIGURATION
NORMALISED OBJECT DISTANCE
20
C
D
2
D
C
X/X.DE
X/TH.DE.QL
Y/Y.DE
Y/PH.DE.QL
A
D
NORMALISED IMAGE DISTANCE
0.0 0.5 1.0 1.5 2.0 2.5 3.0 3.5 4.0 4.5
0 -10 -20 -30 -40 -50 -60 -70 -80 -90
0 -200 -400 -600 -800 -1000

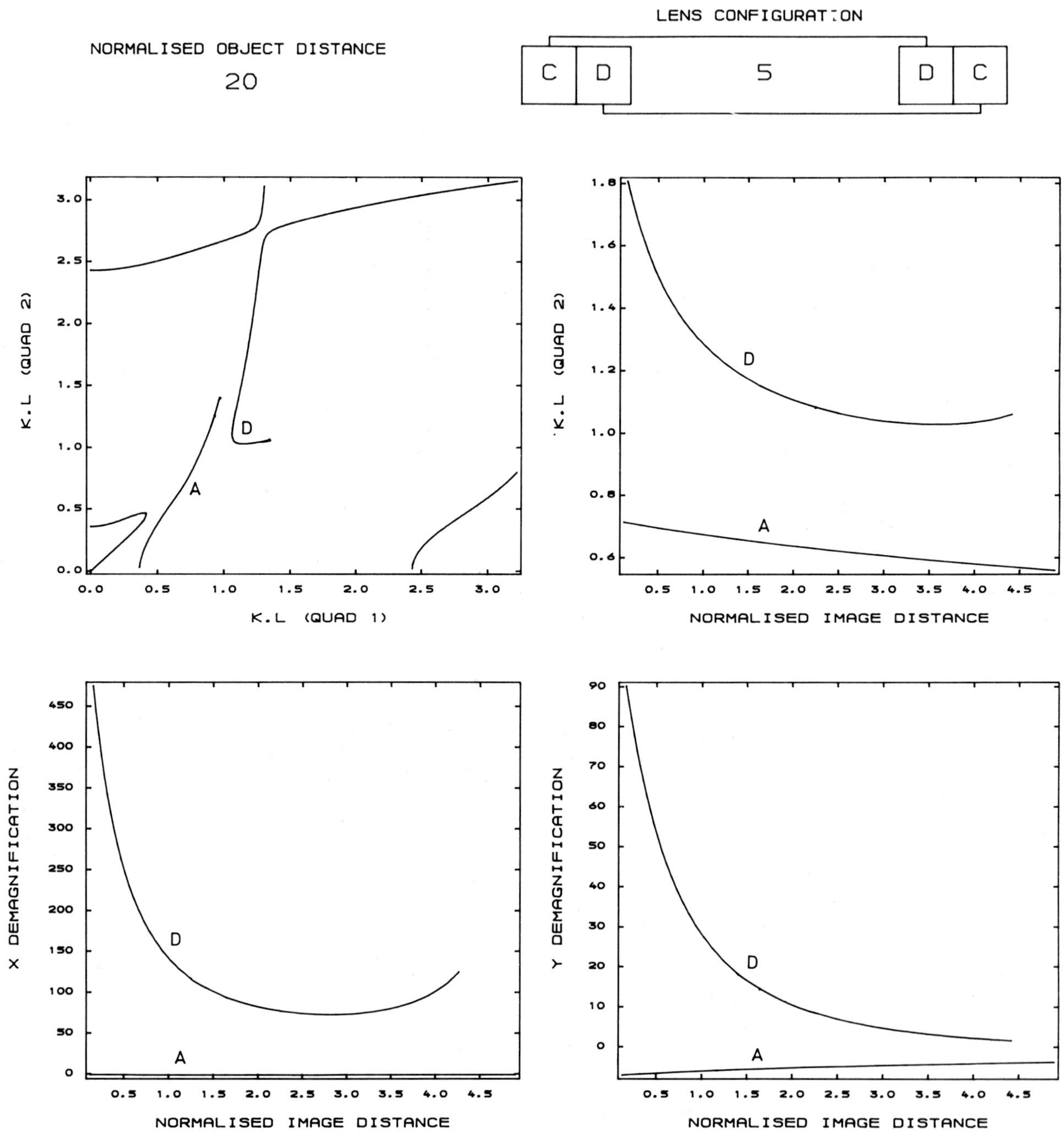
NORMALISED OBJECT DISTANCE
20
LENS CONFIGURATION
C D 5 D C
K.L (QUAD 1)
K.L (QUAD 2)
D
A
NORMALISED IMAGE DISTANCE
K.L (QUAD 2)
D
A
X DEMAGNIFICATION
D
A
NORMALISED IMAGE DISTANCE
Y DEMAGNIFICATION
D
A
NORMALISED IMAGE DISTANCE

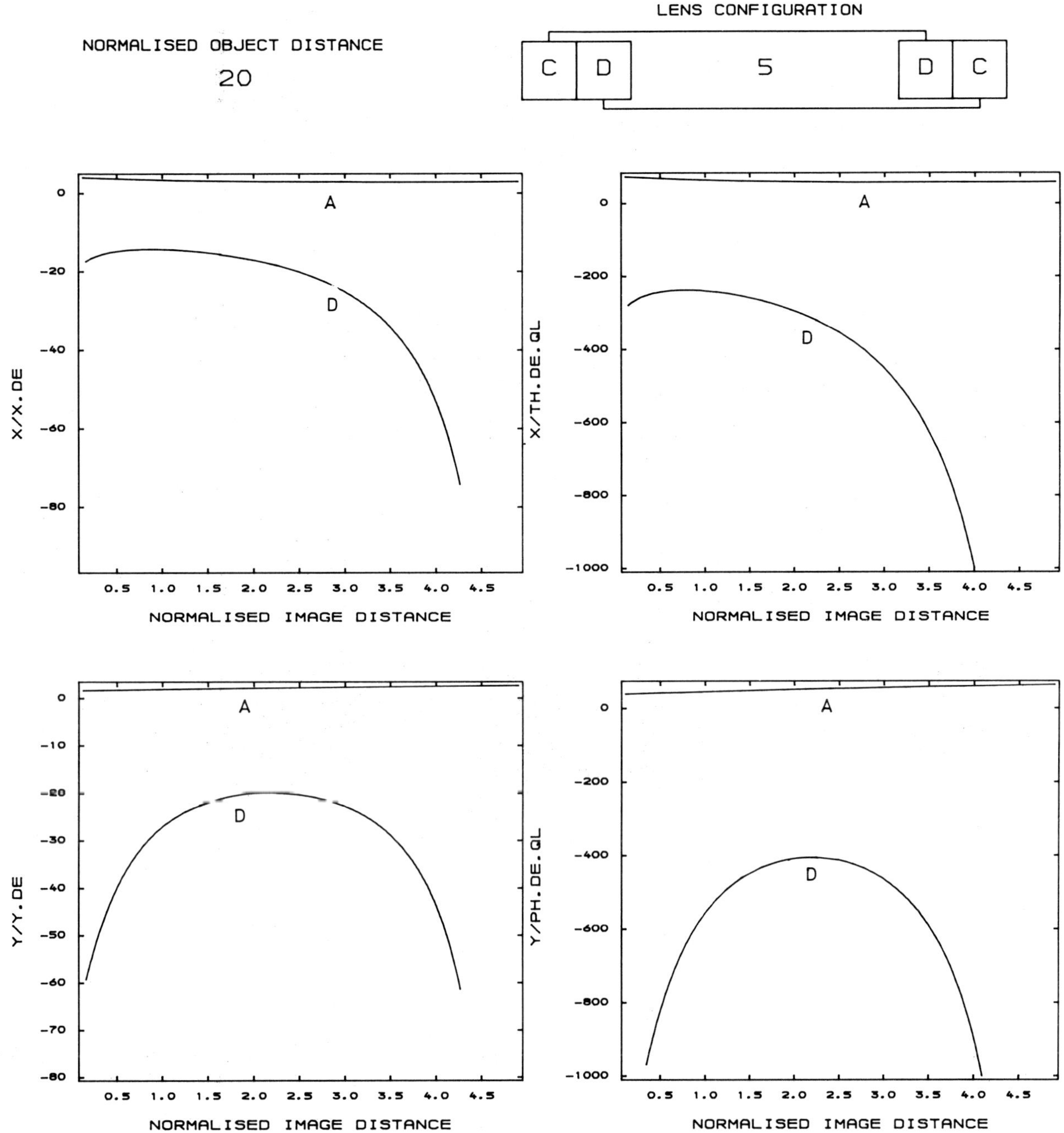
NORMALISED OBJECT DISTANCE
20
LENS CONFIGURATION
C
D
5
D
C
X/X.DE
X/TH.DE.QL
Y/Y.DE
Y/PH.DE.QL
A
D
NORMALISED IMAGE DISTANCE
0.5
1.0
1.5
2.0
2.5
3.0
3.5
4.0
4.5
0
-20
-40
-60
-80
-200
-400
-600
-800
-1000
-10
-30
-50
-70

Lens Configuration 20C0D5D0C

KL (QUAD 1)	KL (QUAD 2)	Normalised image distance	Normalised spherical aberration coefficients			
			$x/\theta^3 QL$	$x/\theta\phi^2 QL$	$y/\phi^3 QL$	$y/\theta^2\phi QL$
0.695	0.693	0.5	−9.2 (3)	−6.27 (4)	−1.18 (4)	−1.17 (4)
0.682	0.672	1.0	−9.0 (3)	−5.63 (4)	−1.28 (4)	−1.13 (4)
0.670	0.652	1.5	−6.0 (3)	−5.09 (4)	−1.17 (4)	−1.26 (4)
0.660	0.635	2.0	−6.9 (3)	−4.68 (4)	−1.32 (4)	−1.17 (4)
0.649	0.619	2.5	−7.3 (3)	−4.52 (4)	−1.24 (4)	−1.21 (4)
0.640	0.604	3.0	−7.4 (3)	−4.28 (4)	−1.39 (4)	−1.17 (4)
0.631	0.591	3.5	−6.5 (3)	−4.40 (4)	−1.51 (4)	−9.9 (3)
0.622	0.578	4.0	−5.9 (3)	−4.29 (4)	−1.30 (4)	−1.14 (4)
1.124	1.497	0.5	1.11 (7)	1.18 (7)	2.26 (7)	5.50 (7)
1.081	1.281	1.0	5.83 (6)	4.23 (6)	4.92 (6)	2.07 (7)
1.061	1.169	1.5	5.04 (6)	2.11 (6)	1.66 (6)	1.26 (7)
1.056	1.102	2.0	5.62 (6)	1.31 (6)	7.74 (5)	1.01 (7)
1.064	1.061	2.5	7.72 (6)	9.69 (5)	4.98 (5)	9.8 (6)
1.086	1.036	3.0	1.32 (7)	7.55 (5)	4.55 (5)	1.13 (7)
1.129	1.026	3.5	2.94 (7)	6.96 (5)	6.11 (5)	1.60 (7)
1.208	1.032	4.0	9.99 (7)	8.19 (5)	1.24 (6)	3.12 (7)

Lens Configuration 20C0D0C0D

KL (QUAD 1)	KL (QUAD 3)	Normalised image distance	Normalised spherical aberration coefficients			
			x/θ^3QL	$x/\theta\phi^2QL$	y/ϕ^3QL	$y/\theta^2\phi QL$
2.148	1.694	0.5	6.48 (7)	7.45 (8)	2.67 (8)	7.05 (7)
1.662	1.608	1.0	5.48 (6)	8.74 (6)	1.88 (6)	2.47 (6)
1.617	1.597	1.5	5.15 (6)	7.31 (6)	2.56 (6)	3.16 (6)
1.603	1.592	2.0	5.69 (6)	7.89 (6)	3.52 (6)	4.19 (6)
1.595	1.588	2.5	6.38 (6)	8.85 (6)	4.54 (6)	5.25 (6)
1.591	1.586	3.0	7.17 (6)	9.89 (6)	5.54 (6)	6.36 (6)
1.588	1.584	3.5	7.94 (6)	1.09 (7)	6.57 (6)	7.45 (6)
1.586	1.583	4.0	8.88 (6)	1.20 (7)	7.56 (6)	8.56 (6)
1.145	1.691	0.4	4.9 (4)	2.26 (6)	−2.15 (6)	−6.0 (4)

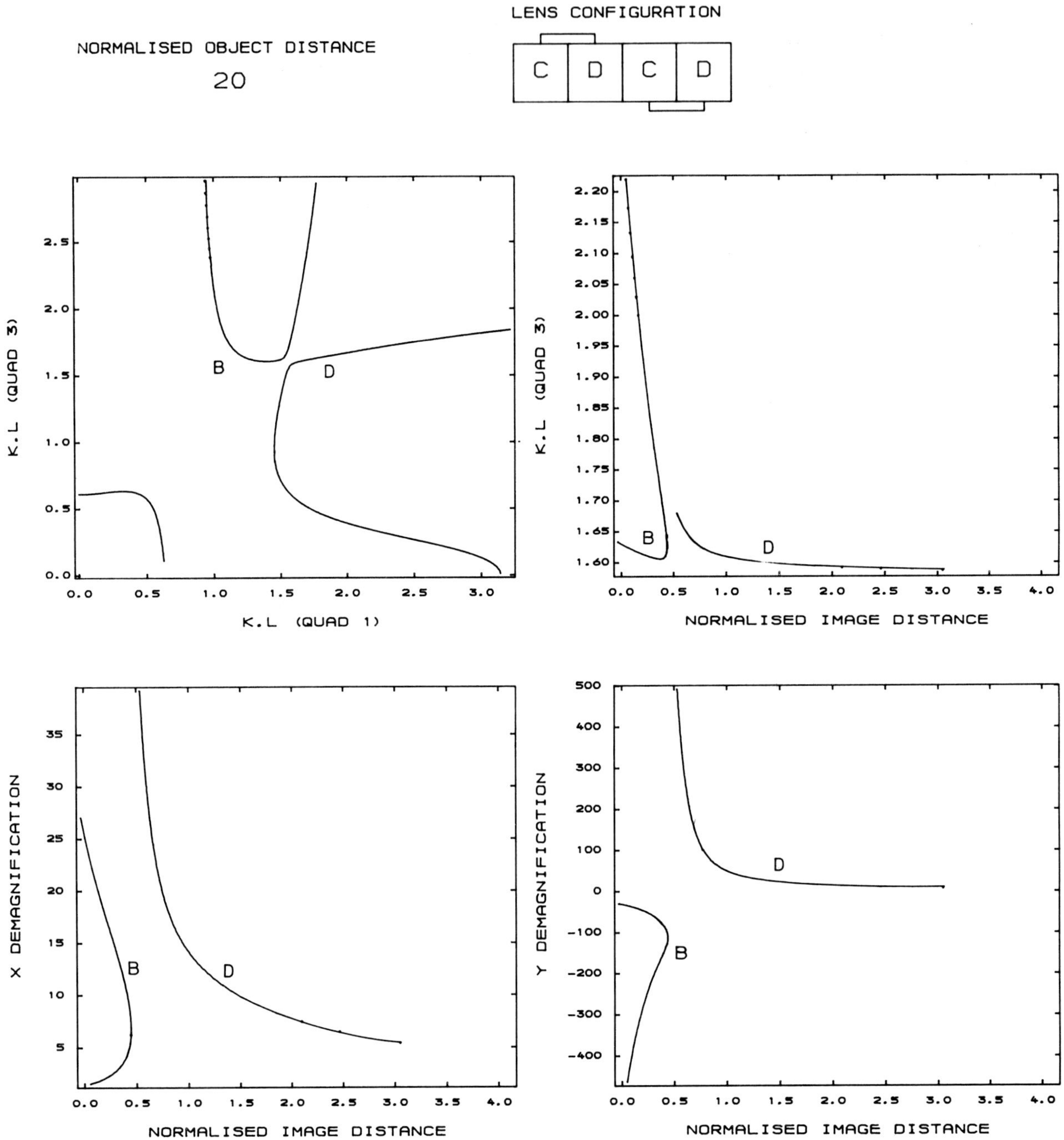

LENS CONFIGURATION
NORMALISED OBJECT DISTANCE
20
C
D
C
D
K.L (QUAD 3)
K.L (QUAD 1)
B
D
NORMALISED IMAGE DISTANCE
X DEMAGNIFICATION
Y DEMAGNIFICATION

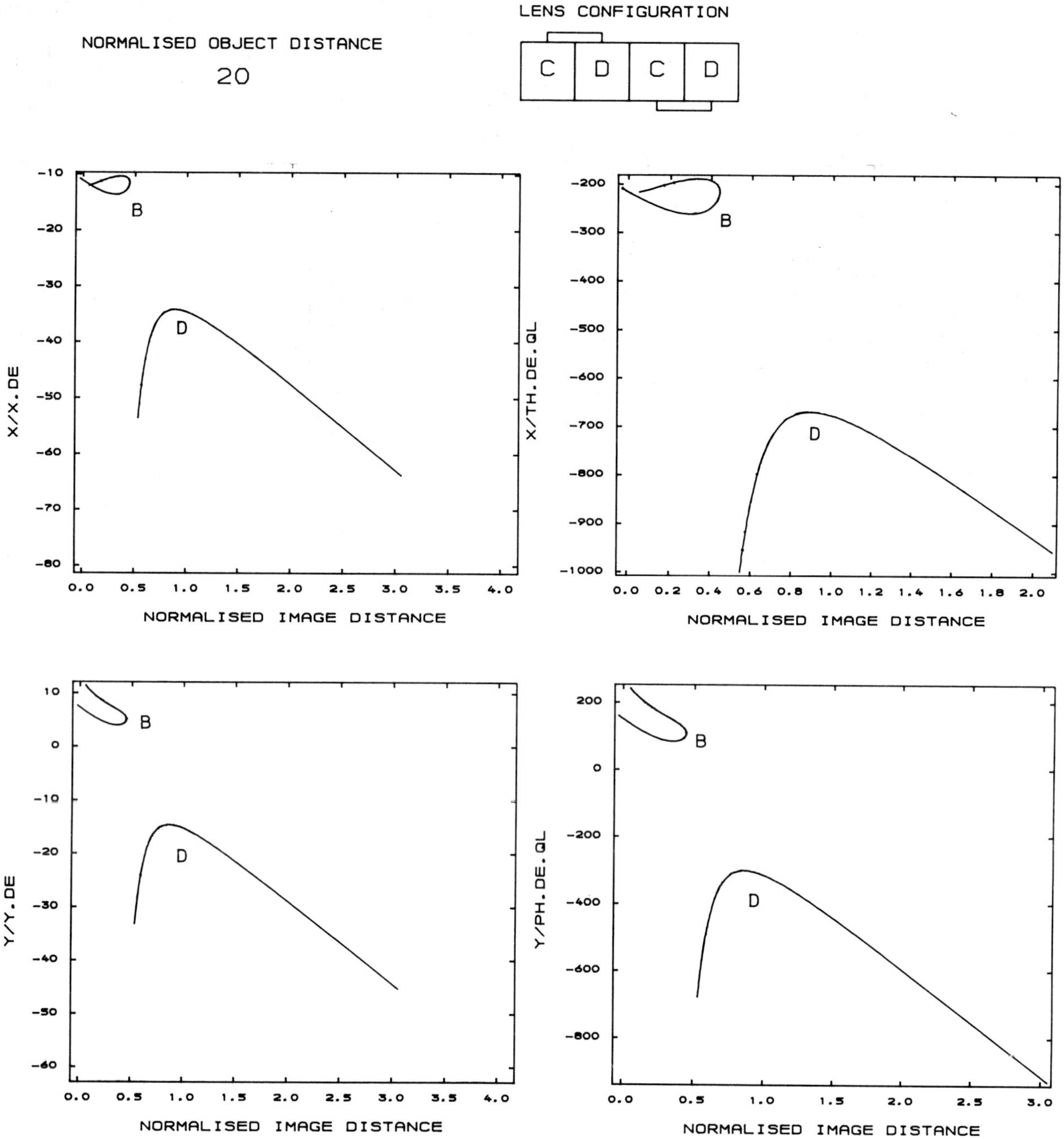
NORMALISED OBJECT DISTANCE
20
LENS CONFIGURATION
C
D
C
D
X/X.DE
X/TH.DE.QL
Y/Y.DE
Y/PH.DE.QL
B
D
NORMALISED IMAGE DISTANCE

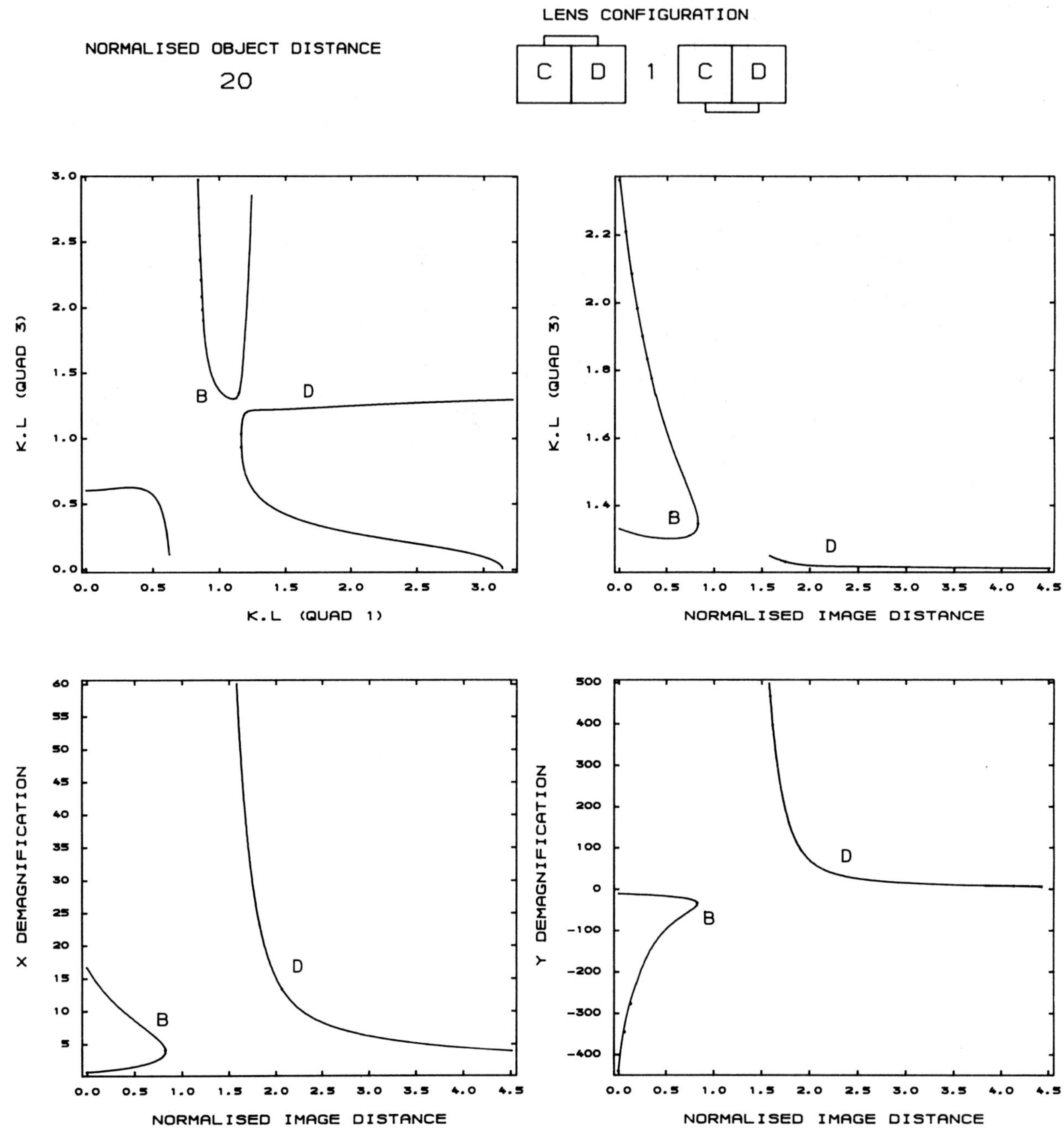
NORMALISED OBJECT DISTANCE
20
LENS CONFIGURATION
C
D
1
C
D
K.L (QUAD 3)
K.L (QUAD 1)
B
D
NORMALISED IMAGE DISTANCE
X DEMAGNIFICATION
Y DEMAGNIFICATION

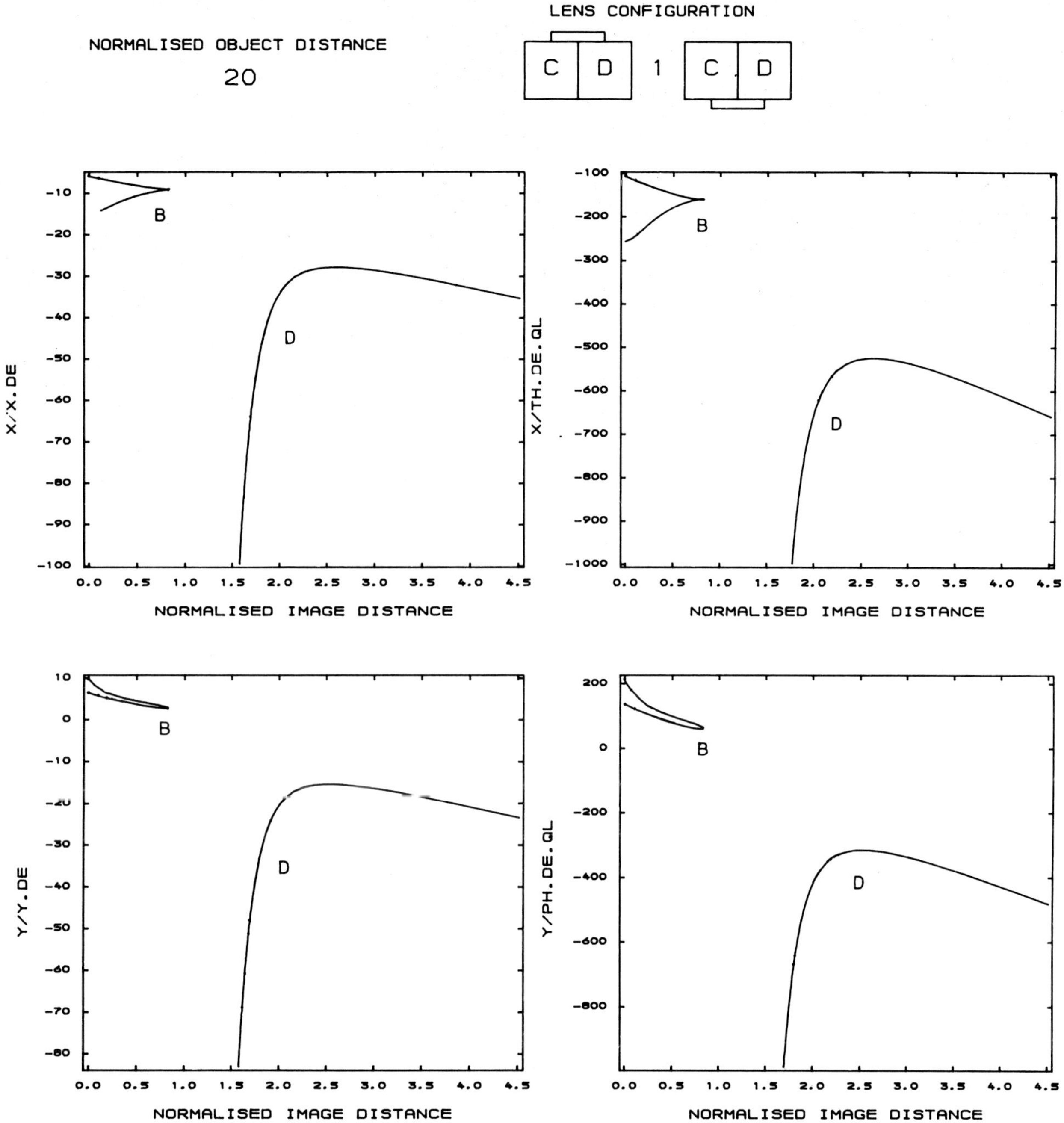
NORMALISED OBJECT DISTANCE
20
LENS CONFIGURATION
C
D
1
C
D
X/X.DE
X/TH.DE.QL
Y/Y.DE
Y/PH.DE.QL
B
D
NORMALISED IMAGE DISTANCE

Lens Configuration 20C0D1C0D

KL (QUAD 1)	KL (QUAD 3)	Normalised image distance	Normalised spherical aberration coefficients			
			x/θ^3QL	$x/\theta\phi^2$QL	y/ϕ^3QL	$y/\theta^2\phi$QL
2.246	1.259	1.5	*	*	*	*
1.450	1.220	2.0	4.32 (6)	1.57 (7)	4.62 (6)	3.31 (6)
1.309	1.217	2.5	1.43 (6)	2.77 (6)	7.64 (5)	8.69 (5)
1.268	1.215	3.0	1.07 (6)	1.64 (6)	5.69 (5)	6.43 (5)
1.248	1.214	3.5	1.00 (6)	1.35 (6)	5.84 (5)	6.15 (5)
1.237	1.213	4.0	9.82 (5)	1.25 (6)	6.33 (5)	6.25 (5)
1.103	1.299	0.5	6.6 (4)	1.47 (5)	−1.09 (5)	−8.6 (4)
0.907	1.614	0.5	1.8 (4)	4.97 (5)	−7.06 (5)	−8.0 (3)

Lens Configuration 20C0D2COD

KL (QUAD 1)	KL (QUAD 3)	Normalised image distance	Normalised spherical aberration coefficients			
			x/θ^3QL	$x/\theta\phi^2$QL	y/ϕ^3QL	$y/\theta^2\phi$QL
2.354	1.090	2.5	*	*	*	*
1.471	1.067	3.0	1.33 (7)	7.93 (7)	2.73 (7)	1.75 (7)
1.237	1.066	3.5	1.70 (6)	5.40 (6)	1.54 (6)	1.59 (6)
1.165	1.068	4.0	8.41 (5)	1.91 (5)	5.67 (5)	6.99 (5)
0.976	1.197	0.5	2.7 (4)	5.6 (4)	−8.0 (4)	−5.3 (4)
0.931	1.213	1.0	1.7 (4)	1.08 (5)	−4.2 (4)	−2.6 (4)
0.882	1.296	1.0	1.3 (4)	1.68 (5)	−5.9 (4)	−1.4 (4)
0.826	1.617	0.5	2.2 (4)	3.14 (5)	−4.4 (5)	−4.0 (3)

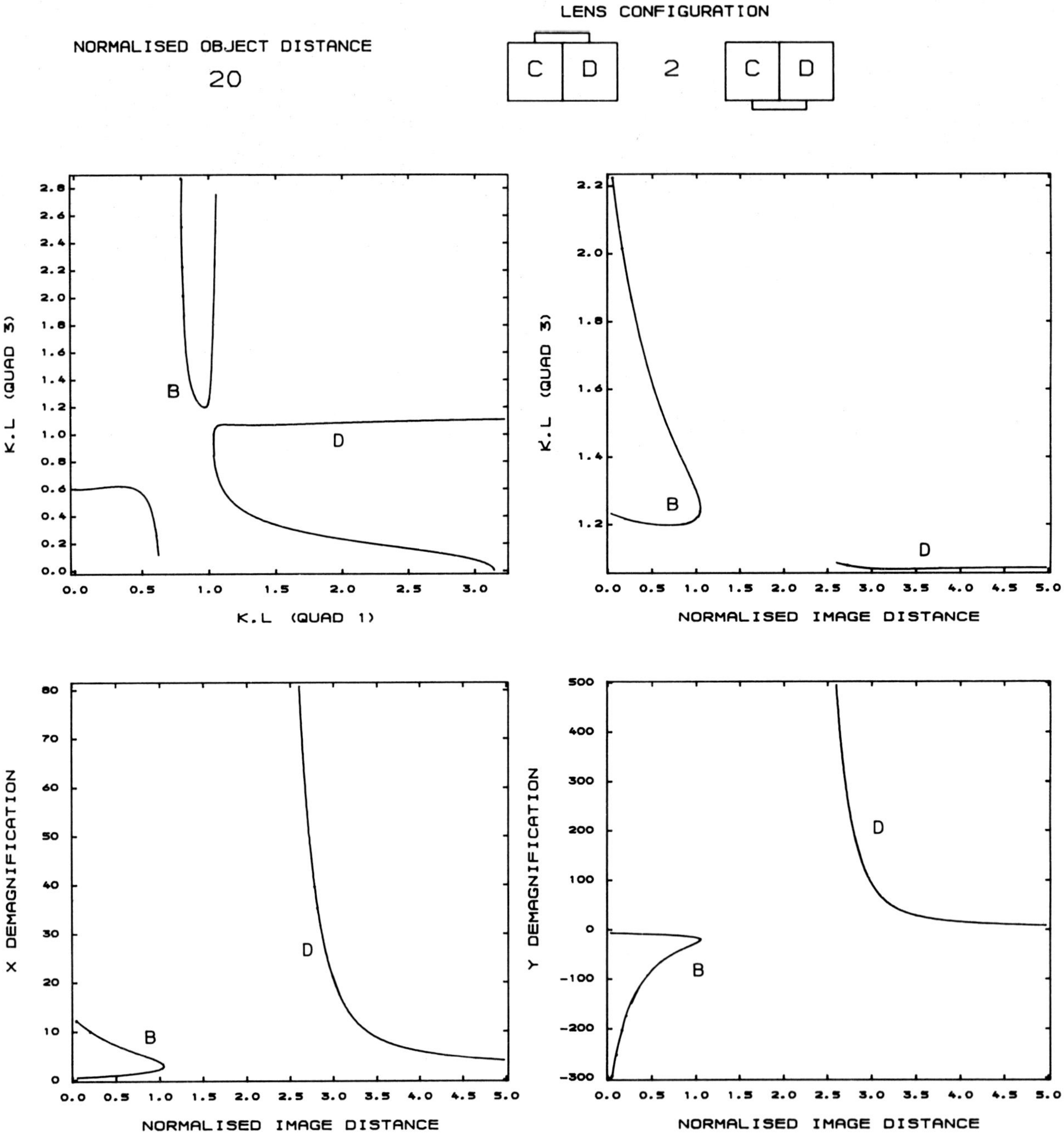
LENS CONFIGURATION
NORMALISED OBJECT DISTANCE
20
C
D
2
C
D
K.L (QUAD 3)
K.L (QUAD 1)
B
D
NORMALISED IMAGE DISTANCE
X DEMAGNIFICATION
Y DEMAGNIFICATION

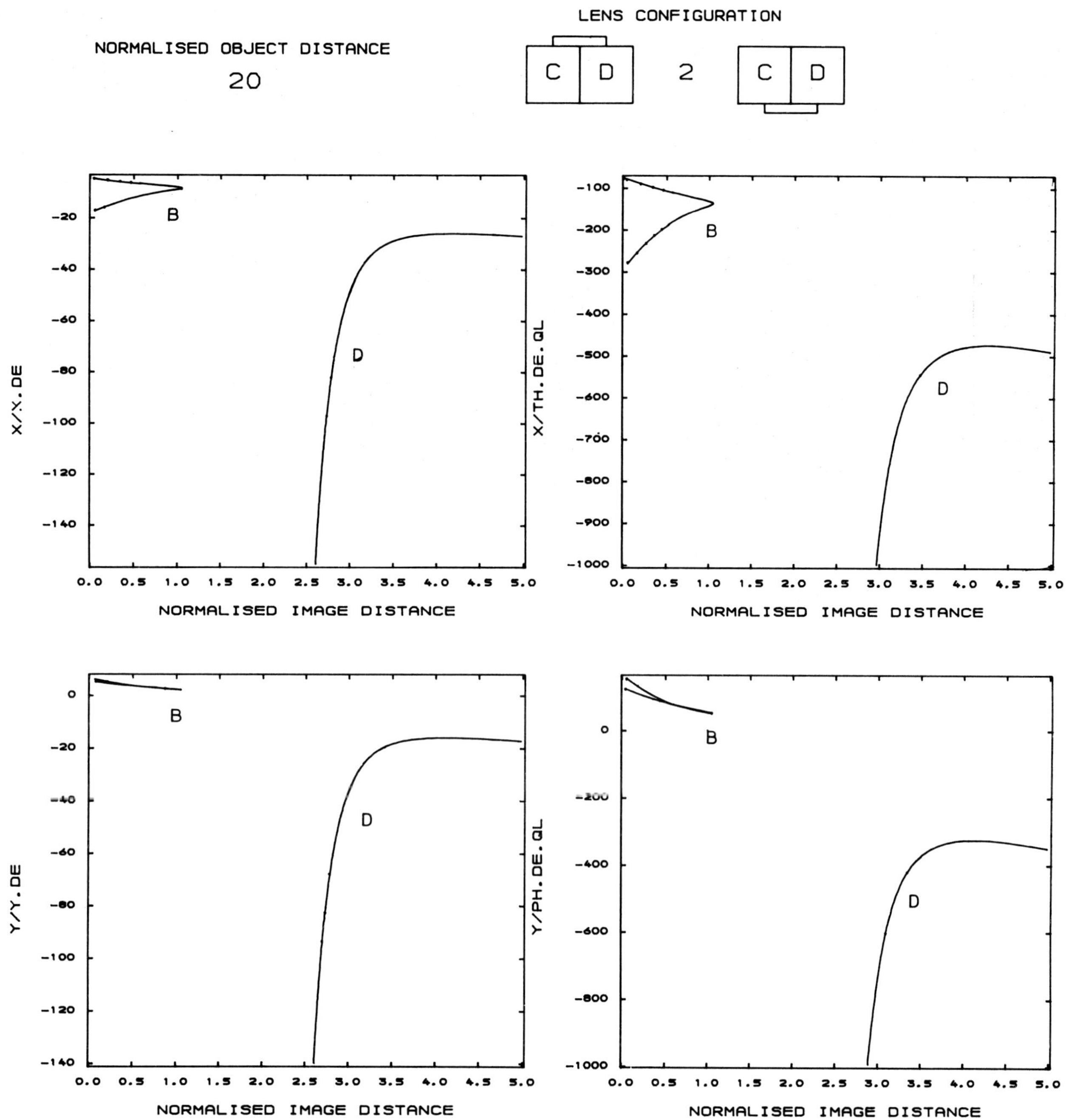

NORMALISED OBJECT DISTANCE
20
LENS CONFIGURATION
C D 2 C D
X/X.DE
X/TH.DE.QL
Y/Y.DE
Y/PH.DE.QL
B
D
NORMALISED IMAGE DISTANCE

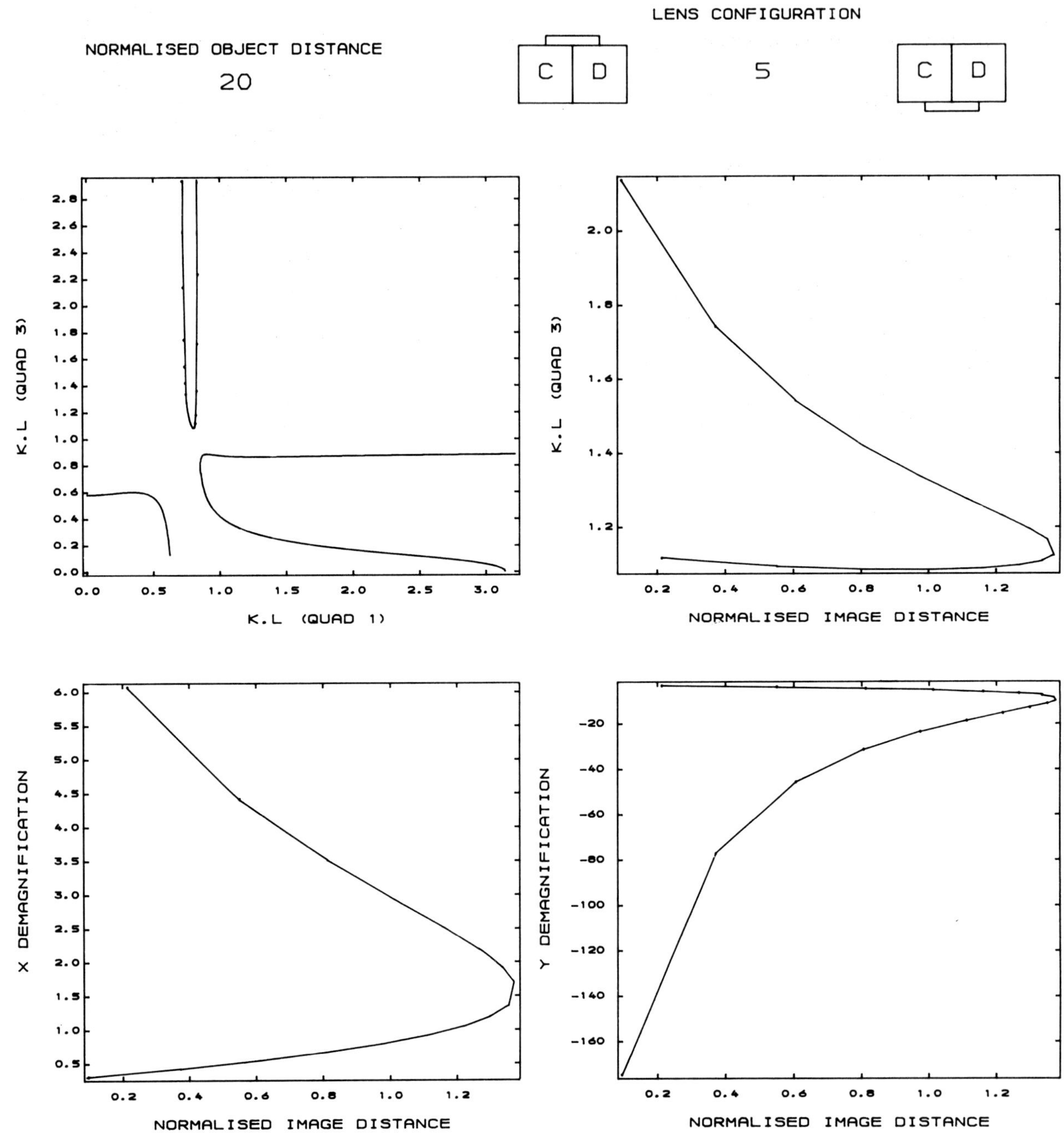
NORMALISED OBJECT DISTANCE
20
LENS CONFIGURATION
C D
5
C D
K.L (QUAD 3)
K.L (QUAD 1)
NORMALISED IMAGE DISTANCE
X DEMAGNIFICATION
Y DEMAGNIFICATION

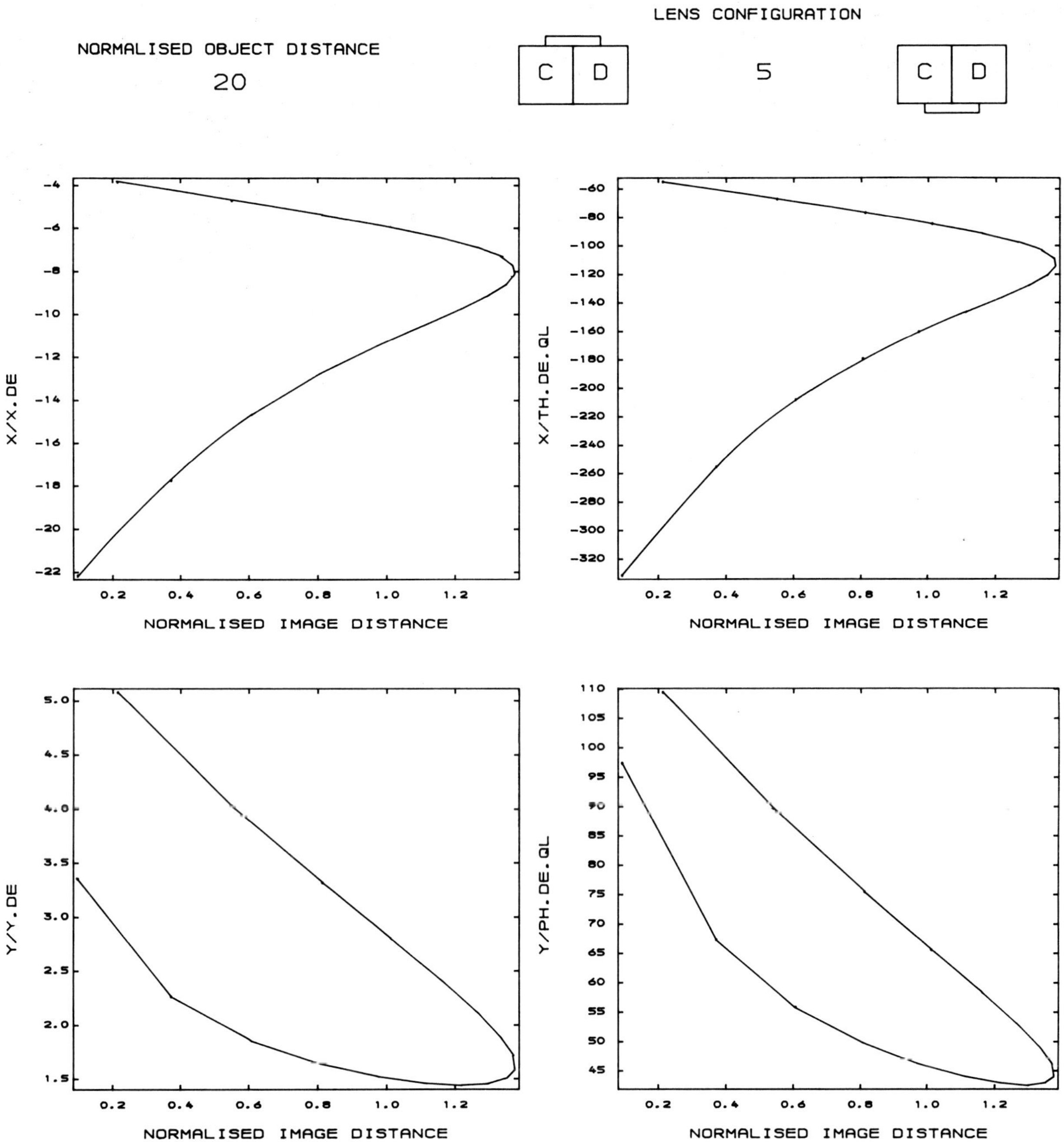
LENS CONFIGURATION
NORMALISED OBJECT DISTANCE
20
C D
5
C D
X/X.DE
X/TH.DE.QL
Y/Y.DE
Y/PH.DE.QL
NORMALISED IMAGE DISTANCE
NORMALISED IMAGE DISTANCE
NORMALISED IMAGE DISTANCE
NORMALISED IMAGE DISTANCE

Lens Configuration 20C0D5C0D

KL (QUAD 1)	KL (QUAD 3)	Normalised image distance	Normalised spherical aberration coefficients			
			x/θ^3QL	$x/\theta\phi^2QL$	y/ϕ^3QL	$y/\theta^2\phi QL$
3.785	0.890	5	*	*	*	*
0.808	1.092	0.5	6.0 (3)	3.1 (4)	−5.5 (4)	−3.6 (4)
0.797	1.080	1.0	8.0 (3)	4.6 (4)	−3.4 (4)	−2.7 (4)
0.743	1.322	1.0	1.9 (4)	1.43 (5)	−3.2 (4)	−6.0 (3)
0.729	1.620	0.5	2.5 (4)	2.39 (5)	−1.47 (5)	−2.0 (3)

Lens Configuration $20\overline{C0D}0\overline{D0C}$

KL (QUAD 1)	KL (QUAD 3)	Normalised image distance	Normalised spherical aberration coefficients			
			x/θ^3QL	$x/\theta\phi^2QL$	y/ϕ^3QL	$y/\theta^2\phi QL$
0.814	0.668	0.5	−4.7 (3)	−4.61 (4)	−2.10 (4)	−1.23 (4)
0.796	0.608	1.0	−5.1 (3)	−4.54 (4)	−2.25 (4)	−1.32 (4)
0.779	0.565	1.5	−6.2 (3)	−4.47 (4)	−1.92 (4)	−1.58 (4)
0.765	0.531	2.0	−5.8 (3)	−4.49 (4)	−2.11 (4)	−1.60 (4)
0.752	0.503	2.5	−5.7 (3)	−4.60 (4)	−2.31 (4)	−1.74 (4)
0.742	0.479	3.0	−5.4 (3)	−4.59 (4)	−1.91 (4)	−1.95 (4)
0.732	0.458	3.5	−5.6 (3)	−4.84 (4)	−2.04 (4)	−1.97 (4)
0.723	0.440	4.0	−5.3 (3)	−4.79 (4)	−1.89 (4)	−2.22 (4)

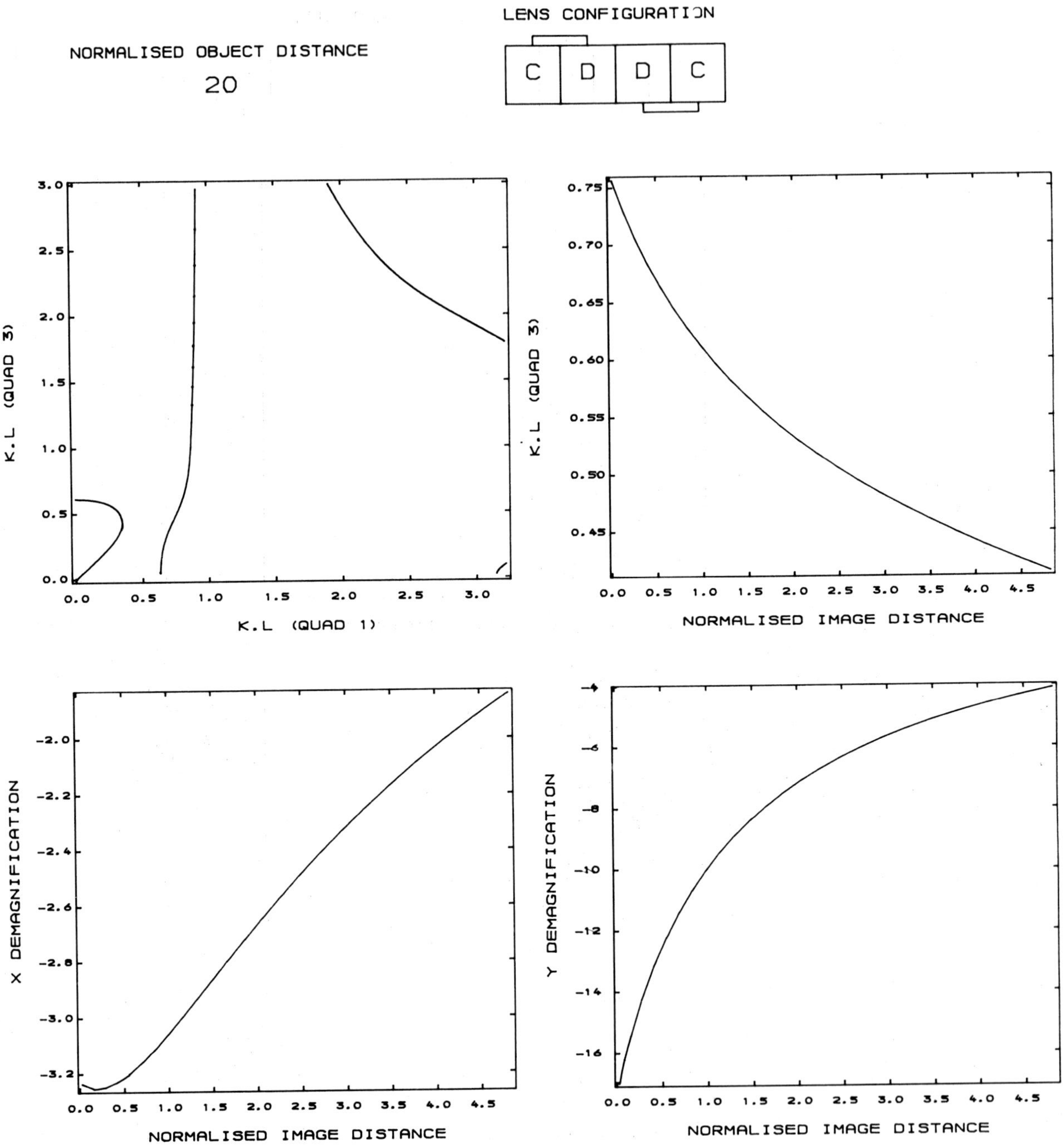

LENS CONFIGURATION
NORMALISED OBJECT DISTANCE
20
C
D
D
C
K.L (QUAD 3)
K.L (QUAD 1)
NORMALISED IMAGE DISTANCE
X DEMAGNIFICATION
Y DEMAGNIFICATION

NORMALISED OBJECT DISTANCE

20

LENS CONFIGURATION

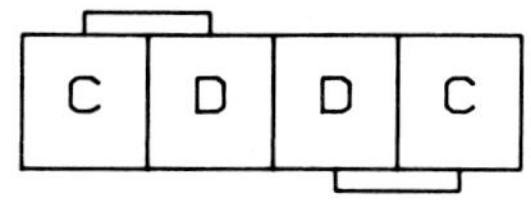

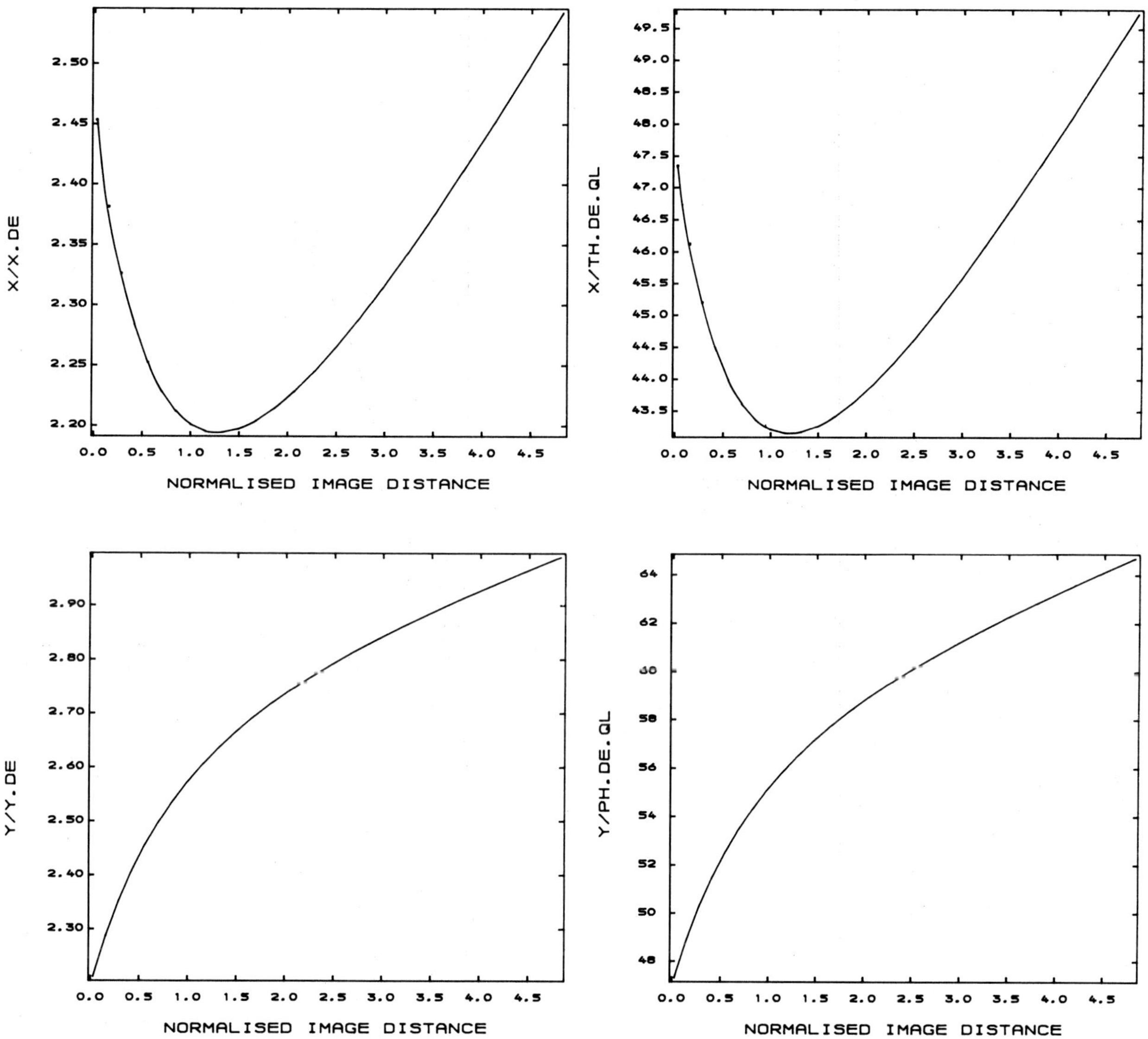

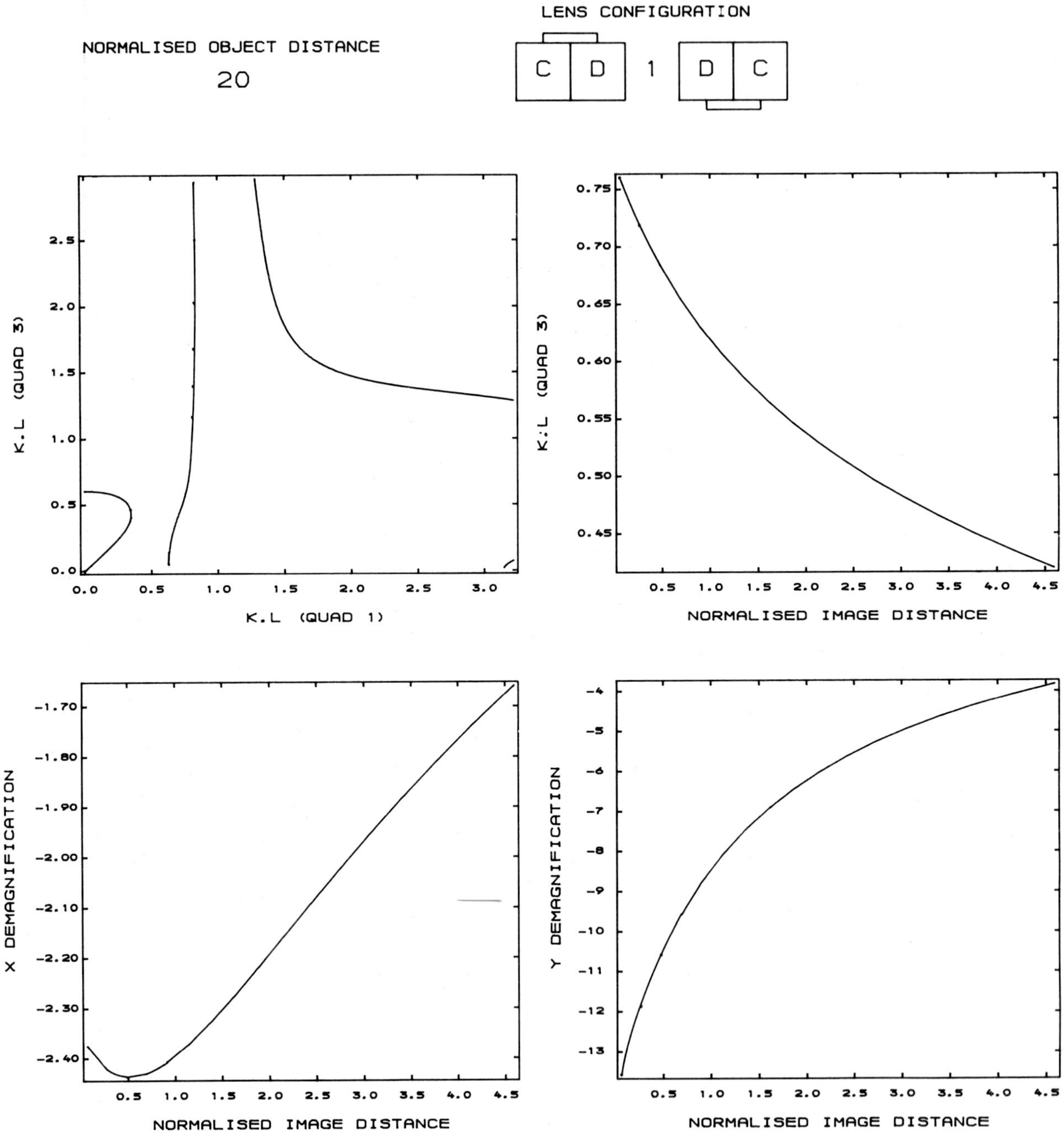

NORMALISED OBJECT DISTANCE
20
LENS CONFIGURATION
C
D
1
D
C
K.L (QUAD 3)
K.L (QUAD 1)
0.0
0.5
1.0
1.5
2.0
2.5
3.0
K.L (QUAD 3)
0.75
0.70
0.65
0.60
0.55
0.50
0.45
NORMALISED IMAGE DISTANCE
0.5
1.0
1.5
2.0
2.5
3.0
3.5
4.0
4.5
X DEMAGNIFICATION
-1.70
-1.80
-1.90
-2.00
-2.10
-2.20
-2.30
-2.40
NORMALISED IMAGE DISTANCE
Y DEMAGNIFICATION
-4
-5
-6
-7
-8
-9
-10
-11
-12
-13
NORMALISED IMAGE DISTANCE

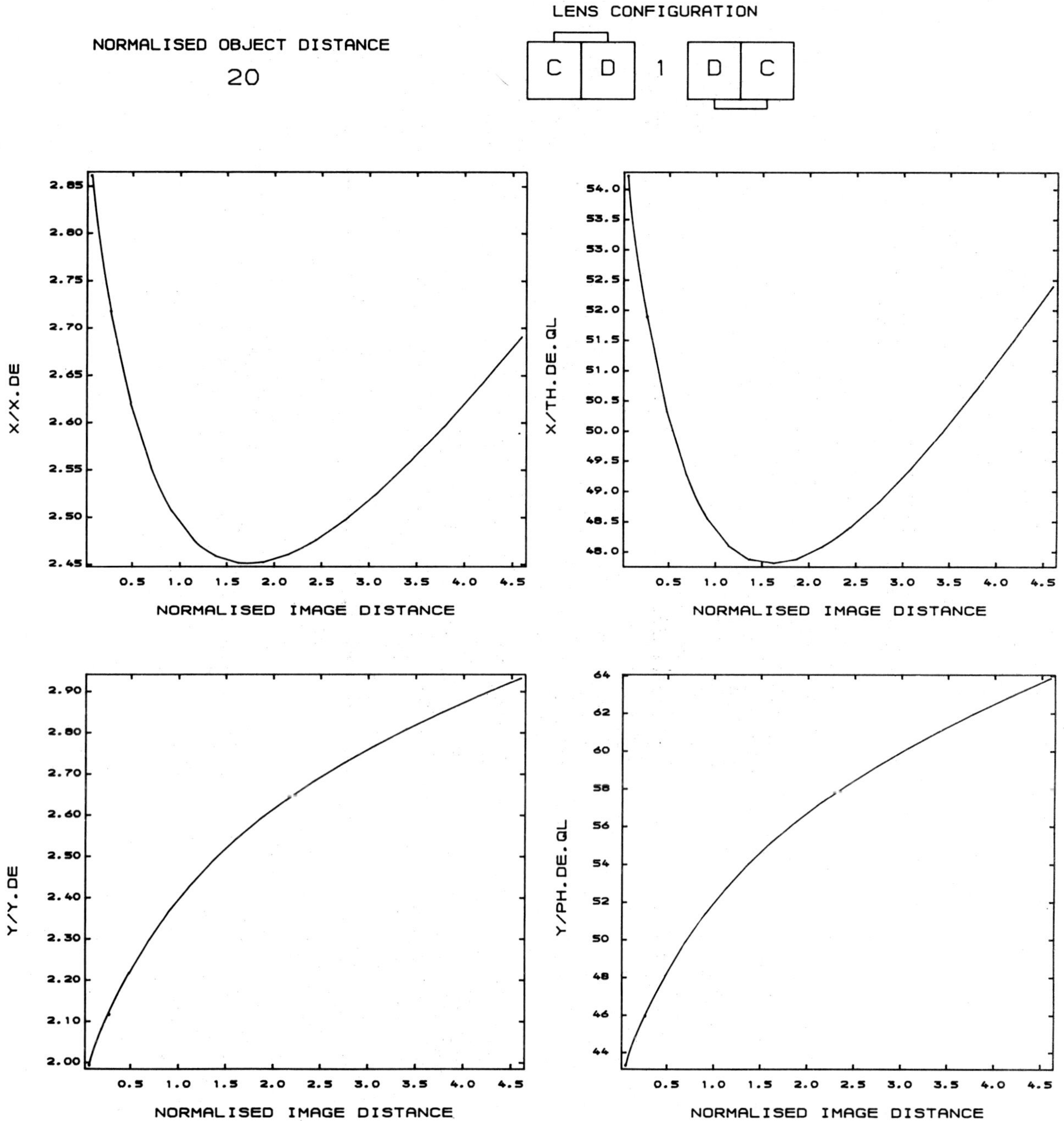
LENS CONFIGURATION
NORMALISED OBJECT DISTANCE
20
C
D
1
D
C
X/X.DE
X/TH.DE.QL
Y/Y.DE
Y/PH.DE.QL
NORMALISED IMAGE DISTANCE
NORMALISED IMAGE DISTANCE
NORMALISED IMAGE DISTANCE
NORMALISED IMAGE DISTANCE

Lens Configuration 20C0D1D0C

KL (QUAD 1)	KL (QUAD 3)	Normalised image distance	Normalised spherical aberration coefficients			
			x/θ^3QL	$x/\theta\phi^2QL$	y/ϕ^3QL	$y/\theta^2\phi QL$
0.814	0.668	0.5	−4.7 (3)	−4.61 (4)	−2.10 (4)	−1.24 (4)
0.796	0.608	1.0	−5.1 (3)	−4.54 (4)	−2.25 (4)	−1.32 (4)
0.779	0.565	1.5	−6.2 (3)	−4.47 (4)	−1.92 (4)	−1.58 (4)
0.765	0.531	2.0	−5.8 (3)	−4.49 (4)	−2.11 (4)	−1.60 (4)
0.752	0.503	2.5	−5.7 (3)	−4.60 (4)	−2.31 (4)	−1.74 (4)
0.742	0.479	3.0	−5.4 (3)	−4.59 (4)	−1.91 (4)	−1.95 (4)
0.732	0.458	3.5	−5.6 (3)	−4.84 (4)	−2.04 (4)	−1.97 (4)
0.723	0.440	4.0	−5.3 (3)	−4.79 (4)	−1.88 (4)	−2.22 (4)

Lens Configuration 20C0D2D0C

KL (QUAD 1)	KL (QUAD 3)	Normalised image distance	Normalised spherical aberration coefficients			
			x/θ^3QL	$x/\theta\phi^2$QL	y/ϕ^3QL	$y/\theta^2\phi$QL
0.746	0.684	0.5	−8.5 (3)	−5.24 (4)	−1.35 (4)	−1.12 (4)
0.737	0.621	1.0	−6.3 (3)	−5.13 (4)	−1.66 (4)	−1.35 (4)
0.729	0.574	1.5	−7.3 (3)	−5.01 (4)	−1.69 (4)	−1.41 (4)
0.721	0.537	2.0	−5.7 (3)	−4.91 (4)	−1.75 (4)	−1.87 (4)
0.713	0.507	2.5	−6.3 (3)	−5.14 (4)	−1.76 (4)	−1.76 (4)
0.707	0.481	3.0	−6.2 (3)	−5.07 (4)	−1.83 (4)	−1.94 (4)
0.700	0.457	3.5	−7.0 (3)	−5.00 (4)	−2.00 (4)	−2.09 (4)
0.694	0.437	4.0	−8.5 (3)	−5.17 (4)	−1.96 (4)	−2.00 (4)
1.211	1.642	0.5	4.83 (7)	4.46 (5)	7.87 (5)	2.41 (7)
1.378	1.374	1.0	1.03 (8)	3.76 (6)	2.17 (6)	9.04 (7)
1.690	1.232	1.5	8.69 (8)	1.66 (8)	2.90 (7)	1.58 (9)
2.433	1.144	2.0	*	*	*	*
3.328	1.081	2.5	*	*	*	*
3.604	1.038	3.0	*	*	*	*
3.702	1.004	3.5	*	*	*	*
3.749	0.977	4.0	*	*	*	*

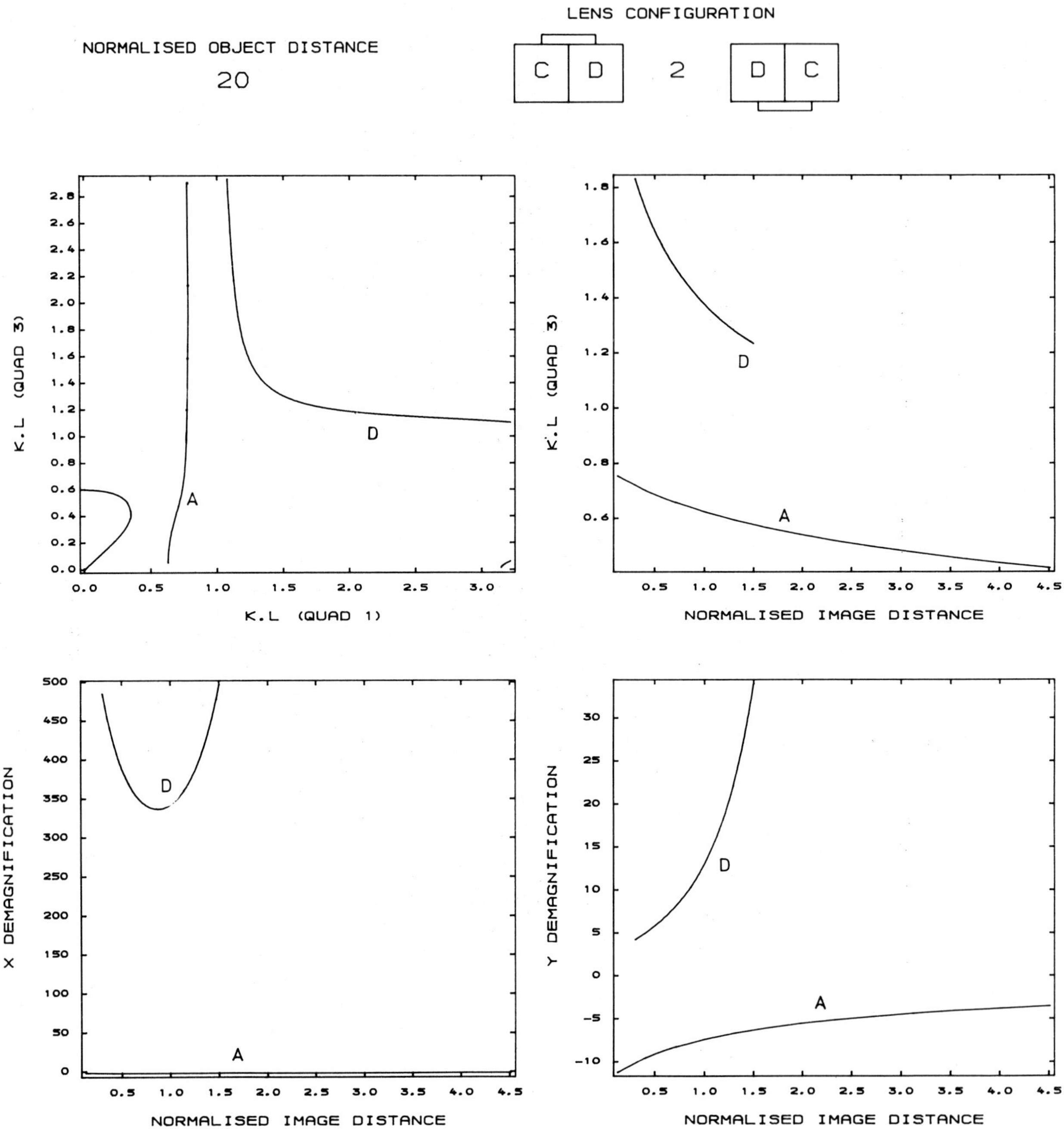

NORMALISED OBJECT DISTANCE
20
LENS CONFIGURATION
C D 2 D C
K.L (QUAD 3)
K.L (QUAD 1)
A
D
NORMALISED IMAGE DISTANCE
X DEMAGNIFICATION
Y DEMAGNIFICATION

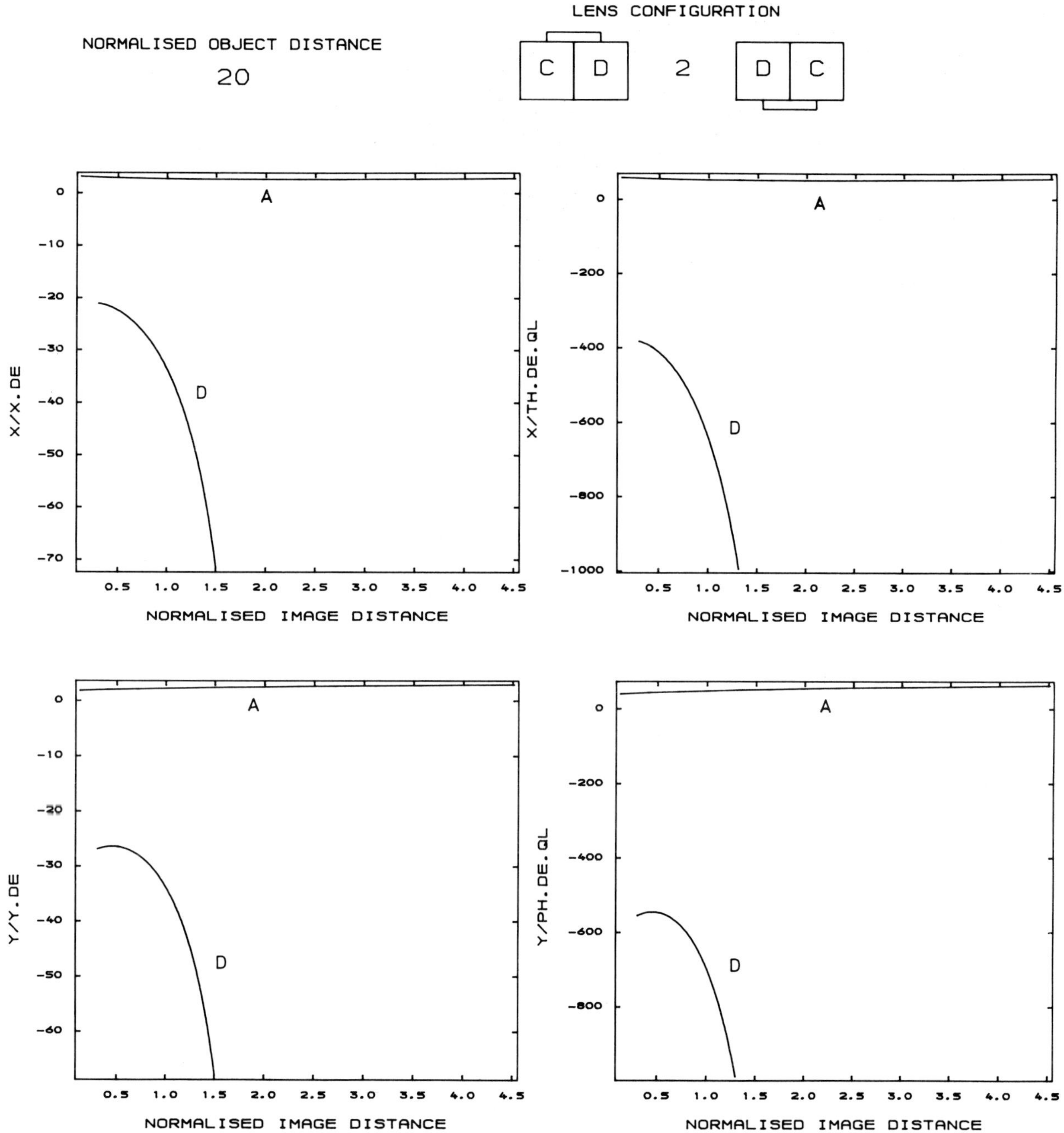

NORMALISED OBJECT DISTANCE
20
LENS CONFIGURATION
C D 2 D C
X/X.DE
X/TH.DE.QL
Y/Y.DE
Y/PH.DE.QL
A
D
NORMALISED IMAGE DISTANCE

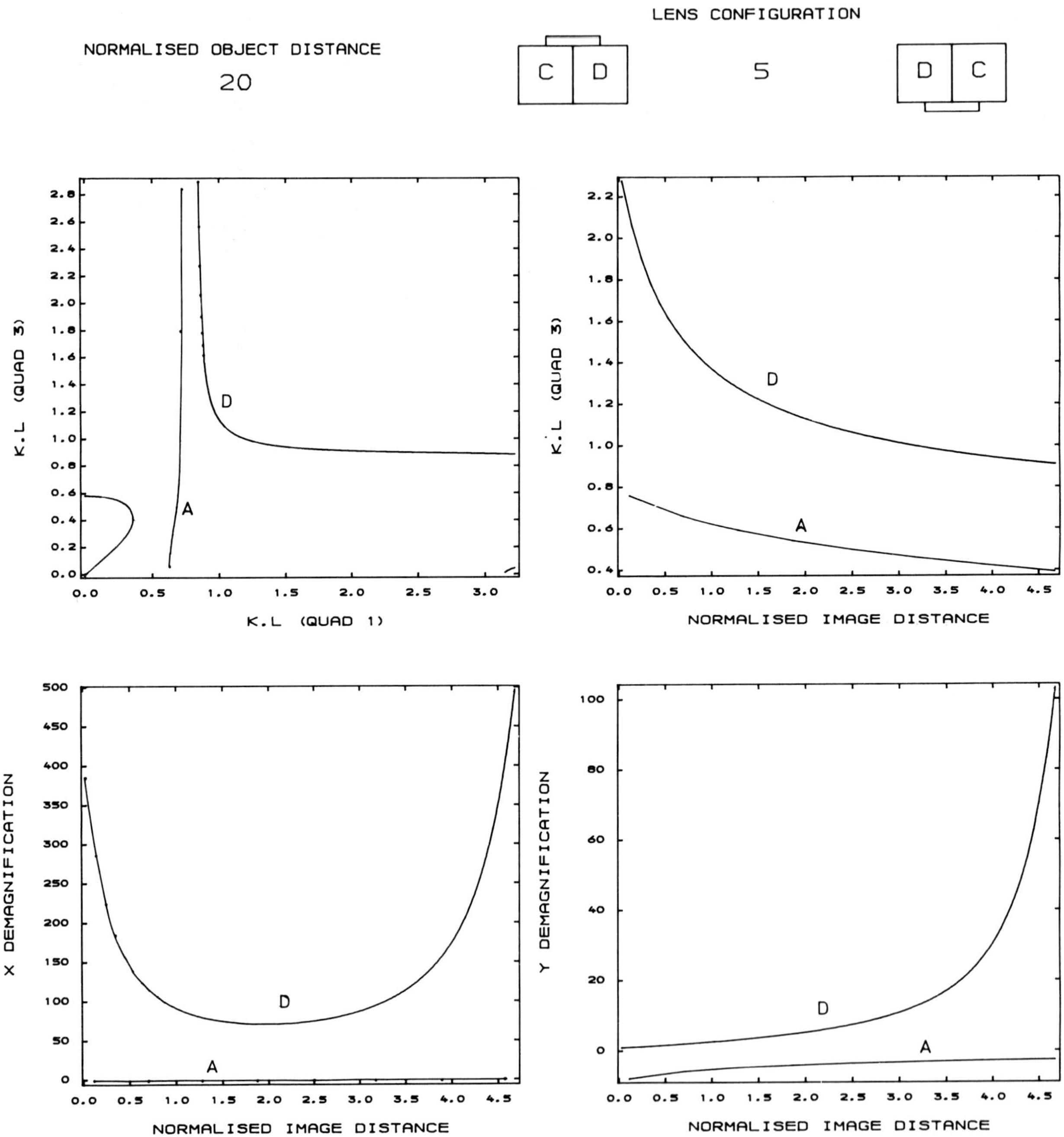
NORMALISED OBJECT DISTANCE
20
LENS CONFIGURATION
C D
5
D C
K.L (QUAD 3)
K.L (QUAD 1)
D
A
NORMALISED IMAGE DISTANCE
X DEMAGNIFICATION
Y DEMAGNIFICATION

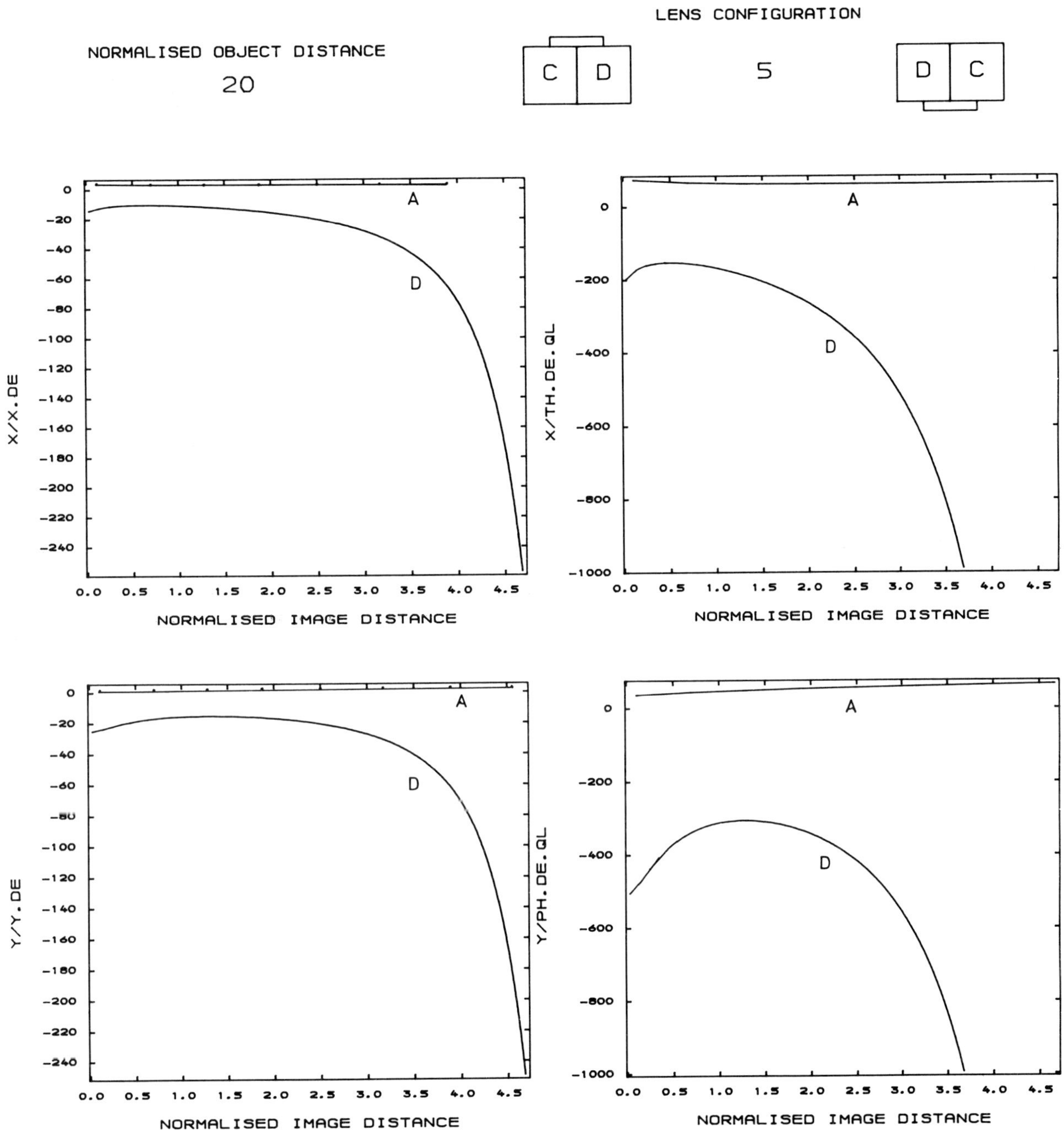

NORMALISED OBJECT DISTANCE
20
LENS CONFIGURATION
C D
5
D C
X/X.DE
X/TH.DE.QL
Y/Y.DE
Y/PH.DE.QL
A
D
NORMALISED IMAGE DISTANCE

Lens Configuration $20\overline{C0D}5\overline{D0C}$

KL (QUAD 1)	KL (QUAD 3)	Normalised image distance	Normalised spherical aberration coefficients			
			x/θ^3QL	$x/\theta\phi^2QL$	y/ϕ^3QL	$y/\theta^2\phi QL$
0.696	0.684	0.5	−8.3 (3)	−6.26 (4)	−1.30 (4)	−1.28 (4)
0.691	0.619	1.0	−9.3 (3)	−5.99 (4)	−1.46 (4)	−1.43 (4)
0.687	0.569	1.5	−9.4 (3)	−5.92 (4)	−1.58 (4)	−1.62 (4)
0.683	0.529	2.0	−1.1 (4)	−5.73 (4)	−1.74 (4)	−1.79 (4)
0.678	0.496	2.5	−7.6 (3)	−5.92 (4)	−1.75 (4)	−1.91 (4)
0.674	0.467	3.0	−9.0 (3)	−5.78 (4)	−1.68 (4)	−2.27 (4)
0.670	0.441	3.5	−1.0 (4)	−5.84 (4)	−1.77 (4)	−2.11 (4)
0.667	0.418	4.0	−7.9 (3)	−5.95 (4)	−2.23 (4)	−2.18 (4)
0.879	1.638	0.5	2.40 (6)	1.6 (4)	1.95 (5)	9.65 (5)
0.912	1.366	1.0	1.81 (6)	3.5 (4)	1.79 (5)	1.17 (6)
0.952	1.219	1.5	2.29 (6)	1.04 (5)	1.82 (5)	2.02 (6)
1.002	1.125	2.0	3.85 (6)	3.10 (5)	2.63 (5)	4.21 (6)
1.068	1.059	2.5	7.95 (6)	1.04 (6)	5.44 (5)	1.04 (7)
1.155	1.009	3.0	2.05 (7)	4.00 (6)	1.59 (6)	3.14 (7)
1.280	0.970	3.5	7.20 (7)	2.07 (7)	6.81 (6)	1.30 (8)
1.472	0.939	4.0	4.14 (8)	2.46 (8)	4.70 (7)	1.11 (9)

Lens Configuration 20C0D0C0D

KL (QUAD 1)	KL (QUAD 2)	Normalised image distance	Normalised spherical aberration coefficients			
			x/θ^3QL	$x/\theta\phi^2QL$	y/ϕ^3QL	$y/\theta^2\phi QL$
0.774	1.144	0.5	−9.8 (3)	−5.38 (4)	−4.97 (4)	−5.35 (4)
0.699	1.096	1.0	−1.18 (4)	−5.95 (4)	−4.16 (4)	−5.59 (4)
0.664	1.057	1.5	−1.52 (4)	−6.13 (4)	−3.85 (4)	−6.23 (4)
0.638	1.025	2.0	−1.77 (4)	−6.59 (4)	−3.68 (4)	−6.63 (4)
0.616	0.998	2.5	−1.94 (4)	−6.82 (4)	−3.46 (4)	−6.97 (4)
0.598	0.975	3.0	−2.35 (4)	−7.22 (4)	−3.31 (4)	−7.11 (4)
0.582	0.955	3.5	−2.95 (4)	−7.41 (4)	−3.83 (4)	−7.30 (4)
0.569	0.937	4.0	−2.10 (4)	−7.96 (4)	−3.59 (4)	−7.46 (4)

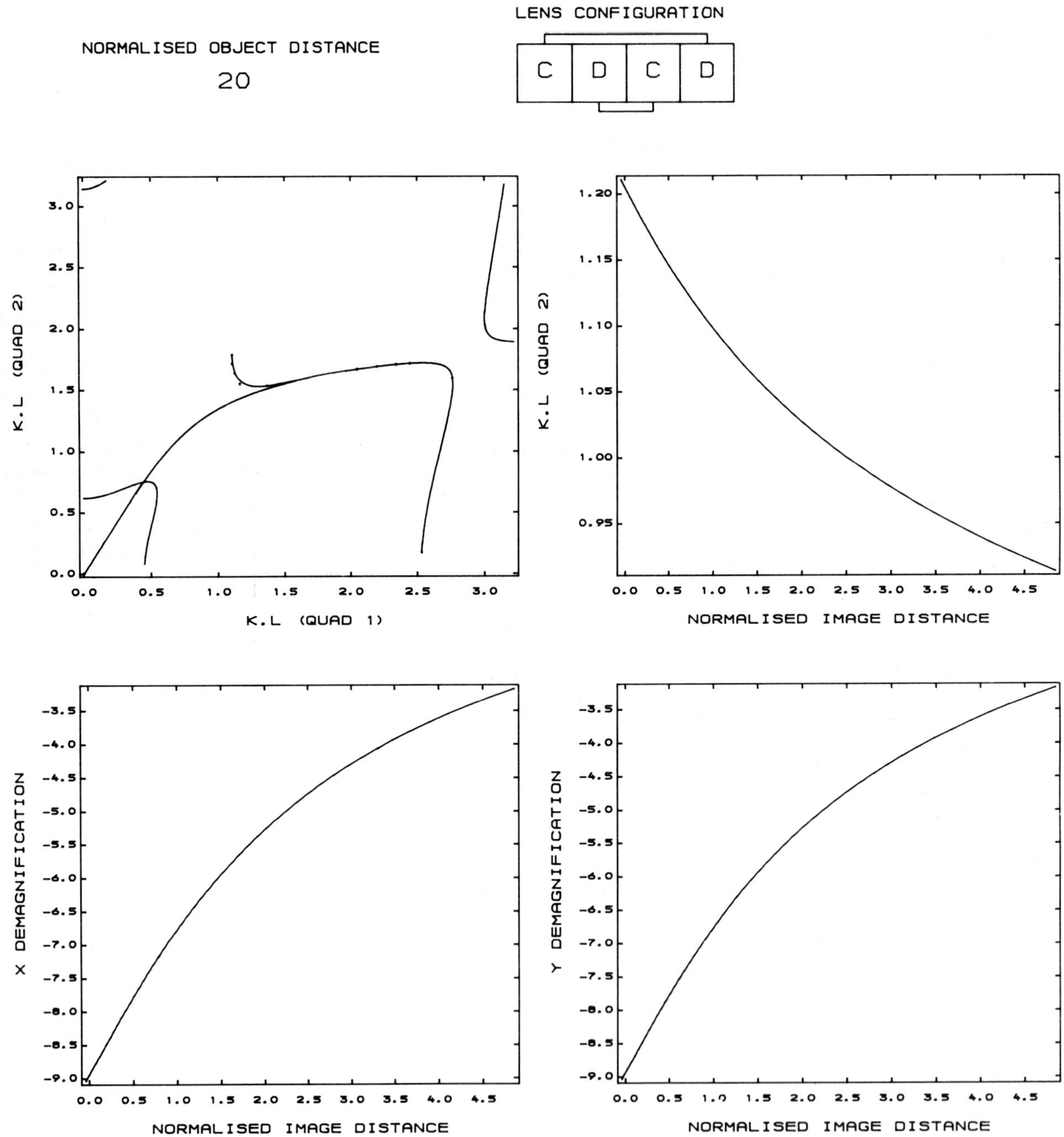
NORMALISED OBJECT DISTANCE
20
LENS CONFIGURATION
C D C D
K.L (QUAD 2)
K.L (QUAD 1)
NORMALISED IMAGE DISTANCE
X DEMAGNIFICATION
Y DEMAGNIFICATION

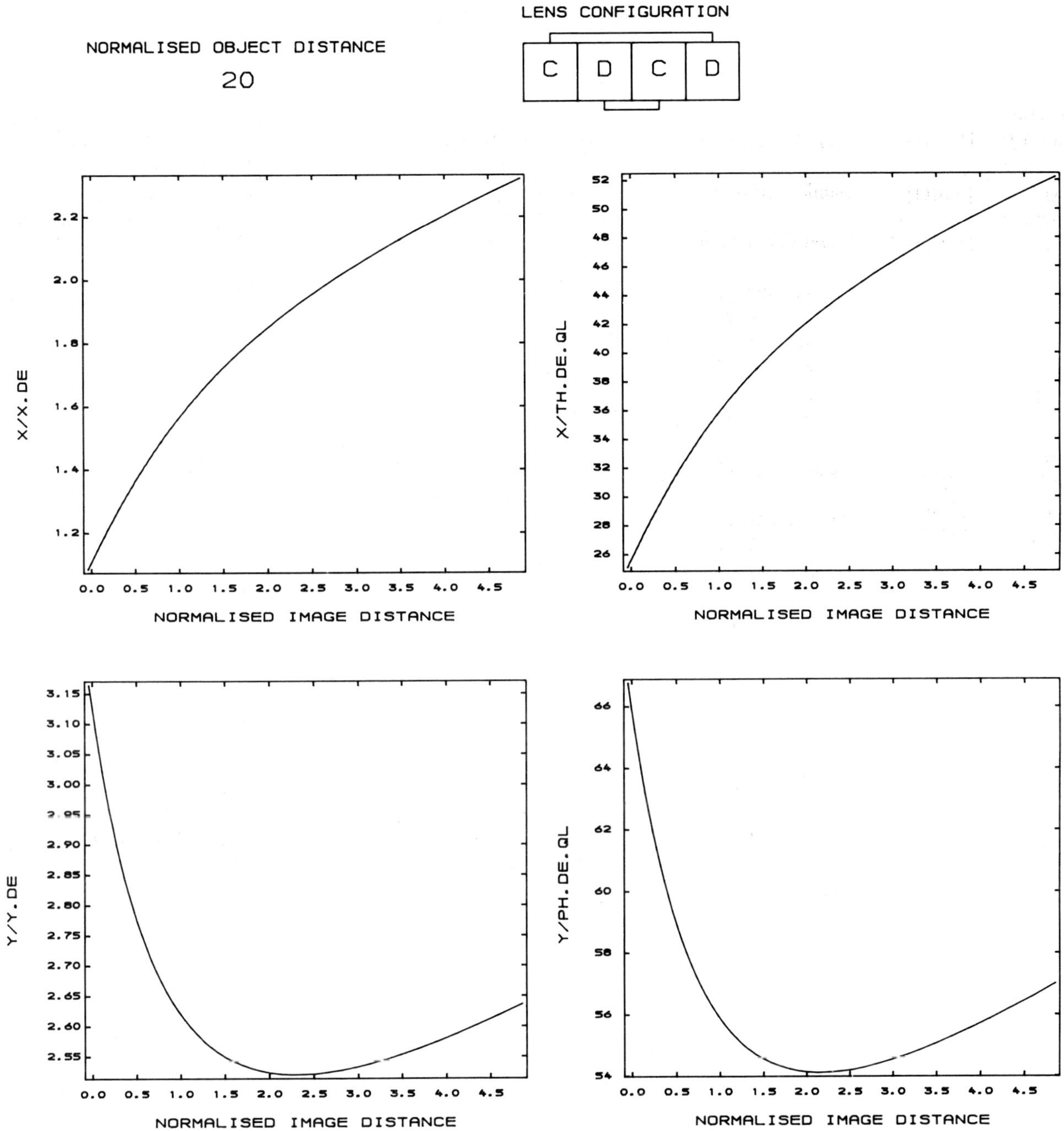
NORMALISED OBJECT DISTANCE
20
LENS CONFIGURATION
C D C D
X/X.DE
X/TH.DE.QL
Y/Y.DE
Y/PH.DE.QL
NORMALISED IMAGE DISTANCE

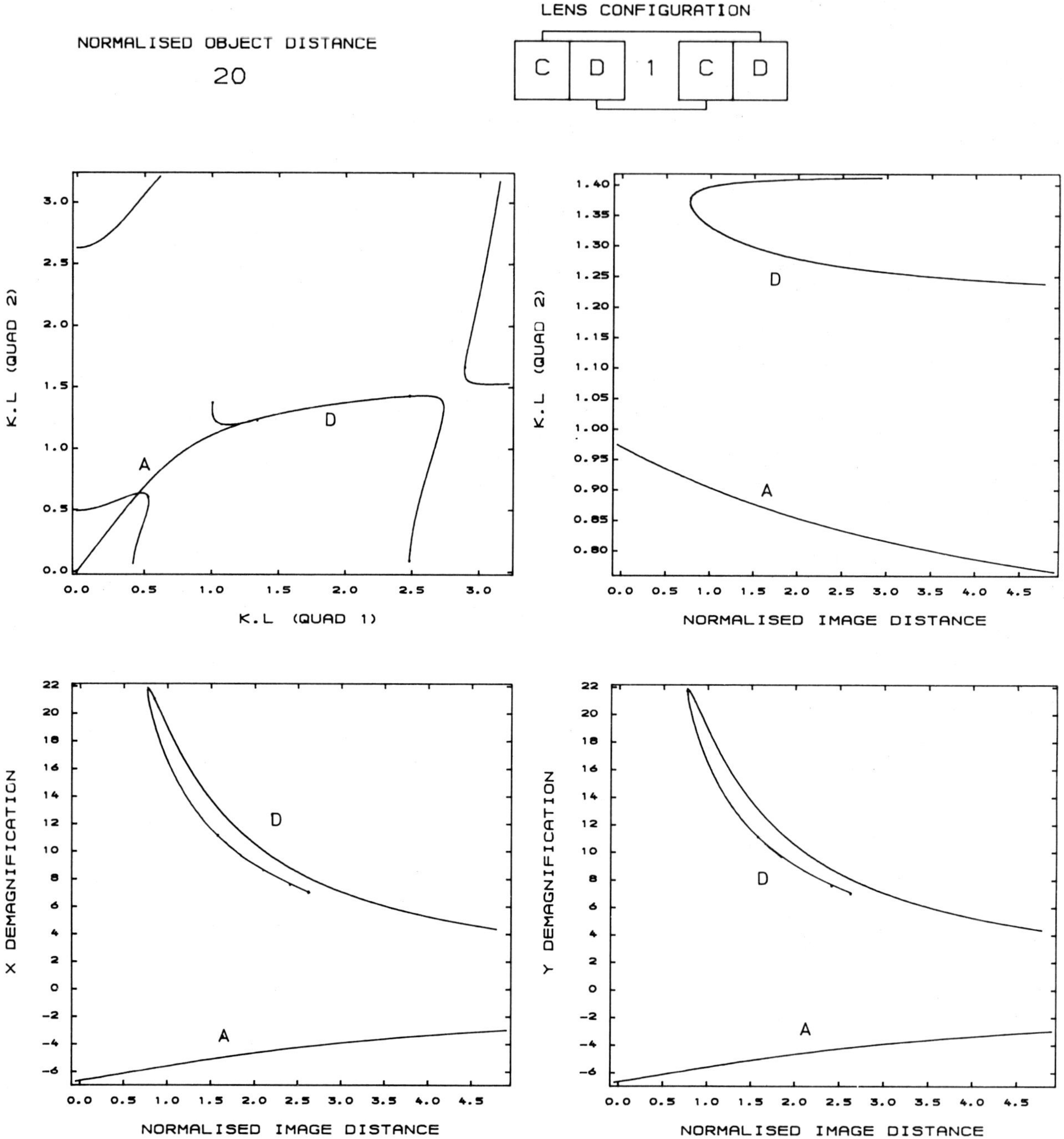

NORMALISED OBJECT DISTANCE
20
LENS CONFIGURATION
C D 1 C D
K.L (QUAD 2)
K.L (QUAD 1)
A
D
K.L (QUAD 2)
NORMALISED IMAGE DISTANCE
D
A
X DEMAGNIFICATION
NORMALISED IMAGE DISTANCE
D
A
Y DEMAGNIFICATION
NORMALISED IMAGE DISTANCE
D
A

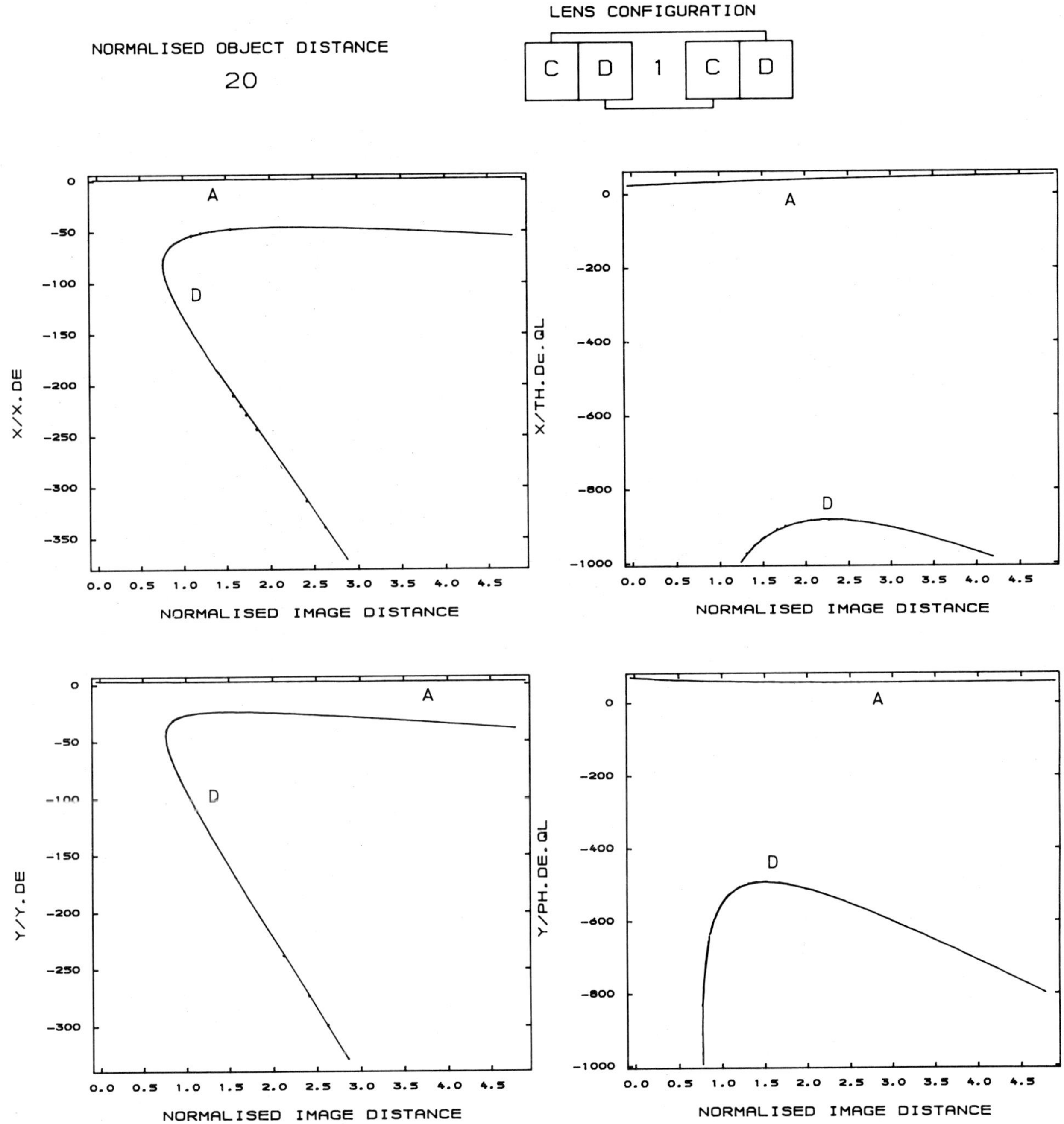
NORMALISED OBJECT DISTANCE
20
LENS CONFIGURATION
C D 1 C D
X/X.DE
X/TH.DE.QL
Y/Y.DE
Y/PH.DE.QL
A
D
NORMALISED IMAGE DISTANCE
0.0 0.5 1.0 1.5 2.0 2.5 3.0 3.5 4.0 4.5
0 -50 -100 -150 -200 -250 -300 -350
0 -200 -400 -600 -800 -1000

Lens Configuration 20C0D1C0D

KL (QUAD 1)	KL (QUAD 2)	Normalised image distance	Normalised spherical aberration coefficients			
			x/θ^3QL	$x/\theta\phi^2QL$	y/ϕ^3QL	$y/\theta^2\phi QL$
0.737	0.932	0.5	−3.7 (3)	−3.14 (4)	−3.27 (4)	−2.87 (4)
0.702	0.901	1.0	−5.1 (3)	−3.23 (4)	−2.65 (4)	−3.07 (4)
0.673	0.874	1.5	−6.7 (3)	−3.37 (4)	−2.60 (4)	−3.12 (4)
0.649	0.851	2.0	−8.3 (3)	−3.46 (4)	−2.25 (4)	−3.43 (4)
0.629	0.831	2.5	−9.2 (3)	−3.67 (4)	−2.27 (4)	−3.55 (4)
0.612	0.814	3.0	−9.7 (3)	−3.74 (4)	−1.88 (4)	−4.05 (4)
0.598	0.798	3.5	−9.8 (3)	−3.92 (4)	−1.79 (4)	−4.00 (4)
0.585	0.785	4.0	−1.1 (4)	−4.23 (4)	−1.95 (4)	−4.15 (4)
1.739	1.330	1.0	1.73 (7)	5.23 (6)	3.97 (6)	5.12 (6)
1.569	1.295	1.5	8.53 (6)	2.89 (6)	2.14 (6)	2.84 (6)
1.485	1.276	2.0	5.82 (6)	2.27 (6)	1.74 (6)	2.21 (6)
1.433	1.263	2.5	4.63 (6)	1.98 (6)	1.61 (6)	1.96 (6)
1.398	1.254	3.0	4.00 (6)	1.85 (6)	1.58 (6)	1.84 (6)
1.373	1.247	3.5	3.61 (6)	1.82 (6)	1.60 (6)	1.81 (6)
1.354	1.242	4.0	3.36 (6)	1.82 (6)	1.67 (6)	1.80 (6)
2.151	1.393	1.0	8.39 (7)	3.29 (7)	4.20 (7)	3.28 (7)
2.228	1.402	1.5	1.37 (8)	6.10 (7)	9.10 (7)	6.11 (7)
2.259	1.406	2.0	1.85 (8)	8.67 (7)	1.38 (8)	8.74 (7)
2.275	1.408	2.5	2.31 (8)	1.11 (8)	1.84 (8)	1.12 (8)
2.286	1.409	3.0	2.76 (8)	1.36 (8)	2.31 (8)	1.38 (8)
†	†	3.5	†	†	†	†
†	†	4.0	†	†	†	†

† Stigmatic focus not calculated with sufficient accuracy to enable spherical aberration coefficients to be estimated.

Lens Configuration 20C0D2C0D

KL (QUAD 1)	KL (QUAD 2)	Normalised image distance	Normalised spherical aberration coefficients			
			x/θ^3QL	$x/\theta\phi^2QL$	y/ϕ^3QL	$y/\theta^2\phi QL$
0.730	0.852	0.5	−3.5 (3)	−2.74 (4)	−3.04 (4)	−2.82 (4)
0.702	0.829	1.0	−4.8 (3)	−2.75 (4)	−2.57 (4)	−2.61 (4)
0.678	0.808	1.5	−4.4 (3)	−2.85 (4)	−2.32 (4)	−2.59 (4)
0.658	0.789	2.0	−4.7 (3)	−2.94 (4)	−2.30 (4)	−2.62 (4)
0.640	0.773	2.5	−6.9 (3)	−2.91 (4)	−2.20 (4)	−2.79 (4)
0.624	0.758	3.0	−8.0 (3)	−3.09 (4)	−1.86 (4)	−2.84 (4)
0.611	0.745	3.5	−7.0 (3)	−3.06 (4)	−2.64 (4)	−2.95 (4)
0.599	0.733	4.0	−6.8 (3)	−3.44 (4)	−2.04 (4)	−3.16 (4)
1.758	1.243	0.5	2.78 (7)	3.44 (6)	1.97 (6)	3.27 (6)
1.497	1.190	1.0	8.88 (6)	1.47 (6)	7.46 (5)	1.41 (6)
1.378	1.160	1.5	4.59 (6)	9.77 (5)	5.42 (5)	9.63 (5)
1.309	1.141	2.0	3.02 (6)	7.91 (5)	4.88 (5)	7.72 (5)
1.265	1.130	2.5	2.28 (6)	7.01 (5)	5.05 (5)	6.71 (5)
1.235	1.118	3.0	1.88 (6)	6.45 (5)	4.94 (5)	6.35 (5)
1.212	1.110	3.5	1.64 (6)	6.30 (5)	5.07 (3)	6.20 (5)
1.195	1.105	4.0	1.50 (6)	6.16 (5)	5.36 (5)	6.02 (5)
2.437	1.336	0.5	1.84 (8)	3.81 (7)	1.00 (8)	3.80 (7)
2.491	1.340	1.0	3.30 (8)	7.57 (7)	2.47 (8)	7.93 (7)
†	†	1.5	†	†	†	†
†	†	2.0	†	†	†	†
†	†	2.5	†	†	†	†
†	†	3.0	†	†	†	†
†	†	3.5	†	†	†	†
†	†	4.0	†	†	†	†

† Stigmatic focus not calculated with sufficient accuracy to enable spherical aberration coefficients to be estimated.

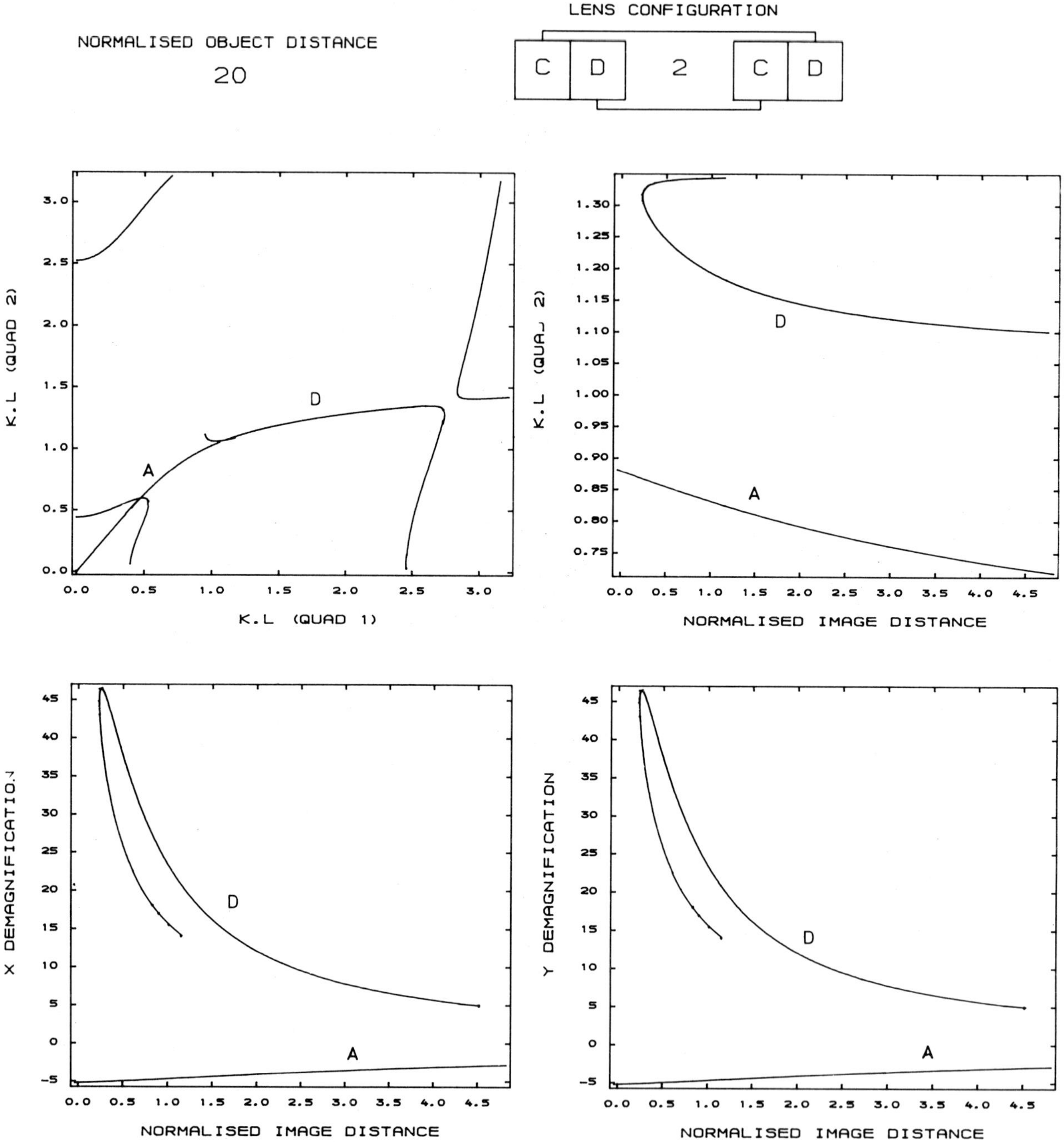
NORMALISED OBJECT DISTANCE
20
LENS CONFIGURATION
C D 2 C D
K.L (QUAD 2)
K.L (QUAD 1)
D
A
K.L (QUAD 2)
NORMALISED IMAGE DISTANCE
D
A
X DEMAGNIFICATION
NORMALISED IMAGE DISTANCE
D
A
Y DEMAGNIFICATION
NORMALISED IMAGE DISTANCE
D
A

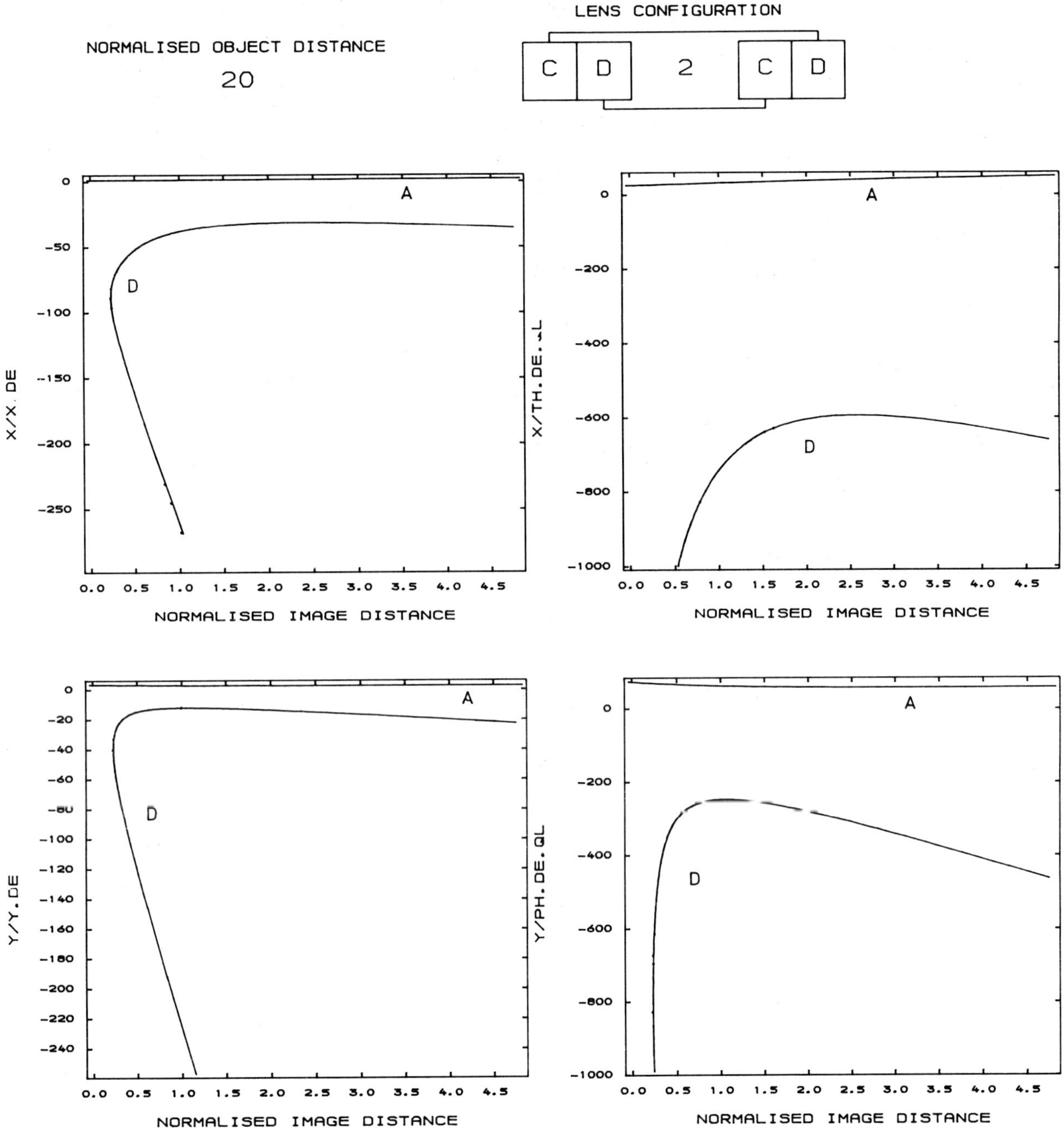
LENS CONFIGURATION
NORMALISED OBJECT DISTANCE
20
C
D
2
C
D
X/X.DE
X/TH.DE.AL
Y/Y.DE
Y/PH.DE.QL
A
D
NORMALISED IMAGE DISTANCE

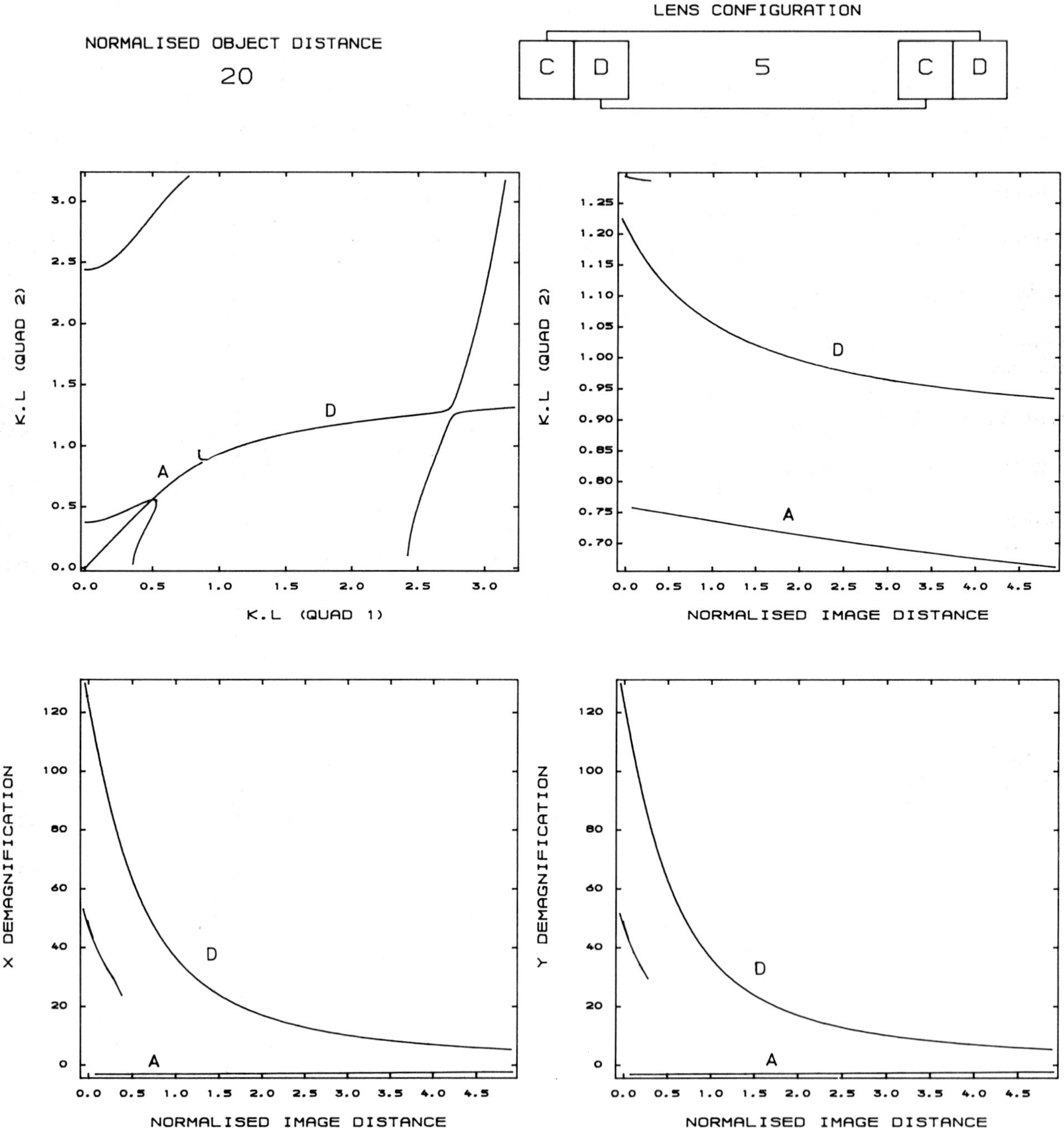

NORMALISED OBJECT DISTANCE
20
LENS CONFIGURATION
C
D
5
C
D
K.L (QUAD 2)
K.L (QUAD 1)
A
D
NORMALISED IMAGE DISTANCE
X DEMAGNIFICATION
Y DEMAGNIFICATION

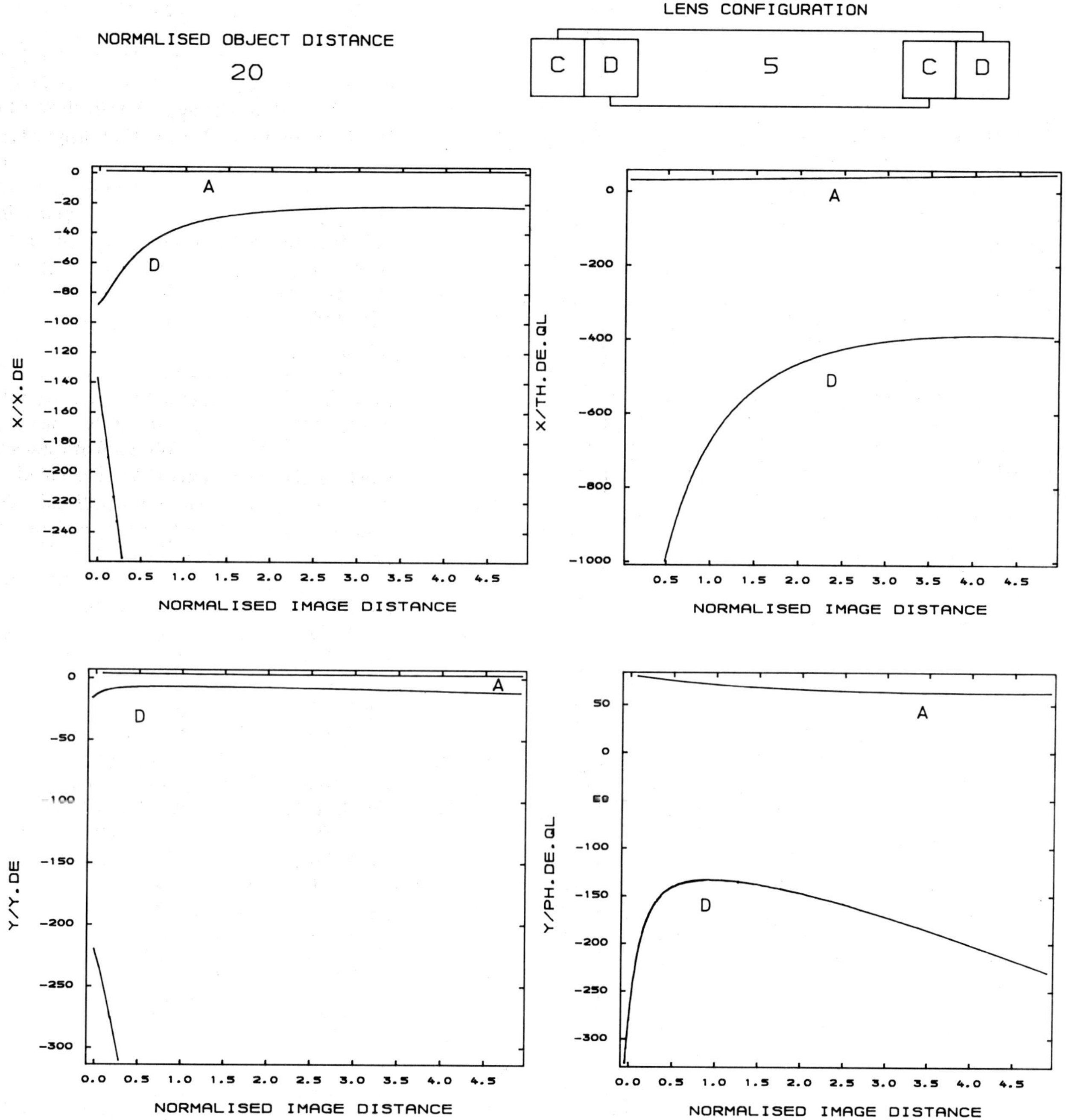
NORMALISED OBJECT DISTANCE
20
LENS CONFIGURATION
C D 5 C D
X/X.DE
X/TH.DE.QL
Y/Y.DE
Y/PH.DE.QL
A
D
NORMALISED IMAGE DISTANCE
NORMALISED IMAGE DISTANCE
NORMALISED IMAGE DISTANCE
NORMALISED IMAGE DISTANCE

Lens Configuration 20C0D5C0D

KL (QUAD 1)	KL (QUAD 2)	Normalised image distance	Normalised spherical aberration coefficients			
			x/θ^3QL	$x/\theta\phi^2$QL	y/ϕ^3QL	$y/\theta^2\phi$QL
0.698	0.745	0.5	−4.1 (3)	−2.94 (4)	−2.85 (4)	−2.94 (4)
0.684	0.734	1.0	−3.6 (3)	−3.02 (4)	−2.02 (4)	−2.78 (4)
0.670	0.722	1.5	−3.9 (3)	−2.83 (4)	−2.27 (4)	−2.79 (4)
0.657	0.711	2.0	−4.5 (5)	−2.77 (4)	−1.91 (4)	−2.91 (4)
0.645	0.701	2.5	−6.1 (3)	−2.68 (4)	−2.41 (4)	−2.38 (4)
0.634	0.691	3.0	−6.5 (3)	−2.65 (4)	−1.69 (4)	−2.68 (4)
0.623	0.682	3.5	−6.3 (3)	−2.74 (4)	−2.12 (4)	−2.52 (4)
0.614	0.673	4.0	−6.6 (3)	−2.79 (4)	−1.88 (4)	−2.70 (4)
1.548	1.108	0.5	4.48 (7)	4.58 (6)	4.76 (5)	4.35 (6)
1.325	1.053	1.0	1.20 (7)	1.95 (6)	2.45 (5)	1.88 (6)
1.210	1.018	1.5	5.10 (6)	1.10 (6)	1.79 (5)	1.08 (6)
1.141	0.993	2.0	2.79 (6)	7.52 (5)	1.55 (5)	7.28 (5)
1.094	0.975	2.5	1.78 (6)	5.66 (5)	1.38 (5)	5.51 (5)
1.061	0.962	3.0	1.27 (6)	4.58 (5)	1.42 (5)	4.49 (5)
1.037	0.951	3.5	9.93 (5)	3.96 (5)	1.39 (5)	3.87 (5)
1.018	0.943	4.0	8.20 (5)	3.51 (5)	1.40 (5)	3.51 (5)
2.870	1.282	0.5	8.90 (8)	2.53 (8)	1.17 (9)	6.93 (7)
2.857	1.281	1.0	1.44 (9)	3.19 (8)	1.68 (9)	9.26 (7)
†	†	1.5	†	†	†	†
†	†	2.0	†	†	†	†
†	†	2.5	†	†	†	†
†	†	3.0	†	†	†	†
†	†	3.5	†	†	†	†
†	†	4.0	†	†	†	†

† Stigmatic focus not calculated with sufficient accuracy to enable spherical aberration coefficients to be estimated.

Appendix 2

Characteristic x-ray energies

This table lists the most prominent characteristic x-ray transition energies encountered in x-ray emission [JO70b]. Although in most cases only one or two of the major emissions are used in the identification of a particular element, weaker transitions have been included to reduce possible misinterpretation between weak lines from major constituents and strong transitions from trace elements.

As an excellent rule-of-thumb guide, the approximate relative intensity ratios of the transitions within an elemental series are (taking $K_{\alpha_1} = 100$):

$K_{\alpha_1} = 100$	$L_{\alpha_1} = 100$	$M_{\alpha_{1,2}} = 100$
$K_{\alpha_2} = 50$	$L_{\alpha_2} = 10$	$M_{\beta} = 60$
$K_{\alpha_{1,2}} = 150$	$L_{\beta_1} = 50$	
$K_{\beta_1} = 15–30$	$L_{\beta_2} = 20$	
$K_{\beta_2} = 1–10$	$L_{\beta_3} = 1–6$	
$K_{\beta_3} = 6–15$	$L_{\beta_4} = 3–5$	
	$L_{\gamma_1} = 1–10$	
	$L_{\gamma_3} = 0.5–2$	
	$L_{\eta} = 1$	
	$L_{l} = 1–3$	

Z	Element	K series					L series										M series			Z
							L_I series			L_{II} series			L_{III} series							
Z	Element	K_{β_3}	K_{β_1}	K_{β_2}	K_{α_1}	K_{α_2}	L_{γ_3}	L_{β_3}	L_{β_4}	L_{γ_1}	L_{β_1}	L_η	L_{β_2}	L_{α_1}	L_{α_2}	L_l	M_β	M_{α_1}	M_{α_2}	Z
1	H																			1
2	He																			2
3	Li				0.052															3
4	Be				0.110															4
5	B				0.185															5
6	C				0.277															6
7	N				0.392															7
8	O				0.525															8
9	F				0.677															9
10	Ne				0.848															10
11	Na		1.067		1.041															11
12	Mg		1.295		1.253															12
13	Al		1.553		1.486	1.486														13
14	Si		1.829		1.740	1.739														14
15	P		2.136		2.013	1.012														15
16	S		2.464		2.307	2.306														16
17	Cl				2.622	2.620														17
18	Ar		3.190		2.957	2.955														18
19	K		3.589		3.313	3.310						0.262				0.260				19
20	Ca		4.012		3.691	3.687				0.350	0.345	0.306		0.341		0.303				20
21	Sc		4.460		4.090	4.085				0.407	0.400	0.353		0.395		0.348				21
22	Ti		4.931		4.510	4.504				0.460	0.458	0.401		0.452		0.395				22
23	V		5.426		4.951	4.944		0.585		0.520	0.519	0.453		0.511		0.446				23
24	Cr		5.946		5.414	5.405		0.654		0.583	0.583	0.510		0.573		0.500				24
25	Mn		6.489		5.898	5.887		0.721		0.652	0.649	0.567		0.637		0.556				25
26	Fe		7.057		6.403	6.390		0.792		0.721	0.718	0.628		0.705		0.615				26
27	Co		7.648		6.929	6.914		0.866		0.794	0.791	0.694		0.776		0.678				27
28	Ni		8.263		7.477	7.460		0.941		0.872	0.869	0.762		0.851		0.743				28
29	Cu		8.904		8.046	8.026		1.023		0.952	0.950	0.832		0.930		0.811				29
30	Zn		9.570	9.656	8.637	8.614		1.107		1.044	1.034	0.906		1.012		0.884				30
31	Ga	10.259	10.263	10.365	9.250	9.223		1.197		1.134	1.125	0.984		1.098		0.957				31
32	Ge	10.976	10.980	11.099	9.885	9.854		1.294	1.286	1.249	1.218	1.068		1.188		1.036				32
33	As	11.718	11.724	11.862	10.542	10.506		1.388		1.360	1.317	1.155		1.282		1.120				33
34	Se	12.437	12.494	12.650	11.220	11.179		1.490		1.477	1.419	1.244		1.379		1.204				34

		K series					L series: L_I series			L_{II} series			L_{III} series				M series			
Z	Element	K_{β_3}	K_{β_1}	K_{β_2}	K_{α_1}	K_{α_2}	L_{γ_3}	L_{β_3}	L_{β_4}	L_{γ_1}	L_{β_1}	L_{η}	L_{β_2}	L_{α_1}	L_{α_2}	L_l	M_{β}	M_{α_1}	M_{α_2}	Z
35	Br	13.282	13.289	13.467	11.922	11.876		1.596			1.526	1.339		1.480		1.293				35
36	Kr	14.102	14.110	14.312	12.648	12.596		1.706	1.697		1.636			1.586						36
37	Rb	14.949	14.959	15.183	13.393	13.333		1.826	1.817		1.752	1.542		1.694	1.692	1.482				37
38	Sr	15.822	15.833	16.082	14.163	14.095		1.947	1.936		1.871	1.649		1.806	1.804	1.582				38
39	Y	16.723	16.735	17.013	14.956	14.880		2.072	2.060		1.995	1.761		1.922	1.920	1.685				39
40	Zr	17.651	17.665	17.967	15.772	15.688		2.201	2.187	2.302	2.124	1.876	2.219	2.042	2.040	1.792				40
41	Nb	18.603	18.619	18.949	16.612	16.518		2.334	2.319	2.461	2.257	1.996	2.367	2.166	2.163	1.902				41
42	Mo	19.587	19.605	19.962	17.476	17.371		2.473	2.455	2.623	2.394	2.120	2.518	2.293	2.289	2.015				42
43	Tc	20.595	20.615	21.002	18.364	18.248					2.536			2.424						43
44	Ru	21.631	21.653	22.070	19.276	19.147		2.763	2.741	2.964	2.683	2.382	2.835	2.558	2.554	2.252				44
45	Rh	22.695	22.720	23.169	20.213	20.070		2.915	2.890	3.143	2.834	2.519	3.001	2.696	2.692	2.376				45
46	Pd	23.787	23.815	24.295	21.174	21.017		3.072	3.045	3.328	2.990	2.660	3.171	2.838	2.833	2.503				46
47	Ag	24.907	24.938	25.452	22.159	21.987	3.749	3.234	3.203	3.519	3.150	2.806	3.347	2.984	2.978	2.633				47
48	Cd	26.057	26.091	26.639	23.170	22.980		3.401	3.367	3.716	3.316	2.956	3.528	3.133	3.126	2.767				48
49	In	27.233	27.271	27.856	24.206	23.998		3.572	3.535	3.920	3.487	3.112	3.713	3.286	3.279	2.904				49
50	Sn	28.439	28.481	29.104	25.267	25.040		3.750	3.708	4.130	3.662	3.272	3.904	3.443	3.435	3.044				50
51	Sb	29.674	29.721	30.388	26.355	26.106		3.932	3.886	4.347	3.843	3.436	4.100	3.604	3.595	3.188				51
52	Te	30.939	30.990	31.698	27.468	27.197		4.120	4.069	4.570	4.029	3.605	4.301	3.769	3.758	3.335				52
53	I	32.234	32.289	33.036	28.607	28.312		4.313	4.257	4.800	4.220	3.780	4.507	3.937	3.925	3.484				53
54	Xe	33.556	33.619	34.408	29.774	29.453								4.109						54
55	Cs	34.913	34.981	35.815	30.968	30.620	5.552	4.716	4.649	5.279	4.619	4.141	4.935	4.286	4.272	3.794				55
56	Ba	36.298	36.372	37.251	32.188	31.812	5.808	4.926	4.851	5.530	4.827	4.330	5.156	4.465	4.450	3.953				56
57	La	37.714	37.795	38.723	33.436	33.028	6.073	5.143	5.061	5.788	5.041	4.524	5.383	4.650	4.633	4.124	0.854			57
58	Ce	39.163	39.251	40.226	34.714	34.273	6.340	5.364	5.276	6.051	5.261	4.731	5.612	4.839	4.822	4.287	0.902			58
59	Pr	40.646	40.741	41.767	36.020	35.544	6.615	5.591	5.497	6.321	5.488	4.935	5.849	5.033	5.013	4.452	0.949			59
60	Nd	42.159	42.264	43.327	37.355	36.841	6.900	5.828	5.721	6.601	5.721	5.145	6.088	5.229	5.207	4.632	0.996			60
61	Pm	43.705	43.818	44.929	38.718	38.165		6.070		6.891	5.960		6.338	5.432	5.407					61
62	Sm	45.281	45.405	46.566	40.111	39.516	7.485	6.317	6.195	7.177	6.205	5.588	6.586	5.635	5.607	4.994	1.100			62
63	Eu	46.896	47.030	48.248	41.535	40.895	7.795	6.570	6.438	7.479	6.455	5.816	6.842	5.845	5.816	5.176	1.153			63
64	Gd	48.547	48.688	49.952	42.989	42.302	8.104	6.830	6.686	7.784	6.712	6.049	7.102	6.056	6.024	5.361	1.209			64
65	Tb	50.221	50.374	51.715	44.474	43.737	8.422	7.095	6.939	8.100	6.977	6.283	7.365	6.272	6.237	5.546	1.266			65
66	Dy	51.949	52.110	53.500	45.991	45.200	8.752	7.369	7.203	8.417	7.246	6.533	7.634	6.494	6.457	5.742	1.325			66
67	Ho	53.702	53.868	55.315	47.539	46.692	9.086	7.650	7.470	8.746	7.524	6.787	7.910	6.719	6.679	5.942	1.383			67
68	Er	55.485	55.672	57.204	49.119	48.213	9.429	7.938	7.744	9.087	7.809	7.057	8.188	6.947	6.904	6.152	1.443			68

		K series					L series										M series			
							L_I series			L_{II} series				L_{III} series						
Z	Element	K_{β_3}	K_{β_1}	K_{β_2}	K_{α_1}	K_{α_2}	L_{γ_3}	L_{β_3}	L_{β_4}	L_{γ_1}	L_{β_1}	L_{η}	L_{β_2}	L_{α_1}	L_{α_2}	L_l	M_{β}	M_{α_1}	M_{α_2}	Z
69	Tm	57.293	57.506	59.085	50.733	49.764	9.778	8.229	8.024	9.424	8.100	7.308	8.467	7.179	7.132	6.341	1.503			69
70	Yb	59.141	59.356	60.974	52.380	51.345	10.141	8.535	8.312	9.778	8.400	7.579	8.757	7.414	7.366	6.544	1.567			70
71	Lu	61.037	61.272	62.956	54.061	52.956	10.509	8.845	8.605	10.142	8.708	7.856	9.047	7.654	7.604	6.752	1.631			71
72	Hf	62.969	63.222	64.969	55.781	54.602	10.889	9.162	8.904	10.514	9.021	8.138	9.346	7.898	7.843	6.958	1.697			72
73	Ta	64.938	65.212	67.001	57.523	56.267	11.276	9.486	9.211	10.893	9.342	8.427	9.650	8.145	8.086	7.172	1.765			73
74	W	66.940	67.233	69.089	59.308	57.972	11.672	9.817	9.524	11.284	9.671	8.723	9.960	8.396	8.334	7.386	1.835	1.775	1.773	74
75	Re	68.983	69.298	71.219	61.130	59.708	12.080	10.158	9.845	11.683	10.008	9.026	10.274	8.651	8.585	7.602	1.906			75
76	Os	71.065	71.401	73.390	62.990	61.476	12.498	10.509	10.174	12.093	10.354	9.335	10.597	8.910	8.840	7.821	1.978			76
77	Ir	73.190	73.548	75.606	64.885	63.276	12.922	10.866	10.509	12.510	10.706	9.649	10.919	9.174	9.098	8.040	2.053	1.980	1.975	77
78	Pt	75.355	75.735	77.864	66.821	65.112	13.359	11.233	10.852	12.940	11.069	9.973	11.249	9.441	9.360	8.267	2.127	2.050	2.046	78
79	Au	77.567	77.971	80.172	68.792	66.978	13.807	11.608	11.203	13.379	11.440	10.307	11.583	9.712	9.626	8.493	2.204	2.123	2.118	79
80	Hg	79.809	80.240	82.530	70.807	68.883	14.262	11.993	11.561	13.828	11.821	10.649	11.922	9.987	9.896	8.720	2.282			80
81	Tl	82.104	82.562	84.933	72.859	70.820	14.734	12.388	11.929	14.289	12.211	10.992	12.270	10.267	10.171	8.952	2.362	2.270	2.265	81
82	Pb	84.436	84.922	87.351	74.956	72.792	15.215	12.791	12.304	14.762	12.612	11.347	12.621	10.550	10.448	9.183	2.442	2.345	2.339	82
83	Bi	86.819	87.328	89.846	77.095	74.802	15.708	13.208	12.689	15.245	13.021	11.710	12.978	10.837	10.729	9.419	2.525	2.422	2.416	83
84	Po	89.231	89.781	92.383	79.279	76.851		13.635	13.083	15.741	13.445		13.338	11.129	11.014	9.662				84
85	At	91.707	92.287	94.974	81.499	78.930		14.065		16.249	13.874			11.425	11.303					85
86	Rn	94.230	94.850	97.622	83.768	81.051		14.509		16.768	14.313			11.725	11.596					86
87	Fr	96.791	97.460	100.307	86.089	83.217		14.973		17.300	14.768		14.448	12.029	11.893					87
88	Ra	99.415	100.113	103.051	88.454	85.419	18.354	15.442	14.745	17.845	15.233	13.661	14.839	12.338	12.194	10.620				88
89	Ac	102.084	102.829	105.849	90.868	87.660		15.929		18.405	15.710			12.650	12.499					89
90	Th	104.813	105.591	108.699	93.334	89.938	19.503	16.423	15.640	18.979	16.199	14.507	15.621	12.967	12.807	11.117	3.145	2.996	2.986	90
91	Pa	107.576	108.409	111.605	95.852	92.271	20.094	16.927	16.101	19.565	16.999	14.944	16.022	13.288	13.120	11.364	3.239	3.082	3.072	91
92	U	110.387	111.281	114.587	98.422	94.649	20.709	17.452	16.573	20.164	17.217	15.397	16.425	13.612	13.437	11.616	3.336	3.170	3.159	92
93	Np		113.725	118.057	100.781	96.844	21.336	17.986	17.058	20.781	17.747	15.874	16.837	13.942	13.757	11.887				93
94	Pu		116.943	120.350	103.300	99.168	21.979	18.537	17.553	21.414	18.291	16.330	17.252	14.276	14.082	12.122				94
95	Am		120.350	123.960	105.949	101.607		19.103	18.060	22.061	18.849		17.673	14.615	14.409	12.381				95
96	Cm		122.733	126.490	108.737	104.168				22.703	19.399		18.096	14.953	14.740					96
97	Bk		126.490	130.484	111.676	106.862				23.389	19.961		18.529	15.304	15.080					97
98	Cf		127.794	133.290	114.778	109.699				24.070	20.557		18.983	15.652	15.418					98

References

[AA72] *Proc. Int. Conf. Modern Trends in Activation Analysis* 1973 *Radioanal. Chem.* **16** 2

[AJ71] F Ajzenberg-Selove 1971 Energy Levels of Light Nuclei $A=16$–17 *Nucl. Phys.* A **166** 1

[AJ76] F Ajzenberg-Selove 1976 Energy Levels of Light Nuclei $A=13$–15 *Nucl. Phys.* A **268** 1

[AJ78] F Ajzenberg-Selove 1978 Energy Levels of Light Nuclei $A=18$–20 *Nucl. Phys.* A **300** 1

[AJ79] F Ajzenberg-Selove 1979 Energy Levels of Light Nuclei $A=5$–10 *Nucl. Phys.* A **320** 1

[AJ80] F Ajzenberg-Selove and C L Busch 1980 Energy Levels of Light Nuclei $A=11$–12 *Nucl. Phys.* A **336** 1

[AN77] H H Anderson and J F Ziegler 1977 Hydrogen: Stopping Powers and Ranges in all Elements *The Stopping and Range of Ions in Matter* vol 3 (New York: Pergamon)

[AR60] G D Archard 1960 *Proceedings of the Second International Symposium on X-ray Microscopy,* Stockholm, p337

[AR71] D D Armstrong and H E Wegner 1971 *Rev. Sci. Instrum.* **42** 40

[AU78] W M Augustyniak, D Betteridge and W L Brown 1978 *Nucl. Instrum. Methods* **149** 669

[BA66] A P Banford 1966 *The Transport of Charged Particle Beams* (London: E and F N Spon)

[BA78] P Bayerl and P Eichinger 1978 *Nucl. Instrum. Methods* **149** 663

[BI74] J R Bird, B L Campbell and P B Price 1974 *At. Energy Rev.* **12** 275

[BI78] J R Bird, B L Campbell and R J Cawley 1978 Lucas Heights *AAEC/E443*

[BI80] J R Bird 1980 *Nucl. Instrum. Methods* **168** 85

[BI82] W Bischof, M Hofert, H R Wilde, B Raith and B Gonsoir 1982 *Nucl. Instrum. Methods* **197** 201

[BL61] D Blanc and A Degeilh 1961 *J. Phys. Radium* **22** 230

[BL74] F Blaschta and M Szilagyi 1974 *Optik* **39** part 5 499

[BL83] G D Blower 1983 *D Phil Thesis* Oxford University

[BO32] B von Borries and E Ruska 1932 *Z. Phys.* **76** 649

[BO77] F Bosch, A El Goresy, B Martin, B Povh, R Nobiling, D Schwarm and K Traxel 1977 *Science* **199** 765

[BO78a] G Bonani, M Suter, H Jung, C Stoller and W Wolfli 1978 *Nucl. Instrum. Methods.* **157** 55

[BO78b] F Bosch, A El Goresy, B Martin, B Povh, R Nobiling, D Schwarm and K Traxel 1978 *Nucl. Instrum. Methods* **149** 665

[BO80a] F Bosch, A El Goresy, W Herth, B Martin, R Nobiling, B Povh, H D Reiss and K Traxel 1980 *Nucl. Sci. Appl.* **1** 3

[BO80b] P Bonjour 1980 *Applied Charged Particle Optics* ed. A Septier, part A (New York: Academic)

[BR65] C M Braams, P Zieske and M J Kofoid 1965 *Rev. Sci. Instrum.* **36** 1411

[BR67] K L Brown 1967 *SLAC-75* (Stanford Linear Accelerator Centre)

[BR70] K L Brown 1970 *SLAC-91* (Stanford Linear Accelerator Centre)

[BR77] D Brune, U Lindh and J Lorenzen 1977 *Nucl. Instrum. Methods* **142** 51

[BR80] K L Brown, D C Carey, Ch Iselin and F Rothacker 1980 *CERN report 80-04*

[BU81] M Budnar, V Ramsak, M Ravnikar, V Valkovic and I Orlic 1981 *Nucl. Instrum. Methods* **181** 255

[CA62] R Castaing and G Slodzian 1962 *J. Microsc.* **1** 395

[CH54] P Chanson and C Magnan 1954 *C. R. Acad. Sci., Paris* **238** 1701

[CH74] Z H Cho, M Singh and A Mohabbatizadeh 1974 *IEEE Trans. Nucl. Sci.* **21** 622

[CH77] W K Chu, M Nicolet, J Mayer 1977 *Backscattering Spectrometry* (New York: Academic)

[CH78] Z H Cho, M Singh and G C Huth 1978 *Ann. New York Acad. Sci.* **306** 223

[CL78] R Clampitt and D K Jefferies 1978 *Nucl. Instrum. Methods* **149** 739

[CL81] E Clayton, D D Cohen and P Duerden 1981 *Nucl. Instrum. Methods* **180** 541–48

[CO65] J K Cobb and R S Cole 1965 *Proc. Int. Symp. Magnet Tech.*, Stanford, p431

[CO70] J A Cookson and F D Pilling 1970 *Harwell report AERE R6300*

[CO72] J A Cookson, A T G Ferguson and F D Pilling 1972 *J. Radioanal. Chem.* **12** 39

[CO73] J A Cooper 1973 *Nucl. Instrum. Methods* **106** 525

[CO76] J A Cookson and F D Pilling 1976 *Phys. Med. Biol.* **21** 965

[CO78] G E Coote, R J Sparks, K P Pohl, J G West and C R Purcell 1978 *Semi-Annual Report INS-R-242* p10 Inst. Nucl. Sci. DSIR Laboratory Lower Hutt, New Zealand

[CO79] J A Cookson 1979 *Nucl. Instrum. and Methods* **165** 477–508

[CO81] J A Cookson 1981 *private communication*

[CR67] A V Crewe, D N Eggenberger, L M Welter and J Wall 1967 *J. Appl. Phys.* **38** 4257

[DA81] R L Dalglish 1981 *Nucl. Instrum. Methods* **191** 191

[DE64] J H M Deltrap 1964 *PhD Thesis* Cambridge University

[DE71] G Deconninck, G Demortier and F Bodart 1972 Chemical Analysis by Charged Particle Bombardment *Proc. Int. Meet.* Namur, Belgium, September 1971 (Lausanne: Elsevier)

[DE81] G Deconninck, F Bodart and E Demoulin 1981 *Nucl. Instrum. Methods* **181** 259–63

[DE82] G Demortier ed 1982 *Proc. Second Int. Conf. Chemical Analysis, Namur 1981 Nucl. Instrum. Methods* **197**

[DH61] D Dhuicq 1961 *PhD Thesis* Third Cycle, Paris

[DH68] D Dhuicq 1968 *PhD Thesis* Orsay, Paris

[DO82] B L Doyle and N D Wing 1982 *Sandia Report SAND 82–2393* Sandia National Laboratories, Albuquerque

[DY65a] A D Dymnikov, T Ya Fishkova and S Ya Yavor 1965 *Sov. Phys.–Tech. Phys.* **10** 340

[DY65b] A D Dymnikov, T Ya Fishkova and S Ya Yavor 1965 *Dokl. Akad. Nauk.* **162** 1265 (Engl. Transl. *Sov. Phys. 'Doklady'* **10** 547)

[DY65c] A D Dymnikov, T Ya Fishkova and S Ya Yavor 1965 *Phys. Lett.* **15** 132

[DY65d] A D Dymnikov, T Ya Fishkova and S Ya Yavor 1965 *Zh. Tekh. Fiz.* **35** 1068 (Engl. Transl. *Sov. Phys.–Tech. Phys.* **10** 822)

[DY65e] A D Dymnikov, T Ya Fishkova and S Ya Yavor 1965 *Nucl. Instrum. Methods* **37** 268

[DY65f] A D Dymnikov, T Ya Fishkova and S Ya Yavor 1965 *Zh. Tekh. Fiz.* **35** 431 (Engl. Transl. *Sov. Phys.–Tech. Phys.* **10** 340)

[DY66] A D Dymnikov, T Ya Fishkova and S Ya Yavor 1966 *Izv. Akad. Nauk.* **30** 739 (Engl. Transl. *Bull. Acad. Sci. USSR, Phys. Ser.* **30** 766)

[DU77] A Durand and P Tanguy 1977 *Nucl. Instrum. Methods* **143** 561

[EN67] H A Enge 1967 *Focusing of Charged Particles* vol 2, ch 4 ed A Septier (New York: Academic)

[EN70] H A Enge and S B Kowalski 1970 *Proc. Third Int. Conf. Magnet Technology* ed G Söhugen *et al* (Hamburg: Deutsches Elektronen-Synchrotron)

[EN78] P M Endt and C Van der Leun 1978 Energy Levels of $A = 21$–44 Nuclei *Nucl. Phys.* **A 310** 1

[FI67] T Ya Fishkova, M Siladi and S Ya Yavor 1967 *Radiotekh. Fiz.* **38** 1311 (Engl. Transl. *Radio Eng. Electron Phys.* **12** 1222

[FI68a] T Ya Fishkova, L A Baranova and S Ya Yavor 1968 *Zh. Tekh. Fiz.* **38** 694 (Engl. Transl. *Sov. Phys.–Tech. Phys.* **13** 520)

[FI68b] T Ya Fishkova, L A Baranova and S Ya Yavor 1968 *Izv. Akad. Nauk.* **32** 1026 (Engl. Transl. *Bull. Acad. Sci. USSR, Phys. Ser.* **32** 951)

[FO68a] A Foster 1968 *Correction of aperture aberrations in magnetic lens spectrometers* Thesis, London

[FO68b] L Fox and D F Mayers 1968 *Computing Methods for Scientists and Engineers* (Oxford: Clarendon)

[FO75] F Folkmann 1975 *J. Phys. E: Sci. Instrum.* **8** 429–43

[FO76a] F Folkmann 1976 *Ion Beam Surface Layer Analysis* vol 2 ed. O Meyer 747–58 (New York: Plenum)

[FO76b] F Folkmann 1976 *Ion Beam Surface Layer Analysis* vol 2 ed. O Meyer 695–718 (New York: Plenum)

[FO78] F Folkmann 1978 *Material Characterisation Using Ion Beams* ed. J P Thomas and A Cachard pp 239–81 (New York: Plenum)

[FU75] Y Fujita and H Matsuda 1975 *Nucl. Instrum. Methods* **123** 495

[FY74] R D Fyvie and D E Lobb 1974 *Nucl. Instrum. Methods* **114** 609

[GA46] D Gabor 1946 *Nature* **158** 198

[GA47] D Gabor 1947 *Nature* **160** 89

[GE76] R V Gentry, T A Cahill, N R Fletcher, H C Kaufmann, L R Medsker, J W Nelson and R G Flocchini 1976 *Phys. Rev. Lett.* **37** 11

[GE82] H-U von Gersch, D Grambole, F Herrman and H-J Thomas 1982 *Exp. Tech. Phys.* **30** 227

[GI73] J V Gilfrich, P G Burkhalter and L S Birks 1973 *Anal. Chem.* **45** 2002

[GO61] R Gomer 1961 *Field Emission and Field Ionisation* (Cambridge Ma: Harvard University Press)

[GO73] F Goulding and J M Jaklevic 1973 *Ann. Rev. Nucl. Sci.* **23** 45

[GR60] P Grivet and A Septier 1960 *Nucl. Instrum. Methods* **6** 126

[GR76] L Grodzins, P Horowitz and J Ryan 1976 *Proc. Conf. Sci. Ind. Appl. Small Accel.* Denton 1975 ed J L Duggan and I L Morgan (New York: IEEE)

[GR78a] P Grivet and A Septier 1978 *Ann. N.Y. Acad. Sci.* **306** 158

[GR78b] L Grodzins, M Aronson and P Horowitz 1978 *Bull. Am. Phys. Soc.* **23** 1035

[GR80] G W Grime 1980 *Unpublished calculations*

[GR81a] T Greenway and G Doucas 1981 *Private communication*

[GR81b] L Grodzins 1981 *Private communication*

[GR82] G W Grime, F Watt, G D Blower, J Takacs and D N Jamieson 1982 *Nucl. Instrum. Methods* **197** 97

[HA66a] P W Hawkes 1966 *Proceedings of Workshop on Electron Microscopy* ANL-7275 p32

[HA66b] P W Hawkes 1966 *Quadrupole Optics* Springer Tracts in Modern Physics vol 42 (Berlin: Springer)

[HA67a] P W Hawkes 1967 *Optik* **25** 439

[HA67b] P W Hawkes 1967 *Optik* **26** 497

[HA67c] P W Hawkes and D F Hardy 1967 *Optik* **25** 450

[HA67d] P W Hawkes and E L Cham 1967 *Optik* **26** 507

[HA67e] D F Hardy 1967 *Thesis* Cambridge University

[HA69] K Halbach 1969 *Nucl. Instrum. Methods* **74** 147

[HA70] P W Hawkes 1970 *Quadrupoles in Electron Lens Design* (New York: Academic)

[HA77] P W Hawkes and U Valdre 1977 *J Phys. E: Sci. Instrum.* **10** 309

[HA79] G R Hanson and B M Siegel 1979 *J. Vac. Sci. Technol.* **16**(6) 1875

[HA80] P W Hawkes 1980 *Advances in Electronics and Electron Physics Supplement 13A* ed. A Septier p45 (New York: Academic)

[HE76a] D Heck 1976 *Kernforschungszentrum Karlsruhe Report KfK 2379* p108

[HE76b] D Heck 1976 *Kernforschungszentrum Karlsruhe Report KfK 2288*

[HE77] D Heck 1977 *Nucl. Instrum. Methods* **143** 423

[HE78a] D Heck 1978 *Kernforschungszentrum Karlsruhe Report KfK 2680* p115

[HE78b] D Heck 1978 *Kernforschungszentrum Karlsruhe Report* 2734

[HE79] D Heck 1979 *Beitr. Elektron. Direkt Oberflaechentech.* **12** 259

[HU80] M J Hurst 1980 *D. Phil. Thesis* Oxford University

[IO79] *Proc. Fourth Int. Conf. Ion Beam Analysis, Aarhus* 25–29 June 1979 *Nucl. Instrum. Methods* **168**

[JA79a] D N Jamieson 1979 *Fourth-year report* School of Physics, University of Melbourne

[JA79b] R A Jarjis and J M Calvert 1979 *Nuclear Cross Section Data for Surface Analysis* (Manchester: University of Manchester)

[JA82] D N Jamieson 1982 *Private communication*

[JO70a] T B Johansson, R Akselsson and S A E Johansson 1970 *Nucl. Instrum. Methods* **84** 141

[JO70b] G G Johnson and E W White 1970 *X-ray Emission Wavelengths and keV Tables for Non-Diffractive Analysis* ASTM Data Series DS46

[JO76] S A E Johansson and T B Johansson 1976 *Nucl. Instrum. Methods* **137** 473–516

[KA68a] K Kanaya, S Okayama and H Mishina 1968 *Bull. Electrotech. Lab.* **32** 583

[KA68b] K Kanaya, S Okayama and H Mishina 1968 *Bull. Electrotech. Lab.* **32** 665

[KA68c] H Kawakatsu, K G Vosburgh and B M Siegel 1968 *J. Appl. Phys.* **39** 255

[KA69] K Kanaya and S Okayama 1969 *Bull. Electrotech. Lab.* **33** 27

[KA77] H C Kaufmann, K R Akselsson and W J Courtney 1977 *Nucl. Instrum. Methods* **142** 251

[KA81] K Kaisla, T Raunemaa and P Paatero 1981 *Nucl. Instrum. Methods* **181** 77

[KE62] V M Kel'man and S Ya Yavor 1962 *Sov. Phys.–Tech. Phys.* **6** 1052

[KO77] H Koops, G Kuck and O Scherzer 1977 *Optik* **48** 225

[KO80] H Koyama-Ito and L Grodzins 1980 *Nucl. Instrum. Methods* **174** 331

[KR79] P Krejcik, R L Dalglish and J C Kelly 1979 *J. Phys. D. Appl. Phys.* **12** 161

[KR80] P Krejcik, J C Kelly and R L Dalglish 1980 *Nucl. Instrum. Methods* **168** 247

[LA65] G P Lawrence, R K Beauchamp and J L McKibben 1965 *Nucl. Instrum. Methods* **32** 357

[LA76] M K Lamvik and T Groves 1976 *Ultramicroscopy* **2** 69

[LE70] G E Lee-Whiting 1970 *Nucl. Instrum. Methods* **83** 232

[LE74] R Levi-Setti 1974 *Scanning Electron Microscopy* ed. O Johari and I Corvin (Chicago: IIT Research Institute) p125

[LE75] S V Lebedev and A I Morozov 1975 *Zh. Tekh. Fiz.* **44** 1547 (Engl. Transl. *Sov. Phys.–Tech. Phys.* **19** 966)

[LE76] C M Lederer, V S Shirley 1976 ed. *Table of Isotopes* (New York: Wiley)

[LE78] G J F Legge 1978 *Proc. Second Austr. Conf. on Nucl. Tech. of Anal.* p18

[LE79a] G F Legge, C D McKenzie and A P. Mazzolini 1979 *J. Microsc.* **117** 185

[LE79b] H W Lefevre and R Booth 1979 *IEEE Trans. Nucl. Sci.* NS26 3115

[LE80a] R Levi-Setti and T R Fox 1980 *Nucl. Instrum. Methods* **168** 139

[LE80b] C Lejeune and J Aubert 1980 *Advances in Electronics and Electron Physics Supplement 13A* ed. A Septier (New York: Academic) p159

[LE80c] R Levi-Setti 1980 *Advances in Electronics and Electron Physics Supplement 13A* ed. A Septier (New York: Academic)

[LE82a] H W Lefevre 1982 *Private communication*

[LE82b] G Legge 1982 *Nucl. Instrum. Methods* **197** 243

[LI62] M S Livingston and J P Blewett 1962 *Particle Accelerators* (New York: McGraw-Hill)

[LI67] H Liebl 1967 *J. Appl. Phys.* **38** 577

[LO72] D E Lobb 1972 *Nucl. Instrum. Methods* **105** 129

[LU71] G J Lutz, R J Boreni, R S Maddock and W W Meinke 1971 *Activation Analysis, A Bibliography* Tech. Note 467 Parts 1 and 2 (Washington DC: National Bureau of Standards)

[LY74] Ya G Lyubchik and T Ya Fishkova 1974 *Zh. Tekh. Fiz.* **44** 2272 (Engl. Transl. *Sov. Phys.–Tech. Phys.* **19** 1403)

[MA62] S S Markowitz and J O Mahony 1962 *Anal. Chem.* **34** 329

[MA66] C Maples, G W Goth and J Cerny 1966 *Nuclear Data 2* nos 5 and 6 p429

[MA71] T Matsuo and H Matsuda 1971 *Int. J. Mass Spectrom. Ion Phys.* **6** 361

[MA72] T Matsuo, H Matsuda H Wollnik 1972 *Nucl. Instrum. Methods* **103** 515

[MA75a] J W Mayer 1975 *New Uses of Ion Accelerators* (New York: Plenum) ch 2B

[MA75b] T Matsuo 1975 *Nucl. Instrum. Methods* **126** 273

[MA77] J W Mayer and E Rimini 1977 *Ion Beam Handbook for Material Analysis* (New York: Academic) chs 2, 4 and 5

[MA80a] J H Malmberg and C F Driscoll 1980 *Phys. Rev. Lett.* **44** 63

[MA80b] C J Maggiore 1980 *Scanning Electronmicrosc.* **1** 439

[MA81a] W Maenhaut, L De Reu, H A Van Rinsvelt, G S Roessler, J W Swanson and M D Williams 1981 *IEEE Trans. Nucl. Sci.* **28** 1386

[MA81b] F W Martin 1981 *Nucl. Instrum. Methods* **189** 93

[MA81c] C J Maggiore 1981 *Nucl. Instrum. Methods* **191** 199

[MA82] F W Martin and R Goloskie 1982 *Nucl. Instrum. Methods* **197** 111

[ME61] W E Meyer *Optik* **18** 69

[ME63] P F Meads 1963 *Thesis* University of California

[ME67] P F Meads 1967 *Rev. Sci. Instrum.* **38** 1752

[ME76] O Meyer, G Linker and F Kappeler 1976 *Ion Beam Surface Layer Analysis* (New York: Plenum)

[MO68] R W Moses 1968 *Thesis* University of Wisconsin

[MO70] R W Moses 1970 *Rev. Sci. Instrum.* **41** 729

[MO71] R W Moses 1971 *Rev. Sci. Instrum.* **42** 828

[MO79] R M Mobley, G Gammel and A W Maschke 1979 *IEEE Trans. Nucl. Sci.* NS-26(3) 3112

[NA82] *Mk 8 Library Application Notes* 1982 (Oxford: Numerical Algorithms Group)

[NO70] L C Northcliffe and R F Schilling 1970 *Nuclear Data Tables A7* 233

[NO75] R Nobiling, Y Givelekoglu, B Povh, D Schwalm and K Traxel 1975 *Nucl. Instrum. Methods* **130** 325

[NO77] R Nobiling, K Traxel, F Bosch, Y Givelekoglu, B Martin, F Povh and D Schwalm 1977 *Nucl. Instrum. Methods* **142** 49

[OR78] J Orloff and L W Swanson 1978 *J. Vac. Sci. Technol.* **15** 845

[OV67] L P Ovsyannikova and S Ya Yavor 1967 *Radiotekh. Elektron.* **12** 489 (Engl. Transl. *Radio Eng. Electron. Phys.* **12** 449)

[OV68] L P Ovsyannikova 1968 *Radiotekh. Elektron.* **13** 1489 (Engl. Transl. *Radio Eng. Electron. Phys.* **13** 1292)

[OV71] L P Ovsyannikova and S Ya Yavor 1971 *Sov. Phys.–Tech. Phys.* **16** 803

[PA76] G Parzen 1976 *Brookhaven National Laboratory report BNL 50536*

[PI68] T B Pierce, P F Peck and D R A Cuff 1968 *Nucl. Instrum. Methods* **67** 1

[PI76] *Proceedings of the International Conference on Particle-Induced X-ray Emission and its Analytical Applications* Lund, 23–26 August 1976 *Nucl. Instrum. Methods* **142**

[PI80] *Proceedings of the Second International Conference on* PIXE *and its Analytical Applications* Lund, June 1980 *Nucl. Instrum. Methods* **181**

[PR69] P B Price and J R Bird 1969 *Nucl. Instrum. Methods* **69** 277

[RA67] A Ralston and H S Wilf 1967 *Mathematical Methods for Digital Computers* (New York: Wiley)

[RE63] E Regenstreif 1963 *Chromatic Aberration in Quadrupole Multiplets* CERN 64–72

[RE67] E Regenstreif 1967 *Focusing of Charged Particles I* ed. A Septier (New York: Academic) ch 2

[RI78] G R Ringo and V E Krohn 1978 *Nucl. Instrum. Methods.* **149** 735

[RU11] E Rutherford 1911 *Phil. Mag.* **21** 669

[SC36] O Scherzer 1936 *Z. Phys.* **101** 593

[SC47] O Scherzer 1947 *Optik* **2** 114

[SC73] K Schmid, H Muller, H Ryssel and I Ruge 1975 *Thin Solid Films* **199** 313

[SC78] R L Schulte and E A Kamykowski *J Radioanal. Chem.* **43** 233

[SE66] A Septier 1966 *Advances in Optical and Electron Microscopy* ed. R Barer and V E Coslett (New York: Academic)

[SE67] A Septier 1967 *Focusing of Charged Particles II* ed. A Septier (NewYork: Academic) ch 3

[SE79] R L Seliger, J W Ward, V Wang and R L Kubena 1979 *Appl. Phys. Lett.* **34** 360

[SH78] R E Shroy, H W Kraner and K W Jones 1978 *Nucl. Instrum. Methods* **157** 163

[SI65] K Siegbahn 1965 *Alpha, Beta and Gamma Ray Spectroscopy* vol 1 ed. K Siegbahn (Amsterdam: North-Holland)

[SI76] M Singh, J Melvin and Z H Cho 1976 *IEEE Trans. Nucl. Sci.* **23** 657

[SI78] G Sidenius 1978 *Nucl. Instrum. Methods* **151** 349

[SI79] P Sigmund 1979 *The Stopping and Range of Ions in Matter* vol 1 (New York: Pergamon)

[SP63] R H Spear, J D Larsen and J D Pearson 1963 *Nucl. Phys.* **41** 353

[ST65] K G Steffen *High Energy Beam Optics* Monographs and Texts in Physics and Astronomy, vol XVII (New York: Wiley)

[SU74] M Suter, W Wolfli, G Bonani, H Jung and C Stoller 1974 ETH Zurich Prog. Rep. 263 *Helv. Phys. Acta.* **48** 511

[SW79] L W Swanson, G A Schwind, A E Bell and J E Brady 1979 *J. Vac. Sci. Technol.* **16** 1864

[SZ72] M Szilagyi, F Blaschta, L P Ovsyannikova and S Ya Yavor 1972 *Electron Microscopy 1972: Institute of Physics Conference Series 14* ed. A M Glauert (London: The Institute of Physics) p84

[SZ73] M Szilagyi, F Blaschta, K Kando, L P Ovsyannikova and S Ya Yavor 1973 *Optik* **38** 425

[SZ74] M Szilagyi, F Blaschta, L P Ovsyannikova and S Ya Yavor 1974 *Optik* **39** 351

[TH67] M G R Thompson 1967 *PhD Thesis* Cambridge University

[TI66] R S Tilbury 1966 *Activation Analysis with Charged Particles* Nat. Acad. Sci. Nat. Res. Council Nuclear Science Series, NAS-NS 3110 (Springfield Va: Clearinghouse for Federal Science Series and Technical Information)

[UE73] M Ueda, K Nagami and H Kurado 1973 *Optik* **37** 528

[UE75] M Ueda 1975 *Optik* **43** 319

[VO67] K G Vosburgh 1967 *PhD Thesis* Cornell University

[WA78] F Watt, L K Fifield, M J Hurst, T J M Symons, C H Zimmerman and K W Allen 1978 *Nucl. Instrum. Methods* **151** 163

[WA81] F Watt, G W Grime, G D Blower and J Takacs 1981 *IEEE Trans. Nucl. Sci.* NS-28.2 1413

[WA82] F Watt, G W Grime, G D Blower, J Takacs and D J T Vaux 1982 *Nucl. Instrum. Methods* **197** 65

[WE75] Wei-Kan Chu 1975 *New Uses of Ion Accelerators* (New York: Plenum) ch 2c

[WI78] H R Wilde, M Roth, C D Uhlhorn and B Gonsoir 1978 *Nucl. Instrum. Methods* **149** 675

[WI80] R G Wilson 1980 *Ion Implantation Technology and Machines* Advances in Electronics and Electron Physics, Supplement 13A ed. A Septier (New York: Academic)

[WI81] H R Wilde, W Bischof, B Raith, C D Uhlhorn and B Gonsoir 1981 *Nucl. Instrum. Methods* **181** 165

[WO72] H Wollnik 1972 *Nucl. Instrum. Methods* **103** 479

[WO73] R Woldseth 1973 *X-ray Energy Spectrometry* (Burlingame, Ca: Kevex Corporation)

[WO75] E A Wolicki 1975 *New Uses of Ion Accelerators* ed. J F Ziegler (New York: Plenum) ch 3

[YA68] S Ya Yavor, T Ya Fishkova and L A Baranova 1968 *Proc. Eur. Reg. Conf. Electron Microsc.* Rome, p169

[YA72] S Ya Yavor, L P. Ovsyannikova and L A Baranova 1972 *Nucl. Instrum. Methods* **99** 103

[ZH69] V V Zhukov, A I Morozov and G Y Shehepkin 1969 *JETP Lett.* **9** 14

[ZI75] J F Ziegler 1975 *New Uses of Ion Accelerators* (New York: Plenum) ch 2A

[ZI76] J F Ziegler 1976 *Ion Surface Layer Analysis* vol 2 ed. E O Mayer (New York: Plenum) pp 759–72

[ZI77] J F Ziegler 1977 *Helium: Stopping Powers and Ranges in all Elemental Matter* The Stopping and Range of Ions in Matter vol 4 (New York: Pergamon)

[ZI79] J F Ziegler 1979 *Stopping Cross Sections For Energetic Ions in All Elements* The Stopping and Range of Ions in Matter vol 5 (New York: Pergamon)

[ZI80] J F Ziegler 1980 *Nucl. Instrum. Methods* **168** 17

Index